Are human infants born with an innate a sense of self? How does that sense of self change as they develop? How do great apes compare to human infants and children in their sense of self? Are they capable of embarrassment? What kind of self-awareness do monkeys have? Do they recognize their own images in mirrors? Do dolphins? Do pigeons? These are some of the many questions addressed in *Self-awareness in Animals and Humans*, a collection of original articles on self-awareness in monkeys, apes, humans, and other species, including dolphins. This volume, which grew out of an interdisciplinary conference on self-awareness, focuses on controversies about how to measure self-awareness, which species are capable of self-awareness and which are not, and why. Several articles focus on the controversial question of whether gorillas, like other great apes and human infants, are capable of mirror self-recognition (MSR) or whether they are anomalously unable to do so. Other articles focus on whether macaque monkeys are capable of MSR. Various contributors present competing theories about which abilities accompany and underlie MSR and which capacities underlie developmentally earlier forms of self-detection in human infants. The focus of the articles is both comparative and developmental: Several contributors explore the value of frameworks from human developmental psychology for comparative studies. In particular, various contributors present differing opinions concerning the relationship between MSR and object permanence, imitation, and theory of mind. This dual focus – comparative and developmental – reflects the interdisciplinary nature of the volume, which brings together biological anthropologists, comparative and developmental psychologists, and cognitive scientists from Japan, France, Spain, Hungary, New Zealand, Scotland, and the United States.

Self-awareness in animals and humans

Detail from "A little culture" by A. Weczerzick; published in *Die Gartenlaube*
(Leipzig: Verlag von Ernst Keil), probably before the turn of the century.

Self-awareness in animals and humans

Developmental perspectives

Edited by

SUE TAYLOR PARKER
Sonoma State University

ROBERT W. MITCHELL
Eastern Kentucky University

AND

MARIA L. BOCCIA
University of North Carolina

CAMBRIDGE
UNIVERSITY PRESS

Published by the Press Syndicate of the University of Cambridge
The Pitt Building, Trumpington Street, Cambridge CB2 1RP
40 West 20th Street, New York, NY 10011-4211, USA
10 Stamford Road, Oakleigh, Melbourne 3166, Australia

First published 1994

Printed in the United States of America

Library of Congress Cataloging-in-Publication Data
Self-awareness in animals and humans : developmental perspectives /
    edited by Sue Taylor Parker, Robert W. Mitchell, Maria Boccia.
        p.   cm.
    Papers presented at a Conference on Self-awareness in Monkeys, Apes, and
Humans, held 1991, at Sonoma State University.
    ISBN 0–521–44108–0
    1. Self-perception – Congresses.   2. Developmental psychology – Congresses.
    3. Psychology, Comparative – Congresses.
    I. Parker, Sue Taylor.   II. Mitchell, Robert W., 1958– .   III. Boccia, Maria.
BF697.5.S43S43   1994
156'.37 – dc20   93–24656
CIP

A catalog record for this book is available from the British Library.

ISBN 0–521–44108–0 hardback

Contents

                                v

Contributors

James R. Anderson, Centre de Primatologie, Université Louis Pasteur, Strasbourg, France

Kim A. Bard, Yerkes Regional Primate Research Laboratory, Emory University, Atlanta, GA

Susan L. Boatright-Horowitz, Dept. of Psychology, Hunter College, City University of New York

Maria L. Boccia, Frank Porter Graham Child Development Center, University of North Carolina, Chapel Hill, NC

Sarah T. Boysen, Dept. of Psychology, Ohio State University, Columbus, and the Yerkes Primate Research Laboratory, Emory University, Atlanta, GA

Kirstan M. Bryan, Dept. of Psychology, Ohio State University, Columbus

Ronald H. Cohn, The Gorilla Foundation, Woodside, CA

Cynthia L. Contie, Institute for Human Resources, Rockville, MD

Deborah Custance, Scottish Primate Group, Psychological Laboratory, St. Andrews University, St. Andrews, Scotland

Siân Evans, Dumond Conservancy for Primates and Tropical Forests, Miami, FL

Suzanne Fegley, Dept. of Psychology, Temple University, Philadelphia, PA

Gordon G. Gallup, Jr., Dept. of Psychology, State University of New York, Albany

György Gergely, Institute of Psychology, Hungarian Academy of Sciences, Budapest

Juan Carlos Gómez, Facultad de Psicologia, Universidad Autónomia, Madrid, Spain

Alison Gopnik, Dept. of Psychology, University of California, Berkeley

Daniel Hart, Dept. of Psychology, Rutgers University, Camden, NJ

William D. Hopkins, Yerkes Regional Primate Research Center, Emory University, Atlanta, GA

Charles W. Hyatt, Yerkes Regional Primate Research Center, Emory University, Atlanta, GA

Shoji Itakura, Dept. of Communication, Oita Prefectural College of Arts and Culture, Oita, Japan

Lindsay E. Law, Dept. of Biological Anthropology, Cambridge University, Cambridge, England

Michael Lewis, Dept. of Pediatric Medicine, Robert Wood Johnson Medical School, University of Medicine and Dentistry, New Brunswick, NJ

Andrew J. Lock, Dept. of Psychology, Massey University, Palmerston, New Zealand

Lori Marino, Dept. of Psychology, State University of New York, Albany

Kenneth Marten, Sea Life Park, Kailua, HI

Andrew N. Meltzoff, Dept. of Psychology, University of Washington, Seattle

Constance Milbrath, Dept. of Psychiatry, University of California, San Francisco

H. Lyn White Miles, Dept. of Anthropology, University of Tennessee, Chattanooga, TN

Robert W. Mitchell, Dept. of Psychology, Eastern Kentucky University, Richmond

Louis J. Moses, Dept. of Psychology, University of Oregon, Eugene

Sue Taylor Parker, Dept. of Anthropology, Sonoma State University, Rohnert Park, CA

Francine G. P. Patterson, The Gorilla Foundation, Woodside, CA

Daniel J. Povinelli, New Iberia Research Center, University of Southwestern Louisiana, New Iberia

Suchi Psarakos, Sea Life Park, Kailua, HI

Diana Reiss, Marine World USA, Vallejo, CA

Traci A. Shreyer, Dept. of Psychology, Ohio State University, Columbus

Karyl B. Swartz, Dept. of Psychology, Lehman College, City University of New York

Robert L. Thompson, Dept. of Psychology, Hunter College, City University of New York

Roger K. R. Thompson, Dept. of Psychology, Franklin and Marshall College, Lancaster, PA

John S. Watson, Dept. of Psychology, University of California, Berkeley

Foreword

Louis J. Moses

Big questions are back on the agenda in the cognitive sciences. We are seeing a renewal of interest in the very nature of subjective experience. Once-suspect notions like consciousness, awareness, meaning, and mind are again being pursued with some vigor. Consciousness, for example, is now variously being "rediscovered" (Searle, 1992), "reconsidered" (Flanagan, 1992), and even "explained" (Dennett, 1991). At the heart of this resurgent enterprise are questions of self and self-awareness. If we are ever to understand psychological experience we will need to grapple with questions concerning the *subject* of that experience. What is a self? Does it have real, tangible status or is it perhaps a mere cognitive or cultural construction that conveniently lends coherence to what would otherwise be quite disparate aspects of experience? What is the structure of the self? Is it unitary or multifaceted? If multifaceted, what are these facets and do they cohere in some organized fashion? What difference does self-awareness make? Does it have verifiable consequences for the psychological and behavioral life of an individual? It is questions of this order that have captured the interest of the various contributors to this volume. Their collective approach to studying the self is an especially fascinating one, focusing on the emergence of a sense of self from two perspectives. One is the perspective of human ontogeny: Do infants enter the world with a ready-made sense of self, or is a self only gradually constructed through some maturational or enculturative process? The other is a comparative perspective: Which species can be said to have a sense of self and what is the evolutionary history of self-awareness? Is a sense of self a primitive phenomenon arising early in evolution and common to many species, or is it a much later arrival, perhaps common only to a handful of primate species?

The empirical study of self-awareness is notoriously difficult, and the ontogenetic and comparative perspectives bring with them their own special challenges. We have no direct way of knowing what the mental life of an infant is like and, a fortiori, the same is true with respect to other, more exotic, forms of life. This of course is no more than a restatement of the well-known "problem of other minds" that pervades any attempt to know the inner, subjective life of another individual. Nevertheless, at least in the case of mature members of our own species, intuitions derived from personal experience might serve us reasonably well; moreover, we can always ask others to tell us about their experience. When it comes to young infants and psychologically distant species, however, the luxury of self-report is no longer an option, and the intuitions on which we are likely to fall back may well lead us dangerously astray. Consequently, whether ontogenetic and phylogenetic questions concerning the self are answerable will ultimately depend on how carefully they are framed and on whether we can devise alternative methodologies allowing us to put the questions to nature in some tractable form.

Certain aspects of self lend themselves more readily to empirical scrutiny than others. William James (1892/1948) observed long ago that we can think of the self in two distinct ways: "the self as knower" and "the self as known." The self as knower or the "I" is the subjective self, the self who thinks, feels, and experiences. The self as known or the "Me" is the objective self, involving those aspects of self that the "I" can reflect upon. The self as known is in other words a conception of self. James was quick to point out that the subjective aspects of self are more difficult to study than the objective aspects and, by and large, subsequent research has borne out his impression. Most contemporary work on the self tries to illuminate the nature of self-awareness indirectly through the study of self-knowledge or self-conception. A fundamental aspect of self-conception that has received more attention than any other, and that forms the centerpiece of the current volume, is self-recognition. It seems reasonable to suppose that any animal with even a minimal conception of self should be capable of recognizing who or what it is. Certainly, if we could somehow demonstrate whether or not an animal recognized itself, we would at least have forced an initial wedge into broader aspects of self-conception.

But how could we know whether an animal was capable of self-recognition? In the early 1970s, a profoundly important methodological breakthrough was made, one that effectively revolutionized the field of self-recognition and self-awareness. Working with chimpanzees, Gordon Gallup (1970) developed a simple yet elegant technique for assessing a certain kind of self-recognition, namely recognition of self in a mirror. After giving his chimpanzees several days' prior exposure to mirrors, Gallup anesthetized the animals and then marked their faces with red dye. When the chimpanzees later awakened, they were again confronted with a mirror and Gallup observed how they reacted to the red mark they saw on the mirror image. He reasoned that if the chimpanzees recognized themselves they would be inclined to touch the corresponding marked region of their own faces significantly more often than they had touched that region prior to having their faces marked. If, on the other hand, they did not recognize themselves then they would show little such "mark-directed" behavior, and might instead treat the image in the mirror as a conspecific or perhaps ignore it altogether. A less invasive, but conceptually similar, variation of the procedure was independently developed at about the same time for use with human infants (Amsterdam, 1972; see also Lewis & Brooks-Gunn, 1979). The mirror task has subsequently proven to be an enormously productive research tool, appearing in numerous studies designed to assess self-recognition in a wide variety of species. Thus far, only humans (by roughly 18 months of age), chimpanzees, and orangutans, have offered unambiguous evidence of self-recognition.

The mirror task serves as a point of departure for many of the contributions in this volume. We have learned a great deal from how animals respond to their mirror images in the past two decades; but we ought to be nervous about sole reliance on a single measure, no matter how fruitful that measure might have been in the past, as an index of any conceptual ability. Much of what is written in the pages that follow can be seen as a response to anxiety of this kind. The response takes different forms – attempts to extend, defend, qualify, and sometimes even deny, the significance of mirror recognition – but, in one way or another, the various chapters represent efforts to confront either methodological or conceptual doubts surrounding the mirror task.

Methodological uneasiness arises because the task is hardly likely to be a psychometrically perfect measure of self-recognition, relying as it does on just one kind of recognition (visual recognition) in a medium (the mirror) that is surely low in terms of ecological significance for most species. Moreover, mirror recognition arguably requires cognitive abilities extraneous to a conception of self, most notably some minimal conception of representational relations (the relation between the self and the mirror image). Consequently, we cannot be entirely confident that an animal failing the test is incapable of any form of self-recognition. One healthy reaction to these methodological qualms that is well represented in this volume involves an attempt to find converging and/or better techniques for assessing emerging conceptions of self. Some of these new methods are close cousins of the mirror task: for example, whether an animal recognizes itself in a videotaped image, in a photograph, or even in its own shadow. Other methods take us farther afield: for instance, recognition of self in the imitative actions of another. The latter appeals as a particularly promising measurement tool because it nicely avoids some of the theoretical and ecological problems that beleaguer the mirror task.

The conceptual brand of doubt is more troubling. How should we interpret successful mirror recognition? What kind of self-conception should we grant to an animal who passes the mirror task? Do we accord the animal an elaborate self-conception similar in kind to that which fully developed members of our own species seem to share? That is, do we want to say that what the animal sees in the mirror is a self with a sense of its own continuity, of its enduring yet finite existence, of its agency and intentionality in moving through the world, of what makes it unique compared to other selves and, at the same time, of its psychological commonality with those other selves? Or do we want to attribute a much more delimited conception of self to the animal, one that involves none of this rich psychology? Perhaps the animal is only aware of its physical self, and perhaps only dimly aware at that. The animal might simply recognize its body in the mirror while having little in the way of a conception of its own mental life.

No simple solution to this interpretive dilemma will be found in the pages of this volume. What we see instead is an impressive effort to resolve the dilemma by means of situating mirror recognition within a wider comparative and epigenetic framework. The epigenetic aspects of this framework require searching for psychological correlates of mirror recognition and tracking its developmental history. That is, we might be better able to judge the significance of mirror recognition if we could locate psychological changes that are concomitant with it and isolate factors that affect its onset. With respect to correlated changes, the kind of variables that will be important should depend on the sophistication of the underlying self-conception. If that conception is merely a rudimentary physical or bodily conception, then self-recognition should have only limited implications for the psychological life of the animal, whereas if it is a more complex conception the ramifications should be widespread.

One potentially discriminating set of factors in this regard is social. If mirror recognition is a symptom of an animal's emerging ability to reflect on its own mental life then, as Gallup (1985) has argued, we might expect to see in the animal's behavior some dawning appreciation of the mental lives of its conspecifics. It is here that research on self-recognition begins to coalesce with recent work on developing "theories of mind." In the last decade or so a whole new subfield in developmental psychology has emerged devoted exclusively to the study of children's conceptions of mental states like belief, desire, and intention (e.g., Flavell, 1988; Moses & Chandler, 1992). A first pass through some of the relevant literature might lead one not to be especially sanguine about Gallup's hypothesis: Whereas infants of around 18 months of age show clear signs of mirror recognition, the prevailing wisdom in writings on theories of mind is that a mature conception of mind does not begin to emerge until well into the preschool period. Any tight coupling between self-recognition and the attribution of mental states to others would seem to be out of the question. On closer inspection, however, the prospects for such a coupling are not so dim after all. The particular conception of mind that is believed by many to be acquired by age 4 or 5 rests on an appreciation of epistemic mental states like knowledge and belief. Wimmer and Perner's (1983) now classic task assessing whether children understand that others sometimes fall prey to, and subsequently act on, *false* beliefs is frequently viewed as a kind of litmus test for such a theory of mind. It turns out, though, that the significance of the false belief task is perhaps even more controversial than that of the mirror task, with some arguing that a genuine understanding of false belief might well be in place considerably earlier (Lewis & Mitchell, in press). In addition, it is not clear that self-recognition should necessarily be related to an understanding of the false beliefs of others. All that need be argued is that an ability to reflect on one's own internal experience might lead one to posit that others too experience subjective, psychological states of mind. That the potentially more advanced step of understanding the precise representational relations between those states of mind and states of the world (e.g., understanding false beliefs) is not taken until sometime later is neither here nor there. If we recast Gallup's hypothesis in these terms, the evidence is now considerably more supportive. For example, at about the same time that infants first recognize themselves in mirrors, they also begin to show a nascent appreciation of the perceptions, desires, and emotions of others (Wellman, 1993). Looked at in this way, tantalizing support for a developmental relation between self-recognition and children's early theory of mind can be seen.

Social factors may also play a role with respect to the other epigenetic aspect of the framework, namely that focused on the origins of self-recognition. Here again there are potential links to the literature on children's theories of mind. An ongoing controversy in that literature concerns what role our first-person experience of mental states might play in the development of an understanding of mental life more generally. According to one prominent view (the "simulation theory"; see Harris, 1991), we have direct, privileged access to our own mental states and we come to infer the mental states of others through a process of imaginative simulation. In this view mental states of self are primary, those of others are derivative. A contrasting and equally prominent view (the "theory theory"; see Gopnik & Meltzoff, *SAAH*10), holds that mental-state conceptions are abstract entities constructed at the same time for self and other on the basis of all of the available evidence. By these lights, while we might have direct access to our own subjective experiences, those experiences can have little meaning until rendered interpretable within some more general theoretical conception.

In the literature on self-recognition, the natural tendency has been to side implicitly with the simulation theory. That is, an ability to attribute mental states to others has typically been seen as a potential offshoot of self-recognition: Once an animal has developed some conception of self, the way is then cleared for a conception of others to be acquired. Nevertheless, consider the alternative possibility: Perhaps conceptions of self and other are constructed simultaneously through evidence available in social interaction. This is certainly not a new idea. We see elements of it in ideas about the "looking-glass self" in the writings of the early self theorists (Cooley, 1902/1964; Mead, 1934/1974). These theorists proposed that the self is born out of reflections in a *social* mirror, that the self is a reflection of how one is viewed by others. One clear prediction of this proposal is that some kind of social experience should be necessary for the development of self-recognition and self-awareness. Findings relating social imitation to self-recognition in this volume and elsewhere (Asendorpf & Baudonnière, 1993) are fully consistent with this view, as are Gallup's early observations that chimpanzees reared in social isolation failed to develop mirror recognition (Gallup, McClure, Hill, & Bundy, 1971). Of course, other explanations might well be found for these effects, but they do illustrate the potentially important role that social factors might play in the development of self-recognition.

If we now look to the comparative side of the framework, clear parallels with its epigenetic counterpart can be found. That is, just as we can use information about related conceptual abilities to gauge the developmental significance of mirror recognition, the same can be done with respect to the significance of its phylogenetic emergence. If mirror recognition is indeed a symptom of a rich conception of self, then those species capable of it ought to show a constellation of other abilities in the arena of mental-state attribution and associated domains like imitation, deception, pretense, and empathy. Although the data are admittedly sketchy, the preliminary evidence is by and large supportive: Those species that pass the mirror task (e.g., chimpanzees) show some of the relevant abilities whereas those that fail (e.g., monkeys) do not (Gallup, 1985; Mitchell, 1993; Povinelli, 1993). Should the data continue to fall out in a generally consistent manner, we might eventually be in a position to use some of these other abilities as markers to help decide ambiguous cases (e.g., gorillas and dolphins) for which the mirror results prove difficult to interpret.

The future of this comparative program rests on what are essentially anthropomorphic assumptions. We currently have a decidedly more solid theoretical and empirical foundation with respect to human ontogeny than we do with respect to how development proceeds in most other species. The comparative approach seeks to rely on well-established theories of human development and well-replicated developmental milestones in examining how the relevant capacities might have evolved in these other species. There are, of course, inherent dangers in anthropomorphic assumptions. Even the most thoroughly tested empirical markers and the most firmly entrenched developmental theories can and do fall by the wayside with disturbing frequency. Consequently we need to be highly cautious in relying on human ontogeny as a theoretical and empirical base.

Even if this foundation were entirely solid, however, extrapolating to other species could still lead us astray in several ways. Suppose that the mirror task does in fact capture a developmental watershed for our own species such that very young infants who fail the task have little or no conception of self. We have no guarantee that the same developmental watershed would be found for other species that show mirror recognition. Nature may not have arranged things in so orderly a fashion. Similarly, we cannot be sure that sharp discontinuity in self-awareness will exist when we compare species that do with those that do not show mirror recognition. Self-recognition – and self-conception more generally – may not have been attained in a single all-or-none step but rather might be a considerably more incremental evolutionary achievement. Finally, we have no assurance that the mirror task even measures the same thing across species (Mitchell, *SAAH*6). It is at least conceivable that the task taps different conceptions in different species, perhaps a simple bodily conception for some and a far more elaborate psychological conception for others.

In the end, however, the critical test of any scientific framework is whether it leads us to new discoveries. For all the potentially hazardous assumptions embedded within it, the comparative approach has an enviable record in this respect, having already generated important advances in our understanding of self-awareness in various species. Besides, anthropomorphic assumptions are ultimately empirically testable, and doing so represents an important part of the comparative-epigenetic program that drives much of the research described in this volume. In what follows we see a state-of-the-art report on some of the most exciting and innovative comparative work being carried out today, work that will be of great interest not only to students of developmental and comparative psychology but to anyone wanting to know the place of the self in contemporary cognitive science. The study of self-awareness, self-conception, and self-recognition carries along with it a whole raft of provocative questions. The collective efforts brought together in this volume offer a rich set of remarkable observations that move us significantly closer to some definitive, empirically based answers.

References

Amsterdam, B. K. (1972). Mirror self-image reactions before age two. *Developmental Psychobiology, 5,* 297–305.

Asendorpf, J. B., & Baudonnière, P. (1993). Self-awareness and other-awareness: Mirror self-recognition and synchronic imitation among unfamiliar peers. *Developmental Psychology, 29,* 88–95.

Cooley, C. H. (1902/1964). *Human nature and the social order.* New Brunswick, NJ: Transaction.

Dennett, D. (1991). *Consciousness explained.* New York: Little, Brown.

Flanagan, O. (1992). *Consciousness reconsidered.* Cambridge, MA: MIT Press.

Flavell, J. H. (1988). The development of children's knowledge about the mind: From cognitive connections to mental representations. In J. W. Astington, P. L. Harris, & D. R. Olson (Eds.). *Developing theories of mind* (pp. 244–267). Cambridge University Press.

Gallup, G. G., Jr. (1970). Chimpanzees: Self-recognition. *Science, 167,* 341–343.

   (1985). Do minds exist in species other than our own? *Neuroscience and Biobehavioral Reviews, 9,* 631–641.

Gallup, G. G., Jr., McClure, M. K., Hill, S. D., & Bundy, R. A. (1971). Capacity for self-recognition in differentially reared chimpanzees. *Psychological Record, 21,* 69–74.

Harris, P. L. (1991). The work of the imagination. In A. Whiten (Ed.), *Natural theories of mind* (pp. 283–304). Oxford: Blackwell Publisher.

James, W. (1892/1948). *Psychology.* Cleveland, OH: World.

Lewis, C., & Mitchell, P. (Eds.) (in press). *Origins of an understanding of mind.* Hillsdale, NJ: Erlbaum.

Lewis, M., & Brooks-Gunn, J. (1979). *Social cognition and the acquisition of self.* New York: Plenum.

Mead, G. H. (1934/1974). *Mind, self, and society.* Chicago: University of Chicago Press.

Mitchell, R. (1993). Mental models of mirror self-recognition: Two theories. *New Ideas in Psychology, 11,* 295–325.

Moses, L. J., & Chandler, M. J. (1992). Traveler's guide to children's theories of mind. *Psychological Inquiry, 3,* 286–301.

Povinelli, D. J. (1993). Reconstructing the evolution of mind. *American Psychologist, 48,* 493–509.

Searle, J. R. (1992). *The rediscovery of the mind.* Cambridge, MA: MIT Press.

Wellman, H. M. (1993). Early understanding of mind: The normal case. In S. Baron-Cohen, H. Tager-Flusberg, and D. J. Cohen (Eds.), *Understanding other minds: Perspectives from autism* (pp. 10–39). Oxford University Press.

Wimmer, H., & Perner, J. (1983). Beliefs about beliefs: Representation and constraining function of wrong beliefs in young children's understanding of deception. *Cognition, 13,* 103–128.

Acknowledgments

The editors would like to thank the following people and agencies for providing financial and logistically support, and for the encouragement that allowed us to organize the Conference on Self-awareness in Monkeys and Apes: the late Brendan O'Regean of the Institute for Noetic Sciences of Sausalito, California; Dr. Dave Fredrickson, director of the Anthropological Studies Center of Sonoma State University; Dr. Robert Karlsrud, dean of Social Sciences of Sonoma State University; and the Conference Center of Sonoma State University. We also thank the Department of Psychology of Eastern Kentucky University and the Department of Psychiatry at the University of Colorado Health Sciences Center for their support. Our freestanding conference was convened on the Sonoma State University campus in August 1991, with the purpose of bringing together a small number of investigators of various aspects of self-awareness in primates and human children for intensive discussion of a variety of theoretical and methodological issues regarding its study. Participants in the three-day conference met to hear invited presentations and commentaries; we are grateful for the lively and informative participation of all who attended. In addition to those published here, we would like to thank the following conference participants and commentators who did not contribute to this volume: Drs. Bernard Baars, Michael Fehling, John Flavell, Kathleen Gibson, Louis Moses, and Carolyn Saarni. We also thank Mr. Jean Baulu, primatologist and head program leader of the Barbados Primate Research Center, for his generous contribution of the photograph of his lithograph of monkeys with a mirror from his original collection of lithographs (see frontispiece), as well as Dr. Michael Schwibbe of the Deutsches Primatenzentrum in Göttingen for providing us with information about the origin of this illustration. We are additionally grateful to our parents and the numerous mentors, colleagues, and editors who have helped to nurture and shape our various approaches to comparative developmental evolutionary studies through the years. Finally we thank Julia Hough, our editor at Cambridge University Press, Michael Gnat, production editor, and Cary Groner, copy-editor, for their Herculean efforts.

Note added in proof

The alert reader will notice inconsistencies among the various chapters with regard to a variety of issues. They are particularly common in reference to the following topics:

1. criteria for MSR;
2. the number of gorillas displaying mirror self-recognition (MSR);
3. the age of onset of MSR in chimpanzees;
4. the developmental relationship between embarrassment and MSR;
5. the relationship between the development of mirror-aided object localization and MSR;
6. the presence or absence of controls in various studies of MSR in human infants;
7. the age of onset of theory of mind (ToM) in human and nonhuman species;
8. the relationship between imitation and mental representation; and
9. the terminology relating to self-generated feedback from the mirror image (i.e., kinesthetic–visual vs. somasthetic–visual vs. proprioceptive–visual).

In some cases, these inconsistencies may have resulted from lack of access to revisions of other chapters in the volume; in others, from the use of different sources, or from differences in memory or interpretation.

Part I

Comparative and developmental approaches to self-awareness

1 Expanding dimensions of the self: Through the looking glass and beyond

Sue Taylor Parker, Robert W. Mitchell, and Maria L. Boccia

... any animal whatever, endowed with well-
marked social instincts, the parental and filial
affections being here included, would inevitably
acquire a moral sense or conscience, as soon as
its intellectual powers had become as well, or
nearly as well developed, as in man.
— Darwin (1871, p. 472)

Introduction

Darwin's conjecture that morality is an epiphenomenon of intelligence is consistent with increasing evidence that self-awareness depends upon cognitive and affective capacities present in only a few species. The connection between self-awareness and morality, of course, is that conscience is a manifestation of self-awareness. If we follow Darwin's lead and compare the manifestations of self-awareness and their development in species closely related to humans, we may be able to begin to trace the evolution of self-awareness. In order to compare kinds and degrees of self-awareness, of course, we need a system for classifying the phenomenon and a method for diagnosing its manifestations. Ideally the classification system will allow us to identify a broad range of self-related phenomena so that we can compare many species and thereby reconstruct the evolutionary history of self-awareness and self-knowledge.

This volume grew out of a 1991 conference on self-awareness in monkeys, apes, and humans at Sonoma State University in Rohnert Park, California. A major goal of the conference was to bring together investigators who espoused opposing viewpoints on a variety of issues relating to self-awareness. Various chapters in this volume reflect quite different, and in some cases divergent, views regarding the following issues:

1. the capacity of various species for self-awareness, especially whether gorillas, macaques, dolphins, and pigeons display mirror self-recognition (MSR);
2. the nature of cognitive correlates of self-awareness, specifically, the developmental relationship between self-awareness and imitation, object permanence, theory of mind (ToM), and other abilities;

3

3. the influence of various rearing conditions on the capacity for MSR;
4. the validity of various methodologies for discovering self-awareness; and
5. the evolution and adaptive significance of MSR.

Another major goal of the conference was to bring developmental and comparative psychologists together with biological anthropologists to discuss the mutual implications of their theories and methodologies for studying the phenomenon of self-awareness across species. In accord with this goal, the conference highlighted the contributions of two pioneers in studies of self-awareness: Gordon G. Gallup, Jr., the comparative psychologist whose methodology has dominated comparative research on self-awareness for the past 20 years; and Michael Lewis, the developmental psychologist whose studies of the development of self-awareness in human infants and children have set the direction of developmental studies in this domain.

Mirror self-recognition as an index of self-awareness

In 1970 Gallup published his landmark paper on mirror self-recognition in chimpanzees, detailing a controlled methodology for verifying whether or not an animal recognizes itself in a mirror (see Gallup, *SAAH*3).[1] In primates, studies of self-awareness as measured by mirror self-recognition have revealed that, in addition to humans, chimpanzees and orangutans – and possibly gorillas – display this ability, whereas monkeys and prosimians do not. A recent study (Hyatt & Hopkins, *SAAH*15) provides preliminary evidence that bonobos (pygmy chimpanzees) also are capable of MSR. The discovery of MSR in great apes has been a watershed in comparative psychology, especially coming toward the end of a period that denied the existence of mind in animals and humans. After two decades of research, however, the results of Gallup's test are raising new questions.

Following 3 days of presentations, many participants in the conference on which this volume is based agreed that MSR is a far more complex phenomenon than was apparent. Important questions are raised by two sorts of ambiguities in the responses of various species to tests for MSR. On one hand, the mark test gives ambiguous positives: Evidence was presented that both human children and great apes experience considerable instability in the development of mirror self-recognition (see, e.g., Miles, *SAAH*16; Mitchell, 1993; also see Guillame, 1926/1971; Zazzo, 1982), sometimes reaching behind the mirror or calling their own image by the name of a sibling or friend after engaging in the self-directed behaviors, especially mark-directed behaviors, that have been considered definitive proof of MSR. Likewise, macaques, which do not spontaneously engage in other self-directed behaviors in response to mirrors, have apparently engaged in mark-directed behaviors (e.g., see Boccia, *SAAH*23; Thompson & Boatright-Horowitz, *SAAH*22).

On the other hand, the mark test gives ambiguous negatives: Evidence was presented suggesting that great apes and human children who show other signs of self-awareness sometimes fail to engage in mark-directed behaviors in response to mirrors, whether out of self-consciousness or embarrassment (e.g., Lewis, conference presentation; Patterson & Cohn, *SAAH*17; Swartz & Evans, *SAAH*11), or lack of self-recognition. This evidence is particularly significant given the repeated failure of gorillas to pass the mark test (Ledbetter & Basen, 1983; Suarez & Gallup, 1981). These puzzling negative results must be reconciled with videotape evidence for MSR in Koko, the language-trained gorilla, as well as reports of MSR in other gorillas (Law & Lock, *SAAH*20; Parker, *SAAH*19; Patterson & Cohn, *SAAH*17; Swartz & Evans, *SAAH*11).

The gorilla question

As has been indicated, conflicting reports regarding the gorilla's capacity for MSR pose an intriguing problem. Observations of spontaneous mirror behaviors of the language-trained gorilla Koko that suggested MSR (Patterson, 1978, 1984) had been discounted by most but not all investigators because of their anecdotal nature (Povinelli, 1987). Nevertheless, because gorillas are close relatives of the other great apes, all of whom display MSR, and because gorillas are like the other great apes in their capacity to learn symbols (e.g., Patterson, 1980) and use tools (e.g., Maple & Hoff, 1982), investigators have been surprised at the negative results of MSR experiments. Gallup (*SAAH*3) and others (Povinelli, *SAAH*18) have attempted to explain the negative results by arguing that gorillas differ from other great apes in their mental abilities.

The conference provided a forum for the presentation and discussion of videotapes of the language-trained gorillas, Koko and Michael, responding to their mirror images in sham-marking tests as well as in spontaneous encounters (see Patterson & Cohn, *SAAH*17). These videotapes convinced most of the participants including Gallup himself that at least one gorilla, Koko, is capable of MSR (see Gallup, *SAAH*3; Povinelli, *SAAH*18).

Given the close phylogenetic relationship of the great apes, it seems clear that the capacity for MSR must have been present in the common ancestor of the clade, and hence that it must either be present in gorillas or have been lost subsequent to gorilla divergence (Parker, 1991; Povinelli, 1987, *SAAH*18). Povinelli concludes that gorillas have lost the capacity for MSR and that Koko's performance reflects induction through special enriched rearing conditions. An alternative interpretation is that the environments of cross-fostered gorillas and naturally fostered gorillas are equally enriched as contrasted with the environments of solitarily reared laboratory animals. Parker (*SAAH*19) expresses this perspective in her chapter, concluding that gorillas have probably retained the capacity for MSR.

Patterson & Cohn (*SAAH*17) note that the shy, almost secretive nature of gorillas may inhibit overt responses to the mark. Although their observations suggest some lines for future investigation, the question of why gorillas perform so poorly on MSR tests remains unresolved. Swartz & Evans (*SAAH*11) report the case of one Florida zoo gorilla, King, who showed MSR, but also note the failure of the gorillas in their group in Gabon to display MSR, and the lower frequency of interest in mirrors displayed by gorillas as compared to chimpanzees.

The macaque question

The conference was also a forum for discussion of the surprising claim that some macaque monkeys pass the face mark test (see Boccia, *SAAH*23; Thompson & Boatright-Horowitz, *SAAH*22). Boccia suggests that the rarity of this phenomenon in macaques reflects the fact that individual variation is greatest at the margins or limits of a species's abilities. Videotaped sequences of macaque responses to face marks raised several interesting questions about the adequacy of the mark test for assessing the kind of self-awareness seen in great apes. For Gallup (*SAAH*3), they raised questions about the adequacy of the methodologies used in these studies; for Boccia (*SAAH*23), about the adequacy of prior methodology for eliciting MSR is macaques. Like several other investigators, Boccia stresses the importance of long-term exposure of monkeys to mirrors, although the same is not true for apes (see Boysen, Bryan, & Shreyer, *SAAH*13; Parker, *SAAH*19).

Questions regarding the adequacy of the mark test for assessing ape like self-awareness arise in this context because macaques fail to display the suite of cognitive abilities that various investigators have associated with self-awareness: imitation, symbol use, role reversal, and deception. Participants agreed with Gallup that the mark test cannot be the sole criterion for self-awareness because of the possibility that the performance is a fluke, or that it is an artifact of training (see Gallup, *SAAH*3; Thompson & Contie, *SAAH*26).

These videotapes also raised questions about the adequacy of the mark test to diagnose different, nonhominoidlike forms of self-awareness. Although the behavior of monkeys in front of mirrors generally indicates that they fail to recognize themselves (Anderson, *SAAH*21), several investigators have suggested that macaques (Boccia, *SAAH*23; Itakura, 1987; Platt & Thompson, 1985), baboons (Benhar et al., 1975), marmosets (Eglash & Snowdon, 1983), and perhaps some other monkeys may have some kind or degree of self-recognition. In particular, Mitchell (*SAAH*6) as well as Parker and Milbrath (*SAAH*7) argue that the purported absence of self-awareness in monkeys may be an artifact of the inadequacy of our current models and methodologies.

Several perspectives on self are proposed that may provide new insights into more primitive forms of self-awareness than those displayed by the great apes. Mitchell (*SAAH*6) distinguishes perceptual, imaginal, and evaluative aspects of self, and proposes that only the first two aspects may be present in monkeys, whereas all three are present in great apes (although the evaluative form occurs in the great apes only in the most rudimentary form). Parker and Milbrath (*SAAH*7) propose comparative use of a social/developmental perspective focusing on the processes by which imitative and role-playing games develop self-awareness.

A comparative developmental perspective reveals that the mirror-related behaviors reported in macaques parallel several behaviors characteristic of human children that precede facial self-recognition (Boccia, *SAAH*23; Lewis & Brooks-Gunn, 1979; Platt & Thompson, 1985; Thompson & Boatright-Horowitz, *SAAH*22). These include social behaviors (facial and postural displays), body movements, body-directed movements (Boccia, *SAAH*23; Itakura, 1987), mirror-mediated reaching (Anderson, 1986), and mirror-mediated threat responses (Eglash & Snowdon, 1983) as well as other social responses to other monkeys (Boccia, *SAAH*23). This suggests that monkeys may display body self-awareness comparable to human children less than a year old.

In conclusion, the macaque studies point to two additional limitations of the mark test: First, data from the mark test do not in themselves indicate the capacities underlying MSR in human children and great apes. Second, because the test results are dichotomous (an animal either passes or fails), the mark test does not tap whatever capacities for self-awareness might exist in species that fail to display MSR. In response to their desires to overcome these limitations, contributors and other investigators have begun looking to alternative means of identifying self-awareness that will reduce ambiguity and give greater insight into the cognitive correlates of MSR, as well as of providing insights into simpler forms of self-awareness that may exist in monkeys and other mammals.

The dolphin question

Quite serendipitously, the conference also became a forum for discussion of self-awareness in dolphins. Given their popular reputation for intelligence, dolphins are natural candidates for studies of MSR. Although dolphins are mammals, their adaptation to aquatic habitats has involved significant divergence from primates and other nonaquatic mammals in many motoric and information processing capacities. These dolphin-specific patterns challenge researchers who wish to study MSR. Two chapters in this volume report efforts to tailor MSR studies to dolphins; both report that some dolphins display behaviors suggestive of MSR, including contingent head and mouth movements and mark-revealing movements (see Marino, Reiss, & Gallup *SAAH*25; Marten & Psarakos, *SAAH*24).

This result is not surprising given that like great apes, dolphins have large brains, a prolonged period of immaturity, the ability to learn symbol systems (Herman, 1980), engage in shifting coalitions, and display tool use and social imitation (Tayler & Saayman, 1973). Reports that dolphins have individual signature whistles (Caldwell & Caldwell, 1965) and that they may use the signature whistles of other animals referentially suggests that they may also use their own signature whistles self-referentially. If so, they would constitute the only nonprimate to use symbols.[2]

The pigeon question

The significance of MSR as an index of self-awareness in chimpanzees has been challenged in a paper by Epstein, Lanza, and Skinner (1981), which reports success in training pigeons to peck at a spot on their bodies that could only be seen with the aid of a mirror (see Gallup, *SAAH*3; Thompson & Contie, *SAAH*26). On the basis of their results, Epstein et al. suggest that MSR is solely an artifact of training. In their chapter, Thompson & Contie (*SAAH*26) report on the results of their replication of this experiment, as well as on spin-off experiments designed to parse out various factors underlying the pigeons' behavior. These investigators have failed to replicate the results reported by Epstein et al. They attribute their failure to elicit the MSR response in pigeons to that species's dependence on narrowly construed matching rules and their inability to understand symmetry and identity relations understood by chimpanzees and children. Thompson & Contie argue that, although monkeys can be trained to use simple identity relations, they do not show the spontaneous propensity for identity relations that chimpanzees do. They end their discussion with a provocative comment on the significance of the difference between spontaneous, internal motivation and enforced, external motivation in regard to test performances in great apes versus monkeys and pigeons.

Robots and the broader comparative question

Philosophers have been arguing the question of whether machines can be designed to have consciousness. Some, like Searle (1988) have argued that they cannot, while others, like Dennett (1991), have argued that they can. One of the participants in our conference, who did not contribute to this volume, is a systems engineer with experience designing robots for the U.S. Navy. Michael Fehling's research suggests that certain constraints operate in all agents capable of acting purposefully, making decisions, and solving problems, whether these agents be animals or robots.

Among the constraints operating on such agents, also noted by Dennett (1991), is the necessity for distinguishing self-generated actions from other-generated actions. One of the intriguing convergences in the conference was that between Fehling's and Watson's models for the development of self-detection in human infants. Watson's approach (*SAAH*8) provides an interesting model for the design features underlying self-detection, which precedes self-recognition and other forms of self-awareness in human development.

Robot design speaks to one of the greatest challenges facing comparative evolutionary psychologists, that is, the need to develop comparative frameworks capable of identifying simpler forms of self in mammals and even in nonmammalian vertebrates and more distantly related life forms. The chapters by Lewis and by Watson provide tantalizing clues to future directions. Whereas Watson distinguishes self-detection from self-recognition, Lewis distinguishes the machine self from the me-self – the idea of the self. Like Dennett (1991), he points to the biological necessity for distinguishing self from nonself at various levels of organization: the cellular level, the proprioceptive level, and so on. The machine self is ubiquitous among animals, while the idea of the self is apparently unique to great apes, humans, and dolphins.

Use of developmental models in comparative research in self-awareness

The conference was convened on the idea that models from developmental psychology constitute one of the most promising sources for comparative models of self-awareness (see, e.g., Anderson, 1984; Gallup, 1979). Various developmental models of self-awareness, for example those published by Lewis and Brooks-Gunn (1979), Baldwin (1894/1903), and Guillaume (1926/1971), have already stimulated both comparative studies and theoretical models of self-awareness. Lewis and Brooks-Gunn's descriptions of contingent play with facial expressions associated with passage of the face mark test in human infants, for example, have stimulated an anecdotal report that zoo gorillas may recognize themselves in mirrors (Parker, *SAAH*19), and developmental studies of self-recognition in chimpanzees (Boysen, Bryan, & Shreyer, *SAAH*13; Custance & Bard, *SAAH*12; Lin, Bard, & Anderson, 1991). Parker (1991) has also used Lewis and Brooks-Gunn's model and Piaget's imitation series as a basis for the hypothesis that MSR in apes is based on the capacity for facial and gestural imitation. Boccia, in contrast, has used Bertenthal and Fischer's (1978) model for the development of mirror self-recognition in human infants in her experimental paper on self-recognition in macaques (Boccia, *SAAH*23). More recently, Cameron and Gallup (1988) have studied the development of shadow self-recognition in human children, and Boysen, Bryan and Shreyer (*SAAH*13) have used their technique to study this phenomenon in chimpanzees. Mitchell (1993) has used Guillaume's idea of kinesthetic–visual matching as a basis for one of his models of MSR. In contrast, Gallup (e.g., 1983) and other comparative psychologists have suggested that MSR in great apes is associated with a capacity for theory of mind.

While most of the studies of self-awareness in primates that draw on developmental models are nondevelopmental, we are pleased to include several studies of the development of self-awareness in great apes in this volume: The two studies of language-trained great apes draw on longitudinal data on single subjects (Miles, *SAAH*16; Patterson & Cohn, *SAAH*17) as does Gómez's (*SAAH*5) discussion of mutual awareness in gorillas. In contrast, the developmental studies of laboratory animals depend upon cross-sectional data on several animals of various ages (Boysen, Bryant, & Shreyer, *SAAH*13; Custance & Bard, *SAAH*12).

Developmental models are useful in comparative studies when they trace the origins of behaviors from their simplest nonverbal beginnings, and when they entail sequential transformations in which each succeeding ability depends upon the preceding ability (Mitchell, 1987; Parker, 1990). Such models are particularly useful from the comparative perspective when they address individual variation and when they propose causal models.

A comprehensive developmental study of self-awareness in human infants that addresses many of the issues conference attendees raised in regard to comparative studies was published in 1979 by one of the main speakers at the conference, Michael Lewis (and his associate, Jeanne Brooks-Gunn). This large-scale, controlled study assessed manifestations of self-recognition in four modalities:

1. recognition of mirror representations of self;
2. recognition of videotape representations of self and others;
3. recognition of pictorial representations of self and others from photographs; and
4. verbal labeling of pictures of self and others.

In the mirror and videotape modalities, they used the following behavioral measures: attention, facial expression, vocalization, and imitation. In the mirror modality they also distinguished mirror- and self-directed (including mark-directed) behaviors, studying reactions to the self and the mother in counterbalanced face-marked and -unmarked conditions. In the videotape modality they also distinguished contingent play. Two controlled studies were done in each modality.

In the mirror study they found that smiling and cooing remained constant across age groups, body-directed behaviors decreased across the 15–18- to the 21–24-month groups, while silly faces and mark-directed behaviors increased from zero to moderate frequencies across the 9–12- to the 15–18-month group, and continued to increase across the 15–18- to the 21–24-month group. The face-marked condition infants displayed increased interest in the mirror image and increased silly and coy behavior as well as face touching; no face touching was seen in infants younger than 15–18 months, however. Some infants in the older age groups failed to touch the face-mark, but commented on it.

In the videotape study infants responded with contingent play (making faces, sticking out their tongues, and playing peekaboo) to live (on-line) images of themselves, while they responded with imitation to delayed images of themselves and to other infants. The incidence of contingent play peaked between 15 and 24 months. The incidence of contingent play in both the mirror and video studies led the investigators to conclude that "contingent play is not a precursor but rather an indicator of self-knowledge" (Lewis & Brooks-Gunn, 1979; p. 109). The incidence of imitation in these studies led them to a similar conclusion about imitation. On the other hand, contingency play and imitation preceded recognition of specific perceptual features.

In the pictorial study, the investigators measured the infants' visual fixation, positive affect, and spontaneous vocalizations in response to pictures of themselves and others of the same and older ages of both genders, including strangers and familiars. In the verbal labeling study, in which they were asked to label pictures of themselves and others, infants used their own names at 15–18 months, but did not use personal pronouns to refer to themselves or others until 21 or 22 months of age.

Over all, Lewis and Brooks-Gunn found a consistent age trend toward self-recognition by all of their measures. They also found a consistent trend toward concurrent recognition of age and gender categories in others. Although they were unable to definitively identify the "causes" of MSR, analysis of cognitive correlates of self-recognition suggested that contingency play was the most significant correlate of self-recognition, and that sixth-stage object permanence, though necessary, was not sufficient. They also concluded that although contingency mediated the emergence of self-awareness, the ability to recognize the self independent of contingency was the next major milestone in the development of self-awareness. Contingency self-recognition was the origin of the existential self, whereas feature self-recognition was the origin of the objective self.

In the 1979 study and in his subsequent work, Lewis (*SAAH2*) has emphasized the developmental relationship between social emotions and self-awareness: In subsequent work, he notes that the self-conscious emotion of embarrassment arises by 2 years of age, coincident with MSR, while the self-evaluative emotions of shame, guilt, pride, and hubris arise at about 3 years of age, coincident with the internalization of standards of conduct (Lewis et al., 1989). Later cognitive achievements in middle childhood and adolescence lead to the development of consensual, and later principled, morality (e.g., Kohlberg, 1984; Piaget, 1965) (see Table 1.1) Developmental studies confirm Darwin's intuition that morality is a product of cognitive processes operating in a social context.

Imitation and the social matrix of self- and other-awareness

Lewis's work highlights the importance of the social context in the development of self- and other-awareness. Although our conference lacked sociologists and social psychologists, six of the presentations focused on social aspects of self-awareness: Hart and Fegley (*SAAH9*) present evidence that MSR in human infants is preceded developmentally by social imitation specifically (as opposed to object imitation); Gopnik and Meltzoff (*SAAH10*) present developmental data from imitation and theory-of-mind research to show that self-awareness and self-concepts develop hand in hand with other-awareness and concepts of others (also see Lewis & Brooks-Gunn, 1979). Parker and Milbrath (*SAAH7*) elaborate a Meadian model to argue that self-awareness

Table 1.1. *Developmental manifestations of self-awareness in humans in relation to Piaget's stages and periods of development*

Age (and Stage)	Subjective self	Objective self
2–3 months (Stage 2 of the sensorimotor period)	Coordination of primary schema	
3–5 months (Stage 3 of the sensorimotor period)	Self-detection	
9–12 months (Stages 4 and 5 of the sensorimotor period)	Contingent body movements in mirror Use of mirror to locate objects	
15–24 months (Stages 5 and 6 of the sensorimotor period)	Self-conscious behavior (embarrassment)[a] Contingent facial movements in mirror Mark-directed behavior in mirror	Verbal self-labeling Pointing at self in mirror
2–3 years (symbolic subperiod of preoperations)	Possessiveness (symbolic)	Personal pronoun labeling Pictorial self-recognition
3–6 years (preoperations)	Shadow self-recognition Self-evaluative behavior (shame and pride) Theory of mind	Self-adornment Authoritarian morality Role reversal
6–11 years (concrete operations)		Consensual morality
11 years– (formal operations)		Principled morality

[a] Reported in Lewis (*SAAH*2); reported as occasionally occurring in 12-month-old infants in Amsterdam (1972) and Lewis et al. (1989).

develops through make-believe and role-playing games in infancy and childhood, and use this model to diagnose the levels of self-awareness in monkeys and great apes. Custance and Bard (*SAAH*12) argue for the importance of social "scaffolding" for the development of imitation, but not for MSR. Itakura (*SAAH*14) reports the results of an experiment on symbolic expression of possession among chimpanzees. The social context of development of self-awareness has important implications for methodologies of comparative studies.

These contributors express notably differing views regarding the role of object permanence, imitation, and theory of mind in the origins of MSR. Boccia (*SAAH*23) follows Fischer and Bertenthal in using object permanence stages as a general index to sensorimotor intelligence without, however, claiming that it is responsible for MSR. Gergely (*SAAH*4) presents a detailed argument in favor of the idea that object permanence is responsible for MSR. He bases his argument on the idea that sixth-stage object permanence involves deductive inferences. Parker (1991) and Mitchell (1993), like Lewis and Brooks-Gunn (1979) consider object permanence to be a necessary but insufficient condition of MSR. Parker (1991) argues for the causal role of trial-and-error (visual–kinesthetic) matching through (Piaget's fourth or fifth-stage) imitation of novel actions, while Mitchell (1993) argues that kinesthetic–visual matching is the causal basis for bodily and facial imitation, MSR, and other cognitive abilities such as planning and pretense. Gopnik and Meltzoff (*SAAH*10) argue that early imitation occurs virtually from birth and hence precedes MSR. They also analogize understanding of the other's and the self's body through imitation with the understanding of other's and self's mind through theory of mind and trace the origins of ToM from imitation.

In contrast to those contributors, who argue that imitation plays a central role in MSR, Custance and Bard (*SAAH*12) argue that imitation is not necessary for – and does not developmentally precede – MSR in chimpanzees. They argue that 6th stage sensorimotor mental representation is necessary for MSR (but they do not subscribe to Piaget's [1952] notion that representation arises out of interiorized imitation). They base their argument on their failure to elicit imitation in a young chimpanzee who had already displayed MSR for some months before he began to display imitation.

Gallup (1983, *SAAH*3) and Marino et al. (*SAAH*25) suggest that MSR implies theory of mind. Gopnik and Meltzoff's review of developmental studies of human children, however, suggests that ToM develops at 4 years of age or later, considerably after MSR (Gopnik & Meltzoff, *SAAH*10), and hence cannot be a correlate of MSR as Gallup suggests. Gergely (*SAAH*4) and Mitchell (1993) elaborate the evidence against Gallup's suggestion that MSR implies theory of mind.

Comparative methodologies in various settings

To the degree that methodologies arise in the service of paradigms, they embody the presuppositions of those paradigms and their models. Each discipline involved in studies of self-awareness has its own paradigm and methodologies. Methodologies in comparative studies have grown out of the behaviorist paradigm, which often presupposes universal learning abilities interacting with external contingencies that can be manipulated by the experimenter in a systematic manner using suitable controls. Comparative psychologists and many developmental psychologists are laboratory scientists who have been trained to do controlled experiments, but the nature of their experimental designs vary with the nature of their theories and models. Other developmental psychologists and comparative psychologists, as well as many anthropological and biological primatologists, have been trained in natural history observation methods and are used to studying spontaneous behaviors.

They often test hypotheses by methods of controlled comparisons rather than direct experimentation (e.g., Martin & Bateson, 1986).

Differences in disciplinary training entail different presuppositions regarding what are reasonable questions, what are reasonable contexts for behaviors, what are reasonable controls, and hence what are significant data. Some of these differences are reflected in the differing interpretations of contributors to this volume.

Accordingly, behaviors pertinent to self-awareness have been studied in several settings using various methodologies:

1. in formally controlled laboratory studies of individually caged and group-caged animals (e.g., Gallup, 1970, 1977; Ledbetter & Basen, 1982; Suarez & Gallup, 1981; Thompson & Boatright-Horowitz, SAAH22);
2. in informally controlled studies of cross-fostered – usually language-trained – great apes (e.g., Miles, SAAH16; Patterson & Cohn, SAAH17);
3. in observational and experimental studies of zoo animals (e.g., Law & Lock, SAAH20; Lethmate & Dücker, 1973; Parker, SAAH19; Swartz & Evans, SAAH11); and
4. in observational and experimental studies of semi–free-ranging animals (Platt & Thompson, 1985).

Each setting and context entails certain advantages and disadvantages.

Studies in laboratory settings allow the greatest degree of rigor; they also entail the greatest perturbation of the animals and therefore produce the least naturalistic data. In addition to experiencing disruptive testing procedures, many laboratory subjects may have been raised in spatially and/or socially restricted environments. The canonical procedure in these studies has been to expose an animal to a mirror for many days while recording its behaviors, anesthetize the animal and mark its brow with an odorless dye, then expose it to a mirror again, recording its behaviors under the marked condition. Mark-directed touching of the dyed area is considered diagnostic of mirror self-recognition (Gallup, 1970).

Studies of cross-fostered animals generally occur in highly enriched environments in human terms, that is, in terms of language training and/or object-manipulation training. When the mark test has been used in this setting it has been controlled with "sham marking," which causes less perturbation of the animal's routine than anesthesia as used in Gallup's procedure. Such studies have the advantage of producing intensive longitudinal data on development across a range of behavioral topics (see Miles, SAAH16; Patterson & Cohn, SAAH17). It is important to remember that the data they provide must be interpreted as outcomes of cross-fostering rather than natural fostering. Whether qualitatively different abilities are engendered by human tutoring or "enculturation" (e.g., Miles, SAAH16; Povinelli, SAAH18; Premack & Premack, 1983; Tomasello, Kruger, & Ratner, 1993) remains to be seen, however.

Studies in zoo settings offer the least opportunity for controlled experiments, but also the least perturbation of the animals. Although zoo animals vary in the size of their environments and in the composition of their social groups, in recent years they have often been housed in quasi-naturalistic breeding groups. Such animals are of special interest in regard to their cognitive performances because they are more similar in their social experiences to their wild-living relatives. Free-ranging animals, of course, offer the closest analogue to wild-living populations.

As various chapters in this volume illustrate, developmental studies of self-awareness in human infants also suggest a variety of methodological approaches to comparative studies of self-awareness in monkeys and apes. These approaches suggest ideas for augmenting the rigorous single measure of Gallup's (1970) face mark test with multiple behavioral measure; for example:

1. other self-directed behaviors (Boysen et al., *SAAH*13; Hyatt & Hopkins, *SAAH*15),
2. embarrassment (Patterson & Cohn, *SAAH*17),
3. contingent play with body movements and facial expressions (Custance & Bard, *SAAH*12; Parker, 1991),
4. verbal self-labeling (Miles, *SAAH*16; Patterson & Cohn, *SAAH*17),
5. picture self-recognition (Itakura, *SAAH*14), and
6. shadow recognition (Boysen et al., *SAAH*13).

Language-trained apes provide an opportunity to apply all of these measures to nonhuman primates (see Miles, *SAAH*16; Patterson & Cohn, *SAAH*17). Multiple measures of MSR are highly desirable for a variety of reasons including avoidance of measurement error inherent in a given measure (Swartz, 1990), and providing a broader context for identifying false negatives and false positives.

These measures have the added advantage that they correlate with such other traditional cognitive and affective developmental measure as stages of object permanence, causality, imitation, and symbolic play, thereby allowing various stages of self-awareness to be placed in a larger cognitive/affective framework. In addition, as spontaneous behaviors, they have the advantage of being observable both inside and outside the laboratory setting.

The distinction between spontaneous behaviors and those elicited through training is crucial: The "same behaviors" may be mediated by profoundly different mechanisms depending on whether they occur spontaneously in high frequencies or only after extensive training. When behaviors are elicited through shaping and external reinforcement they may reveal more about the capacity of the experimenter than the capacity of the subject of the experiment (Thompson & Contie, *SAAH*26).

The question of the validity of various measures and controls in MSR experiments is central to the investigation of self-awareness. Some (e.g., Gallup, *SAAH*3) emphasize control conditions in the marking itself, demanding anesthesia, and single measures rather than repeated measures as the definitive control, while others (e.g., Miles, *SAAH*16; Patterson & Cohn, *SAAH*17) emphasize the importance of avoiding trauma to the subject, and argue that sham marking is an adequate control to ensure that the subject is ignorant of being marked. Others emphasize control conditions for mirror exposure, presenting animals with nonreflective surfaces of the same size and shape of the mirror (e.g., Hyatt & Hopkins, *SAAH*15). Ironically, the attempt to provide a measure of baseline behaviors unaffected by the mirror – by presenting a nonreflective surface – runs into trouble because it fails to provide any salient stimulus at all. The behaviors animals exhibit in this condition represent one tiny sample of all their possible behaviors. Recording their behavior over similar time spans in a variety of social and nonsocial contexts would be a better control. Another alternative would be to parse out various affordances of the mirror (Loveland, 1986) and present them separately. If an investigator is seeking to determine whether contingency testing behaviors occur in response to the animal's own image, then various videotape conditions provide an appropriate control: The responses of animals to live (on-line) videotapes of their own behavior versus their responses to delayed ones, or to those of other animals, provide control conditions that cannot be provided by a nonreflective surface (see, e.g., Law & Lock, *SAAH*20; Lewis & Brooks-Gunn, 1979; Marten & Psarakos, *SAAH*24).

Problems of observational and experimental research design in comparative studies of primate mentality are minor in contrast to problems in comparative studies of dolphins, pigeons, and other nonprimate species that display markedly different sensory systems. Although dolphins have evolved a highly specialized sonar system, they also rely on vision and audition. It is important to remember in this context, moreover, that self-awareness as indicated by MSR, picture self-recognition, verbal self-labeling, and other more complex manifestations, is a multimodal achievement that differs markedly from such simpler phenomena as auditory self-recognition (which in some birds is mediated by a single neuron [Margoliash & Fortune, 1992]) and proprioceptive feedback (see Mitchell, 1993, and *SAAH*6).

Although we believe that research into self-awareness can benefit from the interplay of differing methodologies used in various settings, we also believe that researchers need to pay greater attention to the social contexts and relationships in which self-awareness develops under natural conditions. Quasi-naturalistic and laboratory data often suggest relationships that can be studied only in free-ranging populations. Studies of imitation in human children and ex-captive orangutans, for example, have shown the importance of model selection and action selection, which in turn suggest inadequacies in laboratory research on imitations that fail to take these factors into account (Russon & Galdikas, 1992).

Conclusions: Defining the parameters of the self

Lewis opened the conference on self-awareness by posing the following questions: "What is a self? How do we measure a self? What does a self do?" Several contributors to this volume have addressed his second and third questions, but only two have undertaken to answer Lewis's first question, "What is a self?"

Lewis himself has answered this question (*SAAH*2) by proposing a distinction between two levels of self that are often confused: "the machine self," the greater part of the self, which knows without knowing that it knows, and the "idea of me," the smaller part – included in the machine self – which knows that it knows. Taking a developmental perspective, Lewis points out that "although many features of the self exist early – for example, the newborn must have some form of self-recognition as seen in imitation – the idea of the me is not developed until somewhere in the middle of the second year of life." Lewis's model of the machine self applies not only to very young human infants, but to all life forms, whereas the idea of the me is found only in humans and great apes of a certain age.

Mitchell (*SAAH*6) has undertaken to answer this question by surveying a broad range of concepts of self from scholars in a variety of disciplines, including not only psychology, anthropology, and sociology, but also psychiatry, philosophy, and the cognitive sciences. By chronicling the multiplicity of dimensions of self distinguished by various scholars, he suggests a broad context for thinking about self-awareness and a rich appreciation for the complexity of self-phenomena in our own species. Whether explicitly developmental or not, all the various concepts and dimensions of self he describes project onto the development of self (see, e.g., Damon & Hart, 1988, for a developmental treatment of James's concepts of the subjective and objective self).

This volume represents a tentative step toward answering Lewis's opening questions. It also represents a step toward bringing together investigators from various disciplines to present and debate a variety of models regarding self-awareness and to stimulate new lines of research into self-awareness in animals and humans. We hope that the papers in this volume and the contacts that were forged at the conference will stimulate continuing comparative research on self-awareness that incorporates theories and methods from models of human development within an evolutionary framework.

Notes

1. Throughout this volume, we shall cross-reference papers herein by the abbreviation *SAAH* and the chapter number.
2. Personal identification calls have also been detected in crows; individuals give a typical number of caws in contact calls that apparently uniquely identify them to other crows (Thompson, 1969). Cognitive mechanisms underlying such identity displays are unknown and may differ among species.

References

Anderson, J. R. (1984). The development of self-recognition: A review. *Developmental Psychobiology, 17*(1), 35–49.

(1986). Mirror-mediated finding of hidden food by monkeys (*Macaca tonkenana* and *M. fascicularis*). *Journal of Comparative Psychology, 100,* 237–242.

Baldwin, J. M. (1894/1902). *Social and ethical interpretations of social life.* New York: MacMillan.

Benhar, E., Carlton, P., & Samuel, D. (1975). A search for mirror-image reinforcement and self-recognition in the baboon. In S. Kondo, M. Kawai, & A. Ehara (Eds.), *Contemporary primatology* (pp. 202–208). Tokyo: Japan Science Press.

Bertenthal, B. I., & Fischer, K. W. (1978). Development of self-recognition in the infant. *Developmental Psychology, 14,* 44–50.

Caldwell, M. C., & Caldwell, D. K. (1965). Individualized whistle contours in bottlenosed dolphins (*Tursiops truncatus*). *Nature, 207,* 434–435.

Cameron, P. A., & Gallup, G. G., Jr. (1988). Shadow self-recognition in human infants. *Infant Behavior and Development, 11,* 465–471.

Chevalier-Skolnikoff, S. (1982). A cognitive analysis of facial behavior in Old-World monkeys, apes and human beings. In C. T. Snowdon, C. H. Brown, & M. R. Petersen (Eds.), *Primate communication.* Cambridge University Press.

Damon, W., & Hart, D. (1988). *Self-understanding in childhood and adolescence.* Cambridge University Press.

Darwin, C. (1871). *The descent of man.* New York: Modern Library.

Dennett, D. (1991). *Consciousness explained.* Boston: Little, Brown.

Eglash, A. R., & Snowdon, C. T. (1983). Mirror-image responses in pygmy marmosets. *American Journal of Primatology, 5,* 211–219.

Epstein, R., Lanza, R. P., & Skinner, B. F. (1981). "Self-awareness" in the pigeon, *Science, 212,* 695–696.

Gallup, G. G., Jr. (1970). Chimpanzees: Self-recognition. *Science, 167,* 86–87.

(1977). Self-recognition in primates. *American Psychologist, 32,* 329–338.

(1979). Self-recognition in chimpanzees and man: A developmental and comparative perspective. In M. Lewis & M. Rosenblum (Eds.), *Genesis of behavior, Vol. 2: The Child and its family* (pp. 107–126). New York: Plenum.

(1983). Toward a comparative psychology of mind. In R. L. Mellgren (Ed.), *Animal Cognition and Behavior* (pp. 473–510). New York: North Holland.

Guillame, P. (1926/1971). *Imitation in children.* Chicago: University of Chicago Press.

Herman, L. M. (1980). Cognitive characteristics of dolphins. In L. M. Herman (Ed.), *Cetacean behavior* (pp. 363–429). New York: Wiley.

Itakura, S. (1987). Use of a mirror to direct their responses in Japanese monkeys (*Macaca fuscata fuscata*). *Primates, 28,* 343–352.

Kohlberg, L. (1984). *Essays in moral development, Vol. 2: The psychology of moral development.* San Francisco: Harper & Row.

Ledbetter, D. H., & Basen, J. A. (1982). Failure to demonstrate self-recognition in gorillas. *American Journal of Primatology, 2,* 307–310.

Lethmate, J., & Dücker, G. (1973). Untersuchungen zum Selbsterkennen im Spiegel bei Orang-Utans und einigen anderen Affenarten. *Zeitschrift für Tierpsychologie, 33,* 248–269.

Lewis, M., & Brooks-Gunn, J. (1979). *Social cognition and the acquisition of self.* New York: Plenum.

Lewis, M., Sullivan, M., Stanger, C., & Weiss, M. (1989). Self development and self-conscious emotions. *Child Development, 60,* 148–156.

Lin, A. C., Bard, K. A., & Anderson, J. R. (1992). Development of self-recognition in chimpanzees. *Journal of Comparative Psychology, 106,* 120–127.

Loveland, K. A. (1986). Discovering the affordances of a reflecting surface. *Developmental Review, 6,* 1–24.

Maple, T. L., & Hoff, M. P. (1982). *Gorilla behavior.* New York: Van Nostrand Reinhold.

Margoliash, D., & Fortune, E. S. (1992). Temporal and harmonic combination sensitive neurons in the Zebra finch's Hvc. *Journal of Neuroscience, 12*(11), 4309–4326.

Martin, R., & Bateson, P. (1986). *Measuring Behaviour.* Cambridge University Press.

Mitchell, R. W. (1987). A comparative-developmental approach to understanding imitation. In

P. P. G. Bateson & P. H. Klopfer (Eds.), *Perspectives in ethology* (Vol. 7, pp. 183–215). New York: Plenum.

(1993). Mental models of mirror self-recognition: Two theories. *New Ideas in Psychology, 11*, 295–325.

Parker, S. T. (1987). The origins of symbolic communication: An evolutionary cost/benefit model. In *Symbolism and knowledge*. Geneva: Cahiers de la Fondation Archives Jean Piaget, No. 8.

(1990). Origins of comparative developmental evolutionary studies of primate mental abilities. In S. T. Parker & K. G. Gibson (Eds.), *"Language" and intelligence in monkeys and apes* (pp. 3–64). Cambridge University Press.

(1991). A developmental approach to the origins of self-recognition in great apes and human infants. *Human Evolution, 6*, 435–449.

Patterson, F. (1978). Conversations with a gorilla. *National Geographic, 154*, 438–465.

(1980). Innovative uses of language by gorilla: A case study. In K. E. Nelson (Ed.), *Children's Language* (Vol. 2). New York: Gardner.

(1984). Self-recognition by *Gorilla gorilla gorilla*. *Gorilla, 7*(2), 2–3. Woodside, CA: The Gorilla Foundation.

Piaget, J. (1952). *Origins of intelligence in children*. New York: Norton.

(1965). *The moral judgment of the child*. New York: Free Press.

Platt, M., & Thompson, R. (1985). Mirror responses in a Japanese macaque troop. *Primates, 26*, 300–314.

Povinelli, D. J. (1987). Monkeys, apes, mirrors and minds: The evolution of self-awareness in primates. *Human Evolution, 2*(6), 493–507.

Premack, D., & Premack, A. (1983). *The mind of an ape*. New York: Norton.

Russon, A., & Galdikas, B. (1992). *Imitation in rehabilitant orangutans: Model and action selectivity*. Paper presented at the 14th congress of the International Primatological Society, Strasbourg, France.

Searle, J. (1988). Turing the Chinese room. In T. Singh (Ed.), *Synthesis of science and religion: Critical essays and dialogues*. San Francisco: Bhaktiventa Institute.

Suarez, S. D., & Gallup, G. G. (1981). Self-recognition in chimpanzees and orangutans, but not gorillas. *Journal of Human Evolution, 10*, 175–188.

Swartz, K. (1990). The concept of mind in comparative psychology. *Psychology: Perspectives and Practice* (Vol. 602 of the *New York Academy of Sciences*, pp. 105–111).

Tayler, C. K., & Saayman, G. S. (1973). Imitative behaviour by Indian bottlenose dolphins (*Tursiops aduneus*) in captivity. *Behaviour, 44*, 286–298.

Thompson, N. (1969). Individual identification and temporal patterning in the cawing of crows. *Communications in Behavioral Biology, 3*, 1–5.

Tomasello, M., Kruger, A. C., & Ratner, H. H. (1993). Cultural learning. *Behavioral and Brain Sciences, 16*, 495–552.

Zazzo, R. (1982). The person: Objective approaches. In W. W. Hartup (Ed.), *Review of child development research* (Vol. 6, pp. 247–290). Chicago: Univerity of Chicago Press.

2 Myself and me

Michael Lewis

Introduction

The concept of self has interested the Western mind from the beginning of recorded history. How we approach our understanding of ourselves has varied, both in method and study. Historically, we have been concerned with our relationship with God, and that of ourselves to others, as well as with the idea of identity. In this chapter I have chosen to focus on the general problem of the self. Three questions are addressed:

1. What is a self?
2. How do we measure a self?
3. What does a self do?

In the first, I try to distinguish between many of the current ideas about the self, bringing to bear the idea of levels of self. Of importance here is a particular aspect of the self, perhaps the most restrictive: that of the self as aware of itself. In the second question, I suggest that the measurement of self is highly dependent upon what is meant by self. Here I suggest that the self, aware of its self, can best be measured, especially in nonverbal creatures, through self-recognition. Finally, in addressing the third question, I consider the issue of the social and the feeling self. These three questions address what for me are some of the critical issues in the study of self, and in particular, in the study of its development.

What is a self?

As I sit here at the Hydro Majestic Hotel in the Blue Mountains of Australia, sipping a glass of beer, the late afternoon sun shines and warms my skin. The taste of the beer is cold and sweet. At this moment I have no trouble in recognizing myself. I know where I am and why I am here. I can tell the way I smell, and when I speak I can hear my voice; it sounds like me. The sun's warmth against my skin is comfortable. Sitting here, I can think about myself, I can wonder whether I will find a restaurant tonight. I wonder about my appearance. Is my hair combed properly? As I get up to leave, I pass a mirror; there I see myself, the reflected surface of my being. "Yes, that is me," I say, fixing my windblown hair.

20

I know a great deal about me. One of the things I know is how I look; for example, there is a scar above my left eyebrow. I look familiar to myself, even though I have changed considerably with age. Pictures taken of me 30 years ago look like me. Nevertheless, I know when I look at myself in the mirror I will not look as I did then; my hair and beard are now white, not the brownish yellow they were then. My face will be less smooth, and no matter how much I would like to see a very thin man in the mirror, I know now when I look I will have to tuck in my belly in order to look like anything I wish to be.

I know many people might argue that the concept of self is merely an idea. I, for one, would argue that this idea of self is a particularly powerful one; it is an idea with which I cannot part. It is one around which a good portion of the network of many of my ideas center. This is not to say that what I know about myself is all I know. In fact, this idea of myself is only one part of myself; there are many other parts of which I do not know. There are the activities of body: the joints moving, the blood surging, the action potentials of my muscle movements as well as the calcium exchange along the axons. There are many other things that are part of me of which I do not know. True, I may know of them as ideas, but I cannot find them in myself. I have no knowledge of a large number of my motives – organized, coherent thoughts and ideas that we sometimes call unconscious – and that control large segments of my life. I have no knowledge of how my thoughts occur or why I feel one way or another. Nevertheless, I know that I think and feel even without this knowledge.

The claim has been made that it is possible to know of all things related to the self; for example, the yogi's belief in the control of much of our autonomic nervous system function. Although it might be true that I could know more of some parts of myself if I chose to, it is nonetheless the case that what is known by myself is greater than what I can state I know.

If such facts are true, then, it is fair to suggest a metaphor of myself. I imagine myself to be biological machine that is an evolutionarily fit complex of processes: doing, feeling, thinking, planning, and learning. One aspect of this machine is the idea of me. If I were an Eastern mystic, I would perhaps draw this idea as an eye, like some all-seeing thing. Not being such a metaphysician, I prefer the metaphor of a protoplasmic mass, perhaps resembling the frontal lobes. It is this mass that knows itself and knows it does not know all of itself! The me that recognizes me in the mirror is located in that particular mass. The self, then, is greater than the me, the me being only a small portion of myself. If the metaphor drawn is unsatisfactory, the difference between self and me can be understood from an epistemological point of view. The idea that I know is not the same as the idea that I know I know. The me aspect of the self that I refer to is that which knows it knows.

The failure to make the distinction between these features of self can be blamed for much confusion when studying the issue of development. The distinction between self and me, or between knowing and knowing I know, involves two aspects of self. If we do not confuse *knowing* with *knowing-I-know*, then the argument around the issues of the developmental sequence becomes clearer. To anticipate the argument, although many features of the self exist early – for example, the newborn must have some form of self-recognition, as seen in imitation – the idea of me is not developed until somewhere in the middle of the 2nd year of life. As I will show, there is no incompatibility between my view and those who see early aspects of self. There are earlier aspects of self, but the knower who knows is not one of them because it is likely to emerge only in the middle of the 2nd year.

Two early features of the self are well known and have been given considerable attention: These are self–other differentiation and the conservation of self across time and place. Certainly, by three months, and most likely from birth, the infant can differentiate itself from other. Self–other differentiation also has associated with it a type of recognition. This type of recognition – the self–other differentiation – is part of the hardware of any complex system. This differentiation does not contain, nor is it analogous to, the idea of me. For example, T-cells can recognize and differentiate themselves from foreign protein. The rat does not run into a wall but knows to run around it. The newborn infant recognizes intersensory information. These examples indicate that both simple and complex organisms possess the ability to differentiate self from other. We should not expect this aspect of self to be a differentiating feature when we compare widely different organisms; all organisms have this capacity. The single T-cell must organize the information needed to distinguish self from other in a less complex manner than the multicell creature having a central nervous system. Nevertheless, in both cases, self–other differentiation must exist in order for the organism to survive. This is built into the biological machinery and is not predicated on the idea of me.

Another aspect of the machinery of self has to do with the idea of conservation of the self across time and place. What formal aspects imply conservation? It could entail responding the same way to similar occurrences, or it could mean adapting to them, as in habituation. Conservation of the self is a rather troubling concept. The ability to maintain conservation appears in all creatures. For example, the ability to reach for an object requires some machinery of self, as it requires one to act as if there is another in other space–time. This relationship to the other in space–time has consistency. Habituation, if used as a measure of conservation, is not terribly helpful. Very simple organisms are able to habituate to redundant information. This suggests that in almost all organisms, from one-cell organisms on, biological machinery exists that is capable of maintaining information and utilizing that information in making a response. There is also self-conservation in the process we call identity: "This is me regardless of how I look." This type of conservation, unlike the former, requires more complex processes, those likely to involve the idea of me. Thus, from both a phylogenetic and an ontogenetic perspective, there are different types of conservation.

These two features – self–other and self-conservation – appear relatively early in the child's life. They are based on the biological machinery of the self and do not require the idea of me; they are simple aspects of the functioning of a complex process. All complex creatures, by definition, contain many features (Von Bertalanffy, 1967). These features or parts must be in communication with one another. Moreover, in order for these features to work, they need to be interacting with each other and with the environment. In order to be in interaction with the environment, they need to be able to differentiate self from other. The processes that allow this to occur are multidetermined. They do not require the idea of a me. That these phenomenon, self–other and self-conservation, occur early does not require that the idea of me exists early. Moreover, the idea of me is not needed to affect an organism's actions in the world.

The biological machinery of a self continues to operate until about the middle of the 2nd year of life. It is at this point in ontogeny that the child acquires the idea of me. The acquisition of this idea is accompanied by the development of personal pronouns such as "me" or "mine." It is further seen in the child's behavior toward others. The "terrible 2's" reflect the child's "test of wills." Prior to this point, the child is relatively passive to the adult caregiver's desire. The negativism exhibited by the child at this point would appear to be a reflection of the child's development of the idea of me, in particular the idea that *me does not want to do what you want me to do*. This idea of me is further seen in children's response to distorted images of themselves, as well as in self-recognition. These behaviors – the advent of personal pronouns and the development of will and self-recognition – reflect a new milestone in the child's self-development. At the same time, the biological machinery of self and the processes it controls continue to exert influence on the child's behavior and are further elaborated as the child learns more and interacts with its environment.

In the latter half of the 2nd year of life, the child acquires the idea of me. This concept is an addition to the earlier acquired processes. In this regard, it is interesting to note the speculation on the development of the frontal lobes of the neocortex. It is believed that the frontal lobes achieve some maturity at the same time as the idea of me emerges, indicating a possible link between behavior and brain function and, in particular, a connection between the idea of me and the emergence of a new brain region.

From a phylogenetic point of view there seem to be few creatures capable of acquiring this idea of a me, even though their self systems are well developed. It appears as if the great apes, chimpanzees, orangutans, and gorillas are capable of this idea; other than these animals, there is little support for the belief that any other creature is capable of this achievement (Gallup, 1977). For the human child, visual recognition of the self appears to occur no earlier than 15 months and, except for autistic children and children who do not possess a mental age of 15–18 months, it is achieved by all children by 24 months.

The ontogenetic and phylogenetic coherences found to date support the idea that in order to understand the concept of self, we need to disentangle the common term *self* into at least two aspects. These I have called the *machinery of the self* and the *idea of me*. They have been referred to by other terms, for example, I have referred to the idea of me as objective self-awareness and the machinery of self as subjective self-awareness (Lewis, 1990, 1991, 1992). They both are biological events. The same objective–subjective distinction has been considered by others (see, for example, Duval & Wicklund, 1972). In any consideration of the concept of self, especially in regard to adult humans, it is important to keep in mind that both aspects exist. There is, unbeknownst to us most of the time, an elaborate complex of machinery that controls much of our behavior, learns from experience, has states, and affects our bodies, most likely including what and how we think. The processes are, for the most part, unavailable to us. What is available is the idea of me.

To reiterate, there are two aspects or levels of self. Through much of my daily life, I am confronted by these two aspects. Much of my motor action. although initially planned, is carried out by the machinery of my body, which includes, by definition, self-regulation and self–other differentiation. The same, of course, is true of thinking. A me is necessary to formulate (at least sometimes) what to think about; but a me does not appear to be involved in the processes that actually carry out the task of thinking. Consider this example; we give a subject the problem of adding a 7 to the sum of 7's that precede it (e.g., $7 + 7 = 14 + 7 = 21 + 7 = 28$ etc.). It is clear that a person carrying out this task cannot watch him- or herself do the arithmetic. One aspect of the self, the me, has set up the problem; another aspect, the machinery of self, solves it.

Both of these features of self are the consequences of biological processes. Recent work by LeDoux (1989) also points to specific brain regions that may be responsible for different kinds of self-processes. Working with rats, LeDoux found that even after the removal of the auditory cortex, the animals were able to learn to associate an auditory signal and shock. After a few trials, the rat showed a negative emotional response to the sound even though its auditory cortex had been removed. These findings indicate that the production of a fear state is mediated by subcortical, probably thalamic–amygdala, sensory pathways. Similar findings have been reported in humans, suggesting that states can exist without one part of the self experiencing them. Weiskrantz (1986), among others, has reported on a phenomena called "blindsight." Patients have been found who lack a visual cortex, at least in one hemisphere. When they are asked if they can see an object placed in their blind spot, they report that they cannot see it; that is, they do not experience the visual event. The self reflecting on itself, the me, does not see. When, however, they are asked to reach for it, they show that they have the ability to reach, at least some of the time, for the object. Thus, they can "see" the event, but cannot experience their sight! These findings, as well as the work on split brain, suggest that separate brain regions are responsible for the production and maintenance of both the machinery of self processes and the idea of me (see also, Tulving, 1985, for a similar analysis of memory).

Although many things are meant when we study self systems, one aspect of the system that is unique and may exist only in higher primates and in children past 2 years of age, is the idea of me. This must be differentiated from self as related to the physical machine. If we do not make this distinction we are in danger of making logical errors.

For example, the claim that early infant abilities reflect this idea of me is seen in the work on newborn and early childhood imitation. There is no reason to assume that intersensory integration or early imitation requires the idea of a me, although it certainly involves some aspect of the machinery of a self system. This confusion of the different levels or types of selves is made more difficult by several of the psychoanalytic writers. For example, Kernberg (1976, 1980), Stern (1985), and Lacan (1968) all have claimed that the young infant is capable of experiencing anxiety over its nonexistent self. This is very similar to Rank's (1929) notion of birth anxiety. Freud (1936/1963), in his critique of Rank, pointed out the difficulty of such views. For Freud, anxiety is a signal and, as such, has to be experienced. Only the ego can experience it; "The id cannot be afraid as the ego can, it is not an organization and cannot estimate situations of danger" (p. 80). From our perspective, the dangers of confusing the machinery of self for the idea of me reside in the fact that anxiety over nonexistence, as an adult might experience it, cannot occur to an organism having no idea of a me, nor awareness of its own existence. It is not possible for an organism to be anxious about its existence prior to possessing the capacity to think about itself as existing; that is, prior to having the idea of a me.

This problem of confusing machine self with the idea of me leads to other difficulties as well. The *I am* of intersubjectivity in early infancy (see Emde, 1983; Stern, 1985) does not seem possible from the adult perspective because it too depends on the idea of me. For example, interactive behavior between a 3–6 month-old infant and its mother might be viewed as an early example of intersubjectivity. The ability of family members to share experiences, to match, align, or attune their behavior to each other, might reflect intersubjectivity; it might as well reflect much simpler self-processes. These processes may not involve the idea of me, but instead indicate simple rules of attention getting and holding. For example, it is more efficient to stop moving or talking when someone else is talking if the goal is to pay attention to them. Thus, the action of one member of a dyad may terminate the action of another member for reasons other than those implied by intersubjectivity.

We need to be able to maintain the idea of levels of self, the idea of a machine self – common to newborn creatures and to a wide variety of non-human, even single cell organisms – and the idea of me, which emerges about 2 years of life in the human child, and which is shared by only a few other species. Without careful definition of what we mean by self we will not appreciate the complexity of our concept and will be unable to agree how to measure it.

How do we measure a self?

Measurement issues follow from the constructs we make. If we wish to narrow our definition of self and restrict it to the idea of me, then many purposed forms of measuring self are no longer applicable. Thus, for example, early imitation, intersensory integration, and coordination between infant and mother all cease to be adequate measures of this construct. What we need look for are measures of the idea of me.

The idea of me requires, for the most part, language capacity. If we believe that the emergence of this idea occurs before 2 years of age, we will have difficulty using language as a means of measuring the me. In an adult or older child we could say, "Who are you?" or "Tell me something about yourself," or "Tell me something that you know that others don't know." Alternatively, following R. D. Laing (1970), we could see whether the child understands statements such as "I know, you know, that I know, where you put your teddy bear." As is readily understood, all of these questions imply some idea about me.

Without language, however, the child will have trouble explaining this idea to us. One alternative is to require, without using language, that the child do certain tasks and see if they can be done. If the child understands the task given, it is possible to demonstrate that the child has the idea of me even though it does not have language. Thus, for example, in the work on deception (Lewis, Sullivan, Stanger, & Weiss, 1989), and in the research on theories of mind, Wellman (1990) and others (Moses & Chandler, 1992) have been able to show that the child can intentionally deceive and, as well, place itself in the role of another. In each of these types of studies there is an implicit theory of mind (ToM) that includes the idea of me.

Unfortunately, even these studies require that the children understand complex language although they do not have to produce it. Thus, for example, in the deception studies, children have to understand the experimenters' instructions and therefore cannot be less than 3 years old. By this age, children seem to have the idea of a me. The question, then, is whether this idea is acquired earlier, and if so, how might we measure it. We could still focus on language and argue that the idea of me can be determined by whether the children have acquired their names; after all, we are what we are called. The risk of accepting this as proof is that the child may have been taught to use its name by associating it with a visual array (a photograph of the child) without the idea of me being present (see Putnam, 1981, for a discussion of this problem).

Another language usage, a bit less suspect, is that of personal pronouns such as "me" or "mine." Because parents do not use the label me or mine when referring to the child or teaching it to recognize its picture, the use of these terms by the child is likely to be a reasonable referent to the idea of me. This appears even more the case when we observe children's use of the terms and how they behave when using them. One can see a child saying "mine" as he or she pulls the object away from another child and toward him- or herself. Because moving the object toward oneself does not move the object as far away from the other as possible, the placement of the object next to the body, together with the use of the term "me" or "mine," could reveal the idea of me.

One other procedure can be used to get at the idea of me: self-recognition. The topic of self-recognition in infants and young children has been covered in detail (see Lewis & Brooks-Gunn, 1979a). Following a procedure first described by Gallup (1977) that used reaching for a spot on the face as a sign of visual self-recognition, we instituted a series of studies of the course of infant visual self-recognition. The data from a variety of sources indicate that infants even as young as 2 months, when placed in front of mirrors, will show interest and respond to the mirror image. Children will smile, coo, and try to attract the attention of the child in the mirror. There is no reason for them to believe that the image they see is themselves. Their response to images of other infants is no different than their response to their own image (Lewis & Brooks-Gunn, 1979b). At older ages, when locomotion appears, infants on occasion have been observed going behind the mirror in order to see if they can find the child in the mirror. In addition, they often hit the mirror as if they are trying to touch the other. Somewhere around 15–18 months, it appears to occur to the child that the image in the mirror belongs to itself. They appear to know that the image is themselves. This idea of me is best captured by their use of self-referential behavior. As is well known, the touching of their noses when they look in the mirror seems to reveal that they know it's "me" there.

This ability to use the mirror to reference themselves has often been mistaken for the child's understanding of the reflective property of mirrors. There is ample evidence that although children are able to produce self-referential behavior through the use of the mirror mark technique, they do not know many of the properties of reflected surfaces; for example, they cannot use the mirror to find an object reflected in its surface (Butterworth, 1990). What is important about the self-referential behaviors in the mirror is that they need not be a marker of general knowledge about reflected surfaces, but rather are a marker for the child's knowledge about itself. They are the equivalent of the phrase, "that's me," and reflect the idea of me. This recognition, if put into words, says, "That is me over there, this is me, here."

Self-referential behavior in mirrors is related to the idea of me and not to any elaborate knowledge of reflective surfaces. This idea of me, when it emerges, is restricted; nevertheless, it does emerge. What the idea of me will consist of, what will be elaborated and developed, is related to other emerging cognitive capacities as well as to socialization demands.

I must admit to some discomfort in suggesting that self-referential behavior in mirrors is the sine qua non of this idea of me. For more than 15 years other measures have been considered, but none seems able to separate the self as machine from the idea of me. Because of this, this single measure of self-referential behavior appears to remain as the best index of the construct of me available. However, there are other ways of validating this measurement. One is to determine whether or not coherences exist that are predicted by a theory of the idea of me; for example, whether certain behaviors are likely to occur only if mirror self-referential behavior is present. As will be seen in more detail, there are important social as well as emotional behaviors that occur only when this idea of me develops. These coherences can be used to support the measurement system, especially if overall intellectual ability is taken into account.

There are three areas in which coherence can be found. One of these is in the domain of empathy. Integrated forms of empathy, that is, empathy signaled by both appropriate facial expression and behavior, occur only after self-recognition in mirrors (Halperin, 1989). A second area has to do with self-conscious emotions, in particular, embarrassment. Lewis et al. (1989) have demonstrated a direct relation between mirror self-referential behavior and the appearance of embarrassment. Children do not show embarrassment until such time as they show mirror self-referential behavior. Third is the area of relationships in which the mother–child interactive behavior undergoes change once the child acquires mirror self-referential behavior. These and other co-herences in social and affective behavior suggest that the emergence of the ability to recognize one's self is associated with behavior theoretically related to the idea of me; that is, to a referential self that can pay objective attention to its self.

What does a self (the *idea of me*) do?

Here I wish to discuss:

1. a self that participates in socializing itself,
2. the role of the idea of me in the creation of relationships, and
3. the idea of me and the appearance of self-conscious emotions.

Self socializes itself

Social control and the socialization process are the material for understanding as well as the motive for action. Such a view allows for the active participation of the infant in its own socialization. Environment exerts part of its effect through the structures within the infant. Prior to the development of the idea of me, the social influences are direct, shaping and molding the infant's behavior with minimum participation from the infant itself. During this time the infant can be of relatively little help in the socialization process. In some sense, it cannot join the process because such active participation involves an idea of me; that is, an agent that is trying to understand the rules and generate the goals for itself. Somewhere around the beginning of the second half of the 2nd year of life, the infant itself starts to participate in the socialization process. Until the idea of me emerges, infants are not capable of generating evaluation of their own actions (Kagan, 1981) and therefore are incapable of such motivating emotions as shame and pride in achievement situations (Lewis, 1992). Until the idea of me emerges, the motivating and self-correcting processes related to the child's active participation in social-ization are not supported by the organism itself. This idea will be developed further when we discuss the idea-of-me aspect of self in relation to the emo-tions of shame, guilt, and pride. Although I will go into more detail, it is evident that these self-conscious emotions are critical in the socialization process because only with the rise of pride, for example, can the child's adherence to rules be self-generated and sustained.

The role of the idea of me in relationships

Levels of social relationships. Social relationships require a wide range of abilities that have not been well studied. Hinde (1976, 1979) articulated six dimensions that can be used to characterize a relationship:

1. goal structures,
2. diversity of interactions,
3. degree of reciprocity,
4. meshing of interactions,
5. frequency, and
6. patterning and multidimensional qualities of interactions.

These six features of interaction may be adequate to characterize relation-ships of nonhuman animals, but they are inadequate to describe those of human infants. Young human infants and their mothers, as well as other social creatures, can be shown to have acquired this lower level of relation-ship. Whereas this is one level, human adult relationships require more com-plexity. For example, I find it difficult to accept that a rat and her pups have the same level of relationship as a human mother and her 3-year-old child. Hinde's analysis is helpful because it suggests two further features of a rela-tionship that allow us to conceptualize a higher level for the human child and mother, and to differentiate different levels of relationship. These features include:

1. *cognitive factors*, or those mental processes that allow members of an inter-action to think of the other member as well as of themselves, and
2. something that Hinde calls *penetration* and that I have suggested has to do with ego boundaries.

Notice that interactions alone (which characterize the first six features) are insufficient to describe human relationships. Human relationships require self-awareness (Lewis, 1987). Such a view was suggested by Sullivan (1953), who saw relationships by necessity as the negotiation of at least two selves. If the child does not have the idea of me, there can be no relationship as we characterize it for adults. Without two selves and the knowledge of two selves, there can be no relationship between them (Buber, 1958). Emde (1988) makes reference to the "we" feature of relationships, also pointing to the appearance of the idea of me in the second half of the 2nd year of life.

Interactions and relationships. My model of mature human relationship requires that we consider the developmental changes in relationships rather than seeing no difference between relationships that exist at birth and those of adults. Mature human relationships arise from interaction only after the development of self-awareness. Interactions lead to a relationship through the mediation of cognitive structures, in particular the development of the idea of me (Lewis, 1987). Only after the acquistion of the idea of me can adultlike human relationships occur. Some relationships, for example between a mother rat and her pups, do not need this skill; neither do relationships in early human infancy that occur prior to the acquisition of awareness. The meaning of the term "relationship," therefore, is not the same across ontogeny.

What we mean when we refer to mature human relationships is the level that usually includes what the organism thinks about its self and the other, the desire to share, and the use of empathy to regulate the relationship. Moreover, the issue of ego boundaries, as discussed by Hinde (1979), also needs to be considered. When we consider ego boundaries we need to make reference to the child's growing understanding of privacy, as well as its need to become a "we." Without these skills we may talk about some type of relationship, but not the one we consider between mature humans.

From this point of view, the achievement of adult human relationships has a developmental progression. This progression involves, first, interactions (which may be similar to those shown by all social creatures) and, second, cognitive structures, including the idea of me and such skills as empathy (the ability to place the self in the role of the other). The relationships of 1-year-olds do not contain these cognitive structures and, therefore, may only approximate those of adults. By age 2, most children have identity and the beginning of such skills as empathy (Borke, 1971; Zahn-Waxler & Radke-Yarrow, 1981). Their relationships now approximate more closely those of the mature level. Mahler's concept of individuation is relevant here, for as she has pointed out, only when the child is able to individuate can it be said that a more mature relationship exists (Mahler, Pine, & Bergman, 1975).

Such an analysis raises the question of the nature of the child's relationships prior to the development of the idea of me. To my mind, a relationship prior to the development of the idea of me is nothing more than a set of complex social species-patterned processes imposed by the caregiver, which, through adaptive processes, may exist through the biological endowment of the infant. This imposed (or socialized) complex-patterned system in the beginning of life will give way to a mature relationship in which the child joins and becomes an active outstanding member. The nature of the higher-level relationship is dependent on many factors. These include the nature of socialization practices (the initial interactions imposed on the infant), self-awareness, and the cognitions about the interactions of self and other; that is, the meaning given to them by the selves involved (Bowlby, 1980).

To summarize, socialization determines the pattern of parent–infant interactions. For example, boys are reinforced less for crying than girls, presumably because of the socialization of sex role differences in the expression of emotion (Brooks-Gunn & Lewis, 1982; Malatesta & Haviland, 1982). These interaction patterns provide the social context that facilitates self-development and ultimately the idea of me. The emergence of the idea of me allows for the representation of self and other, and for the representation of the affective "good" or "safe" mother as well as of her actions (Bowlby, 1980). These representations, in turn, help transform the behaviors of interactions into goals and rules. Thus, parental goals and rules generate specific patterns of behavior that, in turn, develop structures (self and representations) enabling the child to reconstruct the parental goals and rules. The socialization of relationships provides the material for the generation of relationships; its influence is on both structures and content.

Self-awareness and emotional life

The proposition we wish to entertain here is that socialization helps create the idea of me. The idea of me, in turn, gives rise to emotional states unavailable before this development. These emotions, called self-conscious emotions, include pride, shame, and guilt (Lewis, 1992). Because these emotions have to do with goals, standards, and rules, the child becomes able to reward or punish itself independent of the parent. The influence of socialization in this regard has to do with:

1. the creation of the idea of me, and
2. the articulation, through action and words, of the goals, rules, and standards the child acquires.

Having done so, the socialization process of the parent is joined by the child who, having developed the idea of me as well as standards, rules, and goals, now serves as an active participant in the evaluation of its own practices.

The development of the idea of me and its evaluation provide the cognitive underpinning for the emergence of emotional states absent in the young child in the 1st year and a half of life. Although the child exhibits the primary emotions including joy, anger, sadness, interest, disgust, and fear, it is not until the acquisition of self-awareness that the child acquires such emotions as embarrassment, envy, empathy, pride, guilt, and shame (Lewis et al., 1989; Lewis, 1992). These latter emotions, often called secondary emotions (as opposed to the primary ones), should be relabeled as self-conscious emotions. Elsewhere (Lewis et al., 1989) I have presented a model of development of

emotional states. The first set of states occurs early and, as states, are independent of the idea of me. The second set of states occurs later and only after
the attainment of the idea of me. Let us consider them, in particular the
emotions of pride and shame.

Pride occurs when the child evaluates its own action against a standard and
finds it successful. The emergence of pride appears only after the idea of me
and is related to achievement motivation (Heckhausen, 1984). Because of
this emotion, the child seeks out action that is likely to lead to this feeling.
Although the infant expresses joy in achievement, we have called this efficacy
(Lewis & Goldberg, 1969); not until the onset of awareness can we see pride.
Happiness is reflected in facial behavior, but pride can be seen in both facial
and bodily action (see Gepphart, 1986).

The emotions of shame and guilt also emerge only after the idea of me
emerges (Lewis, 1992; Zahn-Waxler & Kochanska, 1990). The responses of
these three emotions are quite different, but all have to do with the child's
evaluation of its action against a standard. Failure results in one of these
negative emotions, depending upon socialization factors (Dweck & Leggett,
1988), as well as on dispositional ones (DiBiase & Lewis, in press). Children
will focus on the self's action or the self's action toward another, or they will
focus on themselves. The distinction between self-action and self is one of
global versus specific evaluation. Focus on the self's action – as in "I should
not have said that to Mom" – leads to guilt that is characterized by repara-
tion, that is, action to correct the failure, for example, "I shall not say that
again." Focus on the self leads to shame and to such statements as "I am no
good." Shame is characterized by wanting to hide or disappear. Notice there
is not corrective action save this desire to disappear.

In each of these emotions, the idea of me relative to some goal, standard,
or rule is necessary for the emotional state to occur. Thus, the idea of me is
directly related to the emergence of these emotions; but how does the
emergence of these emotions facilitate the socialization of the child, and how
are they related to social control? As we pointed out earlier, the child joins
in its own socialization. In order for the child to be able to do so, it needs to
develop structures that allow it to generate the social order without the
moment-to-moment control of others, or without the others being present.
This can be done by incorporating the rules, standards, and goals of the socializing others. Incorporation is not merely the storage and organization of
specific parental behaviors (although this is necessary); it requires that the
child represent itself as well as others. This representation is seen as the idea
of me, which also allows for the development of pride, shame, and guilt. The
generation of the idea of me gives rise to representations and motives that
themselves promote socialization.

Summary

I have tried to articulate a special feature of the self system, a feature I have
called the idea of me. Such a concept goes by many names; I have used self-

awareness and objective self-awareness, which might be called self-consciousness. The attempt to delineate and conceptualize features of the self is done to differentiate different aspects of this concept. Many aspects of self, such as self–other differentiation, self-regulation, and the communication between parts of one self, are shared by many self systems. The emergence both phylogenetically and ontogenetically of the idea of me is quite different. As I have tried to show, it gives rise to adaptive functions based on representation and active participation rather than on passive control.

In order to measure and thus differentiate this aspect of self from other, I have argued that self-recognition, the use of personal pronouns, and the ability to understand recursive sentences of the form "I know, you know, I know it's raining" are the most reasonable measures of this construct. Other possible measures should be developed with the idea that they can differentiate this idea of me aspect of self from other features of self systems.

References

Borke, H. (1971). Interpersonal perception of young children: Egocentrism or empathy. *Developmental Psychology, 5,* 263–269.
Bowlby, J. (1980). *Attachment and loss, Vol. 3: Loss, sadness and depression.* New York: Basic.
Brooks-Gunn, J., & Lewis, M. (1982). Affective exchanges between normal and handicapped infants and their mothers. In T. Field & A. Fogel (Eds.), *Emotion and early interaction* (pp. 161–188). Hillsdale, NJ: Erlbaum.
Buber, M. (1958). *I & thou* (2nd ed.). Ronald Gregor Smith (Trans). New York: Scribner.
Butterworth, G. (1990). Origins of self-perception in infancy. In D. Cicchetti & M. Beeghly (Eds.), *The self in transition: Infancy to childhood* (pp. 119–137). Chicago: University of Chicago Press.
DiBiase, R., & Lewis, M. (in press). Temperament and emotional expression: A short-term longitudinal study. *Developmental Psychology.*
Duval, S., & Wicklund, R. A. (1972). *A theory of objective self-awareness.* New York: Academic Press.
Dweck C. S., & Leggett, E. L. (1988). A social-cognitive approach to motivation and personality. *Psychological Review, 95,* 256–273.
Emde, R. N. (1983). The prerepresentational self and its affective core. *Psychoanalytic Study of the Child, 38,* 165–192.
  (1988). Development terminable and interminable II: Recent psychoanalytic theory and therapeutic considerations. *International Journal of Psychoanalysis, 69,* 283–296.
Freud, S. (1936/1963). *The problem of anxiety.* H. A. Bunker (Trans.). New York: Norton.
Gallup, G. G., Jr. (1977). Self-recognition in primates: A comparative approach to the bidirectional properties of consciousness. *American Psychologist, 32,* 329–338.
Gepphart, U. (1986). A coding system for analyzing behavioral expressions of self-evaluative emotions (Technical Manual). Munich: Max Planck Institute for Psychological Research.
Halperin, M. (1989). *Empathy and self-awareness.* Paper presented at the Society for Research in Child Development meeting, Kansas City, MO.
Heckhausen, H. (1984). Emergent achievement behavior: Some early developments. In J. Nicholls (Ed.), *The development of achievement motivation* (pp. 1–32). Greenwich, CT: JAI Press.
Hinde, R. A. (1976). Interactions, relationships, and social structure. *Man, 11,* 1–17.
  (1979). *Toward understanding relationships.* New York: Academic Press.
Kagan, J. (1981). *The second year: The emergence of self-awareness.* Cambridge, MA: Harvard University Press.
Kernberg, O. F. (1976). *Object relations theory and clinical psychoanalysis.* New York: Aronson.
  (1980). *Internal world and external reality: Object relations theory applied.* New York: Aronson.

Lacan, J. (1968). *Language of the self.* Baltimore, MD: Johns Hopkins University Press.

Laing, R. D. (1970). *Knots.* New York: Pantheon.

LeDoux, J. (1989). Cognitive and emotional interactions in the brain. *Cognition and Emotion, 3,* 265–289.

Lewis, M. (1987). Social development in infancy and early childhood. In J. Osofsky (Ed.), *Handbook of infant development* (2nd ed., pp. 419–493). New York: Wiley.

    (1990). The development of intentionality and the role of consciousness. *Psychological Inquiry, 1*(3), 231–248.

    (1991). Ways of knowing: Objective self awareness or consciousness. *Developmental Review, 11* (special issue), 231–243.

    (1992). *Shame, the exposed self.* New York: Free Press.

Lewis, M., & Brooks-Gunn, J. (1979a). Toward a theory of social cognition: The development of self. In I. Užgiris (Ed.), *New directions in child development: Social interaction and communication during infancy* (pp. 1–20). San Francisco: Jossey-Bass.

    (1979b). *Social cognition and the acquistion of self.* New York: Plenum.

Lewis, M., & Goldberg, S. (1969). The acquisition and violation of expectancy: An experimental paradigm. *Journal of Experimental Child Psychology, 7,* 70–80.

Lewis, M., Sullivan, M. W., Stanger, C., & Weiss, M. (1989). Self-development and self-conscious emotions. *Child Development, 60,* 146–156.

Mahler, M. S., Pine, F., & Bergman, A. (1975). *The psychological birth of the infant.* New York: Basic.

Malatesta, C., & Haviland, J. (1982). Learning display rules: The socialization of emotion expression in infancy. *Child Development, 53,* 991–1003.

Moses, J., & Chandler, M. J. (1992). Traveler's guide to children's theories of mind. *Psychological Inquiry, 3,* 285–301.

Putnam, H. (1981). *Reason, truth and history.* Cambridge University Press.

Rank, O. (1929). *The trauma of birth.* London: Kegan Paul.

Stern, D. N. (1985). *The interpersonal world of the infant.* New York: Basic.

Sullivan, H. S. (1953). *The interpersonal theory of psychiatry.* New York: Norton.

Tulving, E. (1985). How many memory systems are there? *American Psychologist, 40,* 385–398.

von Bertalanffy, L. (1967). *Robots, men, and mind.* New York: Brazilles.

Weiskrantz, L. (1986). *Blindsight: A case study and implications.* Oxford University Press.

Wellman, H. M. (1990). *The child's theory of mind.* Cambridge, MA: MIT Press.

Zahn-Waxler, C. J., & Kochanska, G. (1990). The origins of guilt. In R. Thompson (Ed.), *36th annual Nebraska symposium, on motivation: Socioemotional development* (pp. 183–258). Lincoln: University of Nebraska Press.

Zahn-Waxler, C. J., & Radke-Yarrow, M. R. (1981). The development of prosocial behavior: Alternative research strategies. In N. Eisenberg-Berg (Ed.), *The development of prosocial behavior* (pp. 109–138). New York: Academic Press.

3 Self-recognition: Research strategies and experimental design

Gordon G. Gallup, Jr.

Experimental background

The first experimental test of self-recognition in animals was conducted on a group of preadolescent chimpanzees and several species of monkeys (Gallup, 1970). Initially all of the animals acted as if they were seeing other animals when they looked at themselves in the mirror. After a couple of days, however, the chimpanzees (but not the monkeys) began to respond as if they had come to appreciate the dualism implicit in mirrors and now realized that their behavior was the source of the behavior being depicted in the reflection. That is, rather than responding to the mirror as such with species-typical patterns of social behavior, they began to show *self-directed responding* by using the mirror to respond to themselves (e.g., to investigate parts of their bodies that they had not seen before). In an attempt to validate my impressions of what had transpired, I devised a more rigorous, unobtrusive test of self-recognition. After the 10th day of mirror exposure the chimpanzees were placed under anesthesia and removed from their cage. While the animals were unconscious, I applied a bright red, odorless, alcohol-soluble dye (rhodamine-B base) to the uppermost portion of an eyebrow ridge and the top half of the opposite ear. The subjects were then returned to their cages in the absence of the mirror and allowed to recover from anesthetization.

There are three special properties associated with this procedure that have often been ignored or overlooked by other investigators who have attempted to adapt this technique for testing other species:

1. The marks were applied while the animals were under deep anesthesia to ensure that they would have no knowledge about the application of these marks.
2. The dye was carefully chosen to dry quickly and to be free from any residual tactile or olfactory cues.
3. The marks were strategically placed at predetermined points on the chimpanzees' faces that could be seen only in a mirror.

After the animals had recovered from anesthesia, they were given food and water and observed for 30 min to establish a baseline of the number of times a marked portion of the face might be touched spontaneously. Follow-

35

ing this period the mirror was then reintroduced as an explicit test of self-recognition. Upon seeing themselves in the mirror the chimpanzees all reached up and attempted to touch the marks directly while intently watching the reflection. After touching marked portions of the skin, the animals would frequently look at and/or smell their fingers in an apparent effort to identify what was on their faces. In addition to these mark-directed responses there was also an abrupt, threefold increase in the amount of time spent viewing their reflection in the mirror.

As a check on the source of these reactions, several comparable chimpanzees without prior exposure to mirrors were also anesthetized and marked, and then confronted with the mirror. Unlike their mirror-experienced counterparts, when these control animals saw themselves for the first time with red marks on their faces there were no mark-directed responses, no patterns of self-directed behavior, and they acted as if confronted by a chimpanzee they had never seen before. These control data demonstrate that the mark-directed behavior shown by the first group of animals was a consequence of prior mirror experience and not an artifact of anesthetization or due to any residual cues left by the dye. It is also important to note that the use of a pre-exposure baseline period established that the mark-directed responses seen on the test trial by mirror experienced chimpanzees were in fact conditional upon seeing themselves in mirrors.

Before proceeding it would be instructive to examine some implications of the logic behind this strategy. To play the devil's advocate: Maybe when confronting themselves with red marks on their faces the subjects continue to interpret what they see as pertaining to another chimpanzee, but now attempt to alert the other animal to the presence of the marks by touching comparable points on their own faces. It is important to remember, however, that (1) only the chimpanzees with prior mirror experience respond in this manner and (2) on the test trial they often investigate their own fingers after having touched the marks on their faces that can only be seen in the mirror. (If they thought that the marks in the mirror were on the face of another animal, there would be no reason to look at their own fingers.) Alternatively, the skeptic might argue (e.g., Mitchell, 1993) that they seem to respond appropriately because when they see another chimpanzee with marks on its face they think that there may be corresponding marks on themselves. Contrary to this interpretation, however, it is important to note (1) the absence of such behavior on the part of control animals without prior mirror experience and (2) that although chimpanzees do sometimes touch and inspect marks on each other when they are returned to group cages following mark tests, they have never been observed to respond to facial marks on other animals by trying to locate and touch comparable areas on their own faces.

It is critical to emphasize that the "mark test," as it has come to be known, was used originally only to confirm my impressions that arose out of seeing chimpanzees use mirrors to respond to themselves. The mark test is not so much a measure of self-recognition as it is a means of validating impressions that emerge as a result of seeing animals engage in a variety of what would

Table 3.1. *Comparisons between the Gallup and Amsterdam techniques for assessing self-recognition*

	Gallup (1970)	Amsterdam (1972)
Inclusion of data on patterns of self-directed behavior to mirrors prior to the mark test	Yes	No
Unobtrusive application of facial marks	Yes	No
Use of marking material free from tactile and/or olfactory cues	Yes	No
Marks placed on facial features only visible in a mirror	Yes	No
Inclusion of control subjects without prior mirror experience	Yes	No

appear to be mirror-mediated patterns of self-directed behavior. In their zeal to defend the conceptual ability of other species or discount the significance of self-recognition, many people seemed to have missed the essential point of the mark test. For example, Epstein, Lanza, and Skinner (1980) claim that because they could ostensibly simulate mark-directed behavior in pigeons using an extensive training regime, it follows that self-recognition can be subsumed by an operant analysis. However, none of the pigeons that they allegedly trained to peck at dots on themselves ever showed any collateral instances of self-directed behavior. In addition, repeated attempts to replicate the Epstein et al. results have never been successful (see R. K. R. Thompson & Contie, *SAAH*26).

In contrast to the technique I developed in 1970, consider a popular alternative approach for testing human infants that was introduced by Amsterdam (1972). In preparation for a test of mirror self-recognition (MSR), human infants of different ages were brought into the laboratory by their mothers and the mother was instructed to apply a spot of rouge to the side of her child's nose. The infant was then exposed to a mirror to see if it would reach up and touch the rouge. Although superficially similar to the technique I used with chimpanzees, Amsterdam's procedure differs from mine at a number of crucial junctures (see Table 3.1). First, unlike the chimpanzees that were rendered unconscious prior to being marked, the children in Amsterdam's study were fully awake at the time their mother applied rouge to their noses. Second, the application of rouge was not only accompanied by visual and tactile cues as the nose was touched by the mother, but rouge is often scented and could have been a source of telltale olfactory cues. Third, rather than being applied to facial features that could be seen only in a mirror, the rouge was applied to the side of the child's nose which, curiously, is one of the few facial features that can be seen in the absence of a mirror. Prior to the reintroduction of the mirror, the chimpanzees in my experiment had no way of knowing about the presence of their facial marks, whereas in the case of Amsterdam's study all of the infants were unwittingly provided with extensive information about the mark prior to seeing themselves in mirrors. As a consequence, it is impossible to be sure that they were using the mirror to respond to the mark. Finally, people in developmental psychology who work on self-recognition typically collect no data on the way children respond to mirrors prior to conducting the rouge test (e.g., Amsterdam, 1972; Lewis & Brooks-Gunn, 1979; Schulman & Kaplowitz, 1976), nor do they attempt to establish the presence of spontaneous instances of self-directed responding in children before conducting mark tests (see Table 3.1).

Following the original demonstration of self-recognition in chimpanzees there have been many attempts to assess this capacity in other species. The ability to decipher mirrored information about the self correctly has been replicated a number of times with chimpanzees, by a number of different investigators, in a variety of settings (e.g., Calhoun & Thompson, 1988; Lethmate & Dücker, 1973; Lin, Bard, & Anderson, 1992). There are also two independent experimental reports of MSR in orangutans (Lethmate & Dücker, 1973; Suarez & Gallup, 1981). Gorillas, however, appear to be another matter. With the exception of a report by Patterson and Cohn (SAAH17) of self-recognition in a human-reared, sign-language-trained gorilla, others (e.g., Ledbetter & Basen, 1982; D. J. Povinelli, personal communication, 1991; Suarez & Gallup, 1981) have failed to find any evidence that gorillas can recognize themselves in mirrors. Most of the available evidence shows that chimpanzees, orangutans, gorillas, and humans all derived from a common ancestor, and on that basis it seems reasonable to suggest that perhaps the precursor to the human–great ape clade was self-aware (Gallup, 1985; Povinelli, 1987). Furthermore, I have argued that at one point during their evolution gorillas may have been self-aware but, because of changes in their socioecology that failed to put a reproductive premium on the use of introspectively based social strategies, they apparently have lost the capacity for self-awareness (Gallup, 1991). The work of experimental embryologists has shown that environmental alterations early in development can result in the appearance of ancestral morphological traits, and Povinelli (SAAH18) speculates that the same effects may apply to behavioral ontogeny. On the heels of my suggestion that gorillas could have been self-aware during an earlier period of their evolution but subsequently lost that ability, Povinelli reasons that the capacity for self-recognition may still exist as an ancestral/vestigial trait in contemporary gorillas. In support of this position he cites Patterson & Cohn (SAAH17) on the gorilla Koko, who, unlike other gorillas that have been tested to date, shows convincing evidence of being able to recognize herself in a mirror. What makes Koko unique is the presence of early environmental enrichment and intervention in the context of being reared by humans and receiving extensive sign language training, which Povinelli argues may be equivalent to the kind of early environmental perturbation that could trigger the reappearance of ancestral morphological traits during embryonic development. The strength of Povinelli's analysis is that it predicts not only that other gorillas experiencing appropriate early enrichment should show self-recognition, but also that primates outside of the great ape clade (e.g., rhesus monkeys) should fail to show such effects because they did not derive from a self-aware ancestor, as did great apes and humans.

Whereas most of the work on self-recognition in humans has focused on when this capacity emerges developmentally, in the case of animal research most people have been concerned with demonstrating the presence or absence of the capacity in different species. In the human literature the typical finding is that children do not begin to show compelling signs of MSR until they reach 18–24 months of age. Only recently has anyone attempted to track the developmental time course of self-recognition in chimpanzees, and the results show that the onset of this capacity may be maturationally delayed relative to humans, with most of the animals failing to show self-recognition until $2\frac{1}{2}$– 3 years of age (Lin, Bard, & Anderson, 1992). Indeed, there are more recent data (Povinelli et al., in press) showing that the onset of self-recognition in chimpanzees may be even more developmentally delayed than the Lin et al. study suggests.

In the remainder of this chapter I will detail some of the methodological and procedural variables that need to be carefully considered in conducting tests of self-recognition. Upon close inspection it can be shown that many recent claims for self-recognition in other species (including pigeons, dogs, and rhesus monkeys) simply do not meet rigorous criteria of scientific evidence.

Subject variables

When it comes to the choice of different species (or even individuals within a species) to expose to mirrors it is important to have some assurance that the subjects have adequate visual capabilities to see the reflection. It makes little sense to expose individuals to mirrors if they lack the capacity for good visual acuity (as do, e.g., rats) or if they suffer from a serious visual impairment (e.g., cataracts or extreme myopia). It is also important to know something about the subject's current health status and prior medical history. Animals used in infectious disease studies, experimental surgery, or those exposed to neurotoxins ought to be automatically disqualified, unless the objective is to explicitly evaluate these effects using appropriate unaffected controls.

In addition to the animal's medical history it is important to have information concerning its rearing history and prior social experiences. For instance, unlike their group-reared counterparts, chimpanzees reared in social isolation fail to show evidence of self-recognition (Gallup, McClure, Hill, & Bundy, 1971). Not unrelated to this (as noted) is that in humans and chimpanzees the ability to recognize oneself in a mirror is subject to maturational constraints. Thus, subjects that lack normal social experiences or those that have yet to reach an appropriate level of development may fail to show evidence of a capacity that may be present in other members of the same species.

At the very least, to constitute a fair test, any study of MSR is predicated on evidence of initial interest in mirrors. Inasmuch as self-recognition presupposes prior experience with mirrors, animals that fail to show much attention to mirrors hardly constitute adequate subjects for mirror tests. For example, Swartz and Evans (1991) report finding self-recognition in only one out of eleven chimpanzees, and on that basis call into question the methodology I have developed. However, for reasons that may related to their animals' atypical medical history (all of them were maintained in a medical research facility in Africa) many of their animals failed to show much interest in mirrors from the outset. Chimpanzees that show clear evidence of self-recognition typically show an avid interest in the mirror for the first several days of exposure (e.g., Gallup, 1971; Gallup, McClure, Hill, & Bundy, 1970; Suarez & Gallup, 1981).

Mirror exposure variables

Assuming that most of the above criteria concerning subject selection have been satisfied (e.g., healthy, socially normal, visually capable subjects), one of the next questions concerns various parameters of mirror exposure. In most of the earlier studies of self-recognition the mirror was positioned outside the animal's cage and out of reach to prevent it from being broken. However, a case can be made that animals provided with physical as well as visual access to the mirror have an advantage; that is, experience with the inanimate mirror surface ought to pose the question about the source of what is being depicted in the mirror in a more direct way (e.g., Gallup, 1987). In studies that have provided animals with direct access to mirrors it is frequently reported that they will attempt to reach or look behind the mirror in an apparent attempt to localize the "other" animal they see when they confront their own reflection (e.g., Anderson, 1984; Gallup, 1968). An obvious extension of this strategy is to give the subjects hand-held mirrors that allow animals to manipulate and gain more explicit control over what they see in the reflection (e.g., Anderson & Roeder, 1989).

The issue of the duration of mirror exposure has received considerable attention. Chimpanzees frequently begin to show evidence of self-recognition (i.e., patterns of self-directed behavior) within two or three days of mirror exposure (8 hours per day). In one recent and exceptional instance, a 3-year-old female chimpanzee evidenced convincing signs of MSR within about an hour of initial exposure to herself in a mirror (Povinelli, Rulf, Landau, & Bierschwale, in press). At the other extreme are species that fail to exhibit signs of self-recognition despite extended experience with mirrors. We have a pair of rhesus monkeys that have been reared together in front of a mirror since 1978. They get 14 hours of mirror exposure every day, 7 days a week. To date they have received in excess of 65,000 hours of mirror exposure, yet despite repeated attempts to make the identity of the reflection more explicit, neither animal has ever shown any evidence of self-recognition (e.g., Gallup

& Suarez, 1991). The spacing of mirror exposure is also an interesting question. As yet, no one has collected systematic information about the potential effects of massed versus distributed practice on the onset of MSR in chimpanzees (or any other species).

Another important consideration relates to the context in which animals are exposed to mirrors. In the original study with chimpanzees (Gallup, 1970) I kept animals in a small individual cage in an otherwise empty room for the duration of mirror exposure (10 days). The rationale for testing animals by themselves was that minimizing the availability of external stimuli ought to provide for a more focused period of enforced confrontation with the mirror. Animals exposed to mirrors in large cages containing other animals have much more to do and are easily distracted by the activity of cagemates and other extraneous events, and as a consequence may actually spend very little time looking at themselves in the mirror.

On the other hand, animals exposed to mirrors in the presence of one or more familiar cagemates have a distinct advantage, at least in principle (Gallup, Wallnau, & Suarez, 1980). Animals that can recognize one another (individual recognition) are confronted with an intriguing situation when they look at themselves and others in mirrors; that is, seeing the reflection of familiar cagemates ought to pose the question in a much more direct way about the source of the unfamiliar animal (themselves) that they also see in the mirror. Although the use of a paired exposure strategy has been tried without success in several species that do not recognize themselves in mirrors (e.g., Gallup et al., 1980; Ledbetter & Basen, 1982), it would be interesting to see if this strategy would facilitate and/or accelerate the development of MSR in species that do show self-recognition (e.g., chimpanzees).

Several other mirror-exposure variables worthy of note include the use of multiple mirrors to create an array of images (see Anderson, 1984). As yet there are no published reports of animals confronted with distorting mirrors, but in humans there are striking differences between schizophrenics and normal people in the ability to adjust a distorting mirror to achieve an undistorted image of the self (Traub & Orbach, 1964). For animals that have learned to recognize themselves in mirrors it would be interesting to observe their reactions to distorting mirrors. Indeed, distorting mirrors that target a particular part of the image for distortion could be used to corroborate impressions of self-recognition. (That is, would the animal preferentially inspect the corresponding portion of its body that appeared distorted in the mirror?)

The other use of multiple mirrors that is methodologically important relates to the problem of gaze aversion. That direct stares often function as threat responses has led many different species of primates to show an aversion to making eye contact with other animals. To the extent that such species are less likely to directly confront their images in mirrors, gaze aversion represents a potential problem in studies of self-recognition, particularly as facial features are typically targeted for marking. Although the existence of this problem has been acknowledged for some time (e.g., Gallup et al., 1980; Gallup, 1987), a practical solution was only recently introduced by Anderson and Roeder (1990). In a larger study of mirror behavior in capuchins, Anderson and Roeder developed a technique that circumvents the gaze aversion problem. Using a pair of mirrors positioned to form an angle of approximately 60° they were able to create a situation in which the animals could see their mirror images but could not make direct eye contact with them. Despite this clever arrangement, however, none of the capuchins showed any evidence of self-recognition.

Mark test considerations and controls

Contrary to the impression created by some investigators (e.g., Epstein et al., 1980), I have never maintained that the mark test is the sine qua non of self-recognition. Appropriate behavior in response to unobtrusively applied facial marks that can only be seen in a mirror constitutes a means of validating impressions that arise out of seeing animals use mirrors in ways that suggest they realize that their behavior is the source of the behavior depicted in the reflection. In trying to demonstrate self-recognition in other species, some people appear to have lost sight of this and have focused almost exclusively on the mark test (e.g., Boccia, SAAH23; R. L. Thompson & Boatright-Horowitz, SAAH22). A monkey that happens to touch a mark on its face after it has been kept for months in front of mirrors and has been repeatedly water deprived to near dehydration in an attempt to get it to maintain eye contact with the reflection, says nothing about self-recognition (see R. L. Thompson & Boatright-Horowitz, SAAH22). Indeed, in a videotape of this sequence that has been widely cited as "evidence" of self-recognition by monkeys, it is curious to note that there are no spontaneous instances of self-directed behavior, and immediately after passively brushing the mark with its wrist, this monkey proceeded to direct a distinctive social response (scalp retraction and jaw protrusion) toward its reflection in the mirror. Whereas appropriate performance on the mark test in the context of self-directed responding provides fairly compelling evidence of self-recognition, attempts to elicit or engineer mark-directed behavior under conditions in which subjects fail to display spontaneous instances of mirror-mediated patterns of self-directed behavior are fundamentally misguided. (For another critique of these attempts to produce mark-directed behavior see Povinelli, 1991.)

Assuming there is good reason to conduct a mark test (e.g., the animals have shown prior evidence of using mirrors to respond to the self), there are a number of important points to consider. One of the first questions is where to apply the mark(s). In order for positive results to be definitive it is essential that the mark be targeted for some portion of the body that is inaccessible without a mirror. In Amsterdam's (1972) procedure a spot of rouge was placed on the side of a child's nose. As noted previously, most people can see the side of their noses without a mirror, and therefore touching the mark can hardly be construed as compelling evidence of mirror self-recognition.

Another important consideration relates to the color and/or saliency of the mark. The mark test is predicated on the assumption that the animal will be sufficiently motivated or interested to contact and inspect the mark – provided it can localize it in real space. Failing this, the absence of mark-directed responding on the test is inconclusive. (That is, did the animals fail because they did not realize that the marks seen in the mirror were on their faces, or because they simply did not care about superimposed facial marks?)

This raises several important issues. In the original experiment a bright red, odorless, nonirritating, alcohol-soluble dye (rhodamine-B base) was used to mark the uppermost portion of an eyebrow ridge and the top half of the opposite ear on a number of chimpanzees (Gallup, 1970). I chose a red dye not only because it would contrast the chimpanzee's skin and be obvious, but as a color, red also has intrinsic biological significance (e.g., blood, tissue damage, etc.). In some species, however, red may be a poor choice because it fails to stand out against the animal's skin. We encountered this problem with gorillas that have very dark skin, and used a pink mark instead (Suarez & Gallup, 1981).

In the event that animals fail to show mark-directed behavior it is essential to include a control condition that demonstrates not only that the mark could be seen in the mirror, but that it would have been touched and investigated had they been capable of localizing it. There are at least two ways to satisfy this requirement. One means of providing indirect evidence that the mark constitutes a psychologically salient stimulus is to monitor the behavior of cagemates at the conclusion of testing, when the animal is returned to its home cage. If cagemates actively touch and inspect the mark on the target animal, then it is reasonable to suppose that the subject would probably have shown similar interest had it known where the mark was (e.g., Gallup et al., 1980; Povinelli, 1989). One can also apply a control mark to a portion of the animal's body that can be seen directly without a mirror, and then monitor responses to the accessible mark prior to reintroducing the mirror as a test of mark salience (Gallup et al., 1980; Suarez & Gallup, 1981). Neither of these procedures, however, are without tactical drawbacks. The problem is that in order to constitute a fair test of self-recognition one needs independent evidence of mark salience, but the mere act of collecting that evidence may reduce or compromise such salience. That is, when animals are exposed to marks that can be seen directly (either on other animals or on themselves) there is an explicit opportunity for them to discover that the mark is inconsequential, and as a result the animals may habituate to the mark and be less mark responsive on the test trial. There is, however, at least one way to partially counter this problem (Gallup, 1987): To minimize the effects of habituation one could vary the color of control and experimental marks (counterbalanced of course, across animals).

The other obvious issue related to mark salience is that of conducting more than one mark test. For the reasons outlined here, repeated testing occurs at the expense of reduced mark salience; therefore, negative results on subsequent tests become progressively uninterpretable. (That is, was the failure to show mark-directed behavior a consequence of the animal's inability to localize the mark or was it merely due to a growing lack of interest in the now-familiar and inconsequential mark.) To the extent that repeated testing is unavoidable, changing the color of the mark from one test trial to the next would be one way to reduce the problem of diminished mark salience.

These problems are vividly illustrated in a recent study by Swartz and Evans (1991) in which they report self-recognition in only one out eleven chimpanzees. Rather than conduct mark tests to validate apparent instances of self-directed responding, Swartz and Evans adopted a procedure that involved repeated mark testing done independently of whether the animals had shown evidence of self-directed behavior, with some animals receiving as many as five mark tests over a period of 80 hours of mirror exposure. Moreover, not only were the animals marked repeatedly, but following successive mark tests they were returned to cages containing other chimpanzees that were also part of the same study. Thus, in some instances prior to even being tested for the first time, some of the animals had already received extensive exposure to these marks on cagemates, and to the extent that animals were mark tested and/or exposed to others with facial marks before learning to recognize themselves in mirrors, the failure to find "evidence" of self-recognition is hardly surprising. It has certainly been my experience that even under ideal conditions, the chimpanzee's interest in unobtrusively applied facial marks wanes in a relatively short period of time.

There are several other points to consider before conducting a mark test. In order for mark-directed behavior to constitute compelling evidence of MSR it is important to apply the marks unobtrusively. Anesthetization has the advantage of rendering the subject completely unconscious, but there may be medical or ethical reasons (e.g., the use of humans) that preclude this procedure. There are, however, a number of noninvasive alternatives to anesthetization. One would be to apply the marks while the subject is sleeping. The other more widely used alternative is to employ a sham marking procedure (Calhoun, 1983). Under conditions of sham marking, the area of the face or body that is ultimately targeted for marking is repeatedly swabbed with a clear solution. For example, if the dye to be used is alcohol soluble, one would apply alcohol with a cotton swab to the targeted area on several occasions until the animal no longer seemed to notice. Once habituation to the application of alcohol occurs, the mark can be applied in a fairly unobtrusive manner (provided that the animal can be visually distracted so that it does not see the dyed swab).

Videotape as evidence

Nowadays claims concerning self-recognition in different species are being made on the basis of videotape recordings of "critical" behavioral sequences (e.g., Boccia, *SAAH*23; R. L. Thompson & Boatright-Horowitz, *SAAH*22). The use of a video camera to supplement other observational techniques has a lot to recommend it, in that an enduring record is available that can be scored in a variety of different ways by a variety of different people. Behavior on videotape, however, is not necessarily, as some investigators would like to believe, a more compelling form of evidence. Whether the behavior depicted on videotape represents definitive data is a function of

1. the setting in which the tape was made,
2. the conditions that led up to the episode in question, and
3. the presence of appropriate controls.

There is nothing about videotape that guarantees that any of these basic and necessary conditions of experimental design will be satisfied.

Another important question concerning the use of videotape as evidence is that of replication. Simply because the supposedly critical sequence can be played over and over does not mean it can be replicated. It is quite possible, for example, that if I were to mark a monkey on its forehead it might eventually, if left in front of a mirror for a long enough period of time, accidentally touch the mark quite independent of what it saw in the mirror. If I were to edit my tape and show just this sequence, it would appear as if the animal had passed the mark test.

Replication by other independent observers, along with use of appropriate controls, is the cornerstone of the scientific method. Although it is a step in the right direction, even a replication involving the same subject does not suffice. To become a bona fide scientific fact an observation must be replicated by different observers, in different settings, on different subjects. Quite independent of videotape, the issue of replication becomes particularly problematic in instances in which only a single subject is involved. Koko's ability to recognize herself in a mirror (Patterson & Cohn, *SAAH*17) is a case in point. In the absence of evidence of self-recognition in other gorillas, Koko's performance needs, at the very least, to be replicated by other independent investigators.

Another advantage of videotape is that it can be shown to large audiences and even reproduced and shared with other investigators. On the other hand, how one chooses to interpret a particular episode on videotape may be a matter of substantial differences of opinion. R. L. Thompson & Boatright-Horowitz (*SAAH*22) and Boccia (*SAAH*23) have made claims concerning the presence of self-recognition in several pigtailed macaques. In an effort to substantiate those claims, they have produced videotapes of the monkeys during mark tests, and these tapes have been shown to people attending various professional conferences. Nevertheless, the content of their tapes belies their claims. Thompson's videotape shows a pigtailed macaque who had been trained to make eye contact with its own reflection in a mirror under an extremely severe water deprivation schedule (46 hours). The videotape depicts a single instance of the animal brushing the mark on its face with the back of its hand, which is then immediately followed by a social response (jaw thrust) at the mirror. Thompson found no evidence for any self-directed behavior before the test, and needless to say the presence of social responses to the mirror seriously compromises his claim for self-recognition in this particular animal. Despite extended training and extensive exposure to mirrors in other pigtailed macaques, this effect has never been replicated (R. L. Thompson, personal communication, 1991).

In a related study of mirror behavior in a group of pigtailed macaques with a history of extensive exposure to mirrors, Boccia (*SAAH*23) also reports an instance of mark-directed behavior in one of the animals that she tested. However, a videotape of the animal's behavior on the mark test, which she showed at the conference on self-awareness at Sonoma State University in 1991, fails to confirm her claims for mark-directed behavior. On the test trial the animal touched its face on several occasions but not while oriented to the mirror, and it never actually made contact with the mark. Boccia has indicated that she has other tapes that contain more compelling instances, but these have not been made available to me. I have experienced the same failure to be allowed access to tapes in the past.

Clearly, if people are going to make claims based on videotaped sequences, then those tapes should become part of the public domain, and reasonable requests for the tapes should be honored. Not only does the failure to comply with these requests render the claims for self-recognition suspect, it raises fundamental questions about science as an objective, empirical enterprise. In order to qualify as data, objects and events not only need to be observable (directly or indirectly); but they must also be made in such a way as to permit others to confirm or disconfirm their existence. Indeed, it has been clearly stipulated in the *Journal of Comparative Psychology*'s instructions to authors, since I assumed the editorship in 1989, that both manuscript submission as well as publication is predicated on the author's compliance with reasonable requests for access to the original data. Nowadays, in many instances, videotape has become the primary data set.

Videotape as stimulus material

The opportunity to see oneself in real time (i.e., live) on a video monitor is an interesting analog to seeing oneself in a mirror. (Parenthetically, it should be noted that shadows represent another naturally occurring parallel to mirrors [see Cameron & Gallup, 1988].) One of the interesting things about the use of videotape as a stimulus is that it can be presented in a delayed mode. For example, Marten and Psarakos (*SAAH*24) have shown that dolphins show more interest in a live video display of themselves than in watching a tape of themselves in a delayed playback mode. They would like to interpret this difference as evidence that self-awareness in dolphins may be a time-bound or on-line phenomenon. There are, however, far more parsimonious ways of explaining these data. Maybe dolphins simply discover in the delayed mode that they have no control over the image. In the case of a live video image (or that seen in front of a mirror), the observer has complete control over the behavior of the image: The image never responds independent of the observer and always mimics what the observer does. A videotape taken earlier represents the opposite end of this continuum, in that there is a complete dissociation between what is seen on the monitor and what the observer is doing at the moment. In the former case there is a perfect correlation between the behavior of the observer and the behavior of the image on the screen, while in the latter there is no correlation. A typical social encounter falls somewhere between these two extremes: The behavior of one participant serves as a stimulus for the behavior of the other; as a consequence one can predict (albeit imperfectly) what one organism will do based on a knowledge of what the other is doing. In other words, social behavior represents a partial correlation between the behavior of the participants; that is, the correlation between the behavior of the participants will be greater than zero but less than perfect. Given the dolphin's prior social experience with other dolphins that behave in predictable ways, another explanation of Marten's data might be that they simply prefer video images over which they can exercise some control.

It would be interesting to see how Marten and Psarakos's dolphins would respond to a delayed videotape of themselves versus a delayed videotape of another dolphin in the same situation. Under these circumstances there is no prediction/control confound, but it would be important to ensure that the animal's familiarity with the other dolphin on tape be comparable to its familiarity with its own image. If it has never seen itself on a video monitor or in a mirror, it will be seeing the image of a dolphin it has never seen before, and any difference in its responsiveness to that and the tape of a familiar dolphin might be due to greater interest in strange dolphins.

The use of video technology to produce stimulus materials has a number of other interesting research applications. Using a modified version of a technique pioneered by Lewis and Brooks-Gunn (1979) with children, I have outlined an alternative means of assessing self-recognition in dolphins (Gallup, 1979). A dolphin would be given an opportunity to see itself live on an underwater video monitor. Occasionally one could switch to a tape containing the video image of another dolphin in the same situation. Once you were satisfied that the dolphin could distinguish between its own live image on the screen and that of the other dolphin, you could test for self-recognition by the unobtrusive introduction of a visual probe into the background. While watching the videotape of another dolphin, the appearance of an object in the background should not cause the first dolphin to turn around. For example, if you are watching a television program that depicts someone sneaking up on somebody else from behind, you do not typically turn around. However, while watching its own live image on the monitor, a dolphin that can recognize itself ought to turn around in response to the appearance of the probe on the screen in order to confront the novel object directly (see Robinson, Connell, McKenzie, & Day, 1990 for some limitations of this technique).

Alternatives to self-recognition as an index of self-awareness

I have argued on several occasions (e.g., Gallup, 1982, 1983, 1985, 1987, 1988, and 1991) that in addition to being able to recognize themselves in mirrors, organisms that can conceive of themselves ought to be able to use their own experience and knowledge to infer comparable experiences and knowledge states in other organisms. According to the model I developed, species (as well as individuals within a species) that can, and those that cannot, recognize themselves in mirrors ought to behave quite differently on tests of social intelligence. For instance, I would predict that if one were to give both chimpanzees and rhesus monkeys experience with a variety of visual obstructions (e.g., blindfolds or opaque goggles) and then confront them with cagemates wearing blindfolds, they ought to behave quite differently (Gallup, 1985, 1988). Chimpanzees, because of their capacity to become the object of their own attention (as evidenced by MSR), ought to be capable of using their experience with blindfolds to model the impaired visual capacity of a blindfolded companion, whereas rhesus macaques should fail to make the distinction. A growing number of papers provide dramatic evidence in support of these predictions (Cheney & Seyfarth, 1990; Povinelli & deBlois, 1992; Povinelli, Nelson, & Boysen, 1990; Povinelli, Parks, & Novak, 1991). As a consequence, I have recently argued (Gallup, 1991), in the context of the continued debate about which species can or cannot recognize themselves in mirrors, that it is time for people to put their mirrors down and begin to conduct some of these other tests of social intelligence (e.g., visual perspective taking). I would predict, for example, that most gorillas (at least those lacking the environmental benefits enjoyed by Koko) would behave like rhesus monkeys rather than chimpanzees on these kind of tests. On the other hand, if Povinelli is right about environmental triggers for the production of vestigial psychological traits, Koko should approximate her chimpanzee counterparts on such tasks.

There is much more to being self-aware than merely recognizing yourself in a mirror.

References

Amsterdam, B. (1972). Mirror self-image reactions before age two. *Developmental Psychobiology,* 5, 297–305.

Anderson, J. R. (1984). Monkeys with mirrors: Some questions for primate psychology. *International Journal of Primatology,* 5, 81–98.

Anderson, J. R., & Roeder, J. (1989). Responses of capuchin monkeys (*Cebus apella*) to different conditions of mirror-image stimulation. *Primates,* 30, 581–587.

Calhoun, S. (1983). *The question of contingent image self-recognition in apes and monkeys.* Unpublished doctoral dissertation, City University of New York.

Calhoun, S., & Thompson, R. L. (1988). Long-term retention of self-recognition by chimpanzees. *American Journal of Primatology,* 15, 361–365.

Cameron, P. A., & Gallup, G. G., Jr. (1988). Shadow recognition in human infants. *Infant Behavior and Development,* 11, 465–471.

Cheney, D. L., & Seyfarth, R. M. (1990). Attending to behavior versus attending to knowledge: Examining monkeys' attribution of mental states. *Animal Behaviour, 40,* 742–753.

Epstein, R., Lanza, R. P., & Skinner, B. F. (1981). "Self-awareness" in the pigeon. *Science, 211,* 695–696.

Gallup, G. G., Jr. (1968). Mirror-image stimulation. *Psychological Bulletin, 70,* 782–793.

(1970). Chimpanzees: Self-recognition. *Science, 167,* 341–343.

(1979). Self-awareness in primates. *American Scientist, 67,* 417–421.

(1982). Self-awareness and the emergence of mind in primates. *American Journal of Primatology, 2,* 237–248.

(1983). Toward a comparative psychology of mind. In R. L. Mellgren (Ed.), *Animal cognition and behavior* (pp. 473–510). New York: North Holland.

(1985). Do minds exist in species other than our own? *Neuroscience and Biobehavioral Reviews, 9,* 631–641.

(1987). Self-awareness. In G. Mitchell (Ed.), *Comparative primate biology, Vol. 2, Part B: Behavior, cognition, and motivation* (pp. 3–16). New York: Liss.

(1988). Toward a taxonomy of mind in primates. *Behavioral and Brain Sciences, 11,* 255–256.

(1991). Toward a comparative psychology of self-awareness: Species limitations and cognitive consequences. In G. R. Goethals & J. Strauss (Eds.), *The self: An interdisciplinary approach* (pp. 121–135). New York: Springer-Verlag.

Gallup, G. G., Jr., McClure, M. K., Hill, S. D., & Bundy, R. A. (1971). Capacity for self-recognition in differentially reared chimpanzees. *The Psychological Record, 21,* 69–74.

Gallup, G. G., Jr., & Suarez, S. D. (1991). Social responding to mirrors in rhesus monkeys (*Macaca mulatta*): Effects of temporary mirror removal. *Journal of Comparative Psychology, 105,* 376–379.

Gallup, G. G., Jr., Wallnau, L. B., & Suarez, S. D. (1980). Failure to find self-recognition in mother–infant and infant–infant rhesus monkey pairs. *Folia Primatologica, 33,* 210–219.

Ledbetter, D. H., & Basen, J. A. (1982). Failure to demonstrate self-recognition in gorillas. *American Journal of Primatology, 2,* 307–310.

Lethmate, J., & Dücker, G. (1973). Untersuchungen zum Selbsterkennen im Spiegel bei Orang-utans und einigen anderen Aftenarten. *Zeitschrift fur Tierpsychologie, 33,* 248–269.

Lewis, M., & Brooks-Gunn, J. (1979). *Social cognition and the acquisition of self.* New York: Plenum.

Lin, A. C., Bard, K. A., & Anderson, J. R. (1992). Development of self-recognition in chimpanzees (*Pan troglodytes*). *Journal of Comparative Psychology, 106,* 120–127.

Mitchell, R. W. (1993). Mental models of mirror self-recognition: Two theories. *New Ideas in Psychology, 11,* 295–325.

Povinelli, D. J. (1987). Monkeys, apes, mirrors, and minds: The evolution of self-awareness in primates. *Human Evolution, 2,* 493–507.

(1989). Failure to find self-recognition in Asian elephants (*Elephas maximus*) in contrast to their use of mirror cues to discover hidden food. *Journal of Comparative Psychology, 103,* 122–131.

(1991). *Social intelligence in monkeys and apes.* Unpublished doctoral dissertation, Yale University.

Povinelli, D. J., & deBlois, S. (1992). Young children's (*Homo sapiens*) understanding of knowledge formation in themselves and others. *Journal of Comparative Psychology, 106,* 228–238.

Povinelli, D. J., Nelson, K. E., & Boysen, S. T. (1990). Inferences about guessing and knowing by chimpanzees (*Pan troglodytes*). *Journal of Comparative Psychology, 104,* 203–210.

(1992). Comprehension of social role reversal by chimpanzees: Evidence of empathy? *Animal Behavior, 43,* 633–640

Povinelli, D. J., Parks, K. A., & Novak, M. A. (1991). Do rhesus monkeys (*Macaca mulatta*) attribute knowledge and ignorance to others? *Journal of Comparative Psychology, 105,* 318–325.

(1992). Role reversal by rhesus monkeys, but no evidence of empathy. *Animal Behavior, 44,* 269–281.

Povinelli, D. J., Rulf, A. B., Landau, K., & Bierschwale, D. (in press). Self-recognition in chimpanzees: Distribution, ontogeny, and patterns of emergence. *Journal of Comparative Psychology*.

Robinson, J. A., Connell, S., McKenzie, B. E., & Day, R. H. (1990) Do infants use their own images to locate objects reflected in a mirror? *Child Development, 61*, 1558–1568.

Schulman, A. H., & Kaplowitz, C. (1977). Mirror-image response during the first two years of life. *Developmental Psychobiology, 10*, 133–142.

Suarez, S. D., & Gallup, G. G., Jr. (1981). Self-recognition in chimpanzees and orangutans, but not gorillas. *Journal of Human Evolution, 10*, 175–188.

(1986). Social responding to mirrors in rhesus macaques (*Macaca mulatta*): Effects of changing mirror location. *American Journal of Primatology, 11*, 239–244.

Swartz, K. B., & Evans, S. (1991). Not all chimpanzees (*Pan troglodytes*) show self-recognition. *Primates, 32*, 483–496.

Traub, A. C., & Orbach, J. (1964). Psychophysical studies of body-image: 1. The adjustable body-distorting mirror. *Archives of General Psychiatry, 11*, 53–66.

4   From self-recognition to theory of mind

György Gergely

Gordon Gallup's significant contribution to the psychological study of the self is twofold: He introduced an objective empirical method to examine self-recognition in nonverbal organisms through the mirror self-recognition (MSR) test, and he developed a challenging theoretical account that postulates an inherent link between the ability for MSR and the capacity to infer mental states in others, that is, for having a theory of mind (ToM).

Briefly, Gallup's model states that

1. MSR implies self-awareness;
2. this "self-awareness is tantamount to being aware of being aware" (Gallup, 1991, p. 123); and
3. being aware of one's own "mental states and their relation to various external events" allows one "to gain inferential access" to the mental states of others (Gallup, 1991, p. 123).

Thus, because Gallup's model attributes to the organism showing MSR a level of self-awareness that satisfies the representational preconditions for inferring beliefs in others, he predicts a developmental synchrony between the appearance of MSR and "introspectively based social strategies," such as intentional deception, that implies a ToM.

Evidence demonstrating that organisms that seem to possess a ToM – such as young children (Astington, Harris, & Olson, 1988) and chimpanzees (Povinelli, Nelson, & Boysen, 1990; Woodruff & Premack, 1979) – also show MSR (Gallup & Suarez, 1986), is clearly in line with this prediction. In contrast, however, recent data on normal and autistic children suggest that the corollary prediction, that organisms showing MSR will also exhibit a ToM, is empirically false, as there are organisms that show MSR but, to different degrees, lack the capacity for a ToM (see Mitchell, 1993). Thus, normal 3-year-olds, who clearly exhibit MSR (Lewis & Brooks-Gunn, 1979), have been shown to fail on ToM tasks that involve attribution of false beliefs to others (Perner, 1991; Perner, Leekam, & Wimmer, 1987; Wimmer & Perner, 1983), whereas childhood autistics, who seem to lack a ToM (Baron-Cohen, Leslie, & Frith, 1985; Perner, Frith, Leslie, & Leekam, 1989), can, when of the appropriate mental age, recognize themselves in the mirror (Dawson & McKissick, 1984; Ferrari & Matthews, 1983; Neuman & Hill, 1978).

I shall argue on the basis of such evidence that Gallup's conceptual analysis overestimates the cognitive prerequisites minimally necessary for MSR, which, as I shall attempt to show, are but a first (though highly significant) step in the rather complex developmental sequence leading to the eventual establishment of a ToM. In addition, I shall propose an alternative hypothesis about the nature of the underlying cognitive mechanism whose appearance allows apes and humans to develop MSR.

Evidence for mirror self-recognition without a theory of mind

Based on a review of the child literature, Gallup and Suarez (1986) argue that the synchrony prediction of the Gallup model is supported by the available evidence, concluding that "it is now well established that by the end of their second year, children can infer mental states of others" (p. 20). This summary conclusion, however, seems unwarranted in the light of converging evidence from the ToM literature (see e.g., Astington et al., 1988) which indicates a complex developmental sequence leading to the establishment of a ToM only around 4 years of age. For example, the ability to attribute false beliefs to others that is a precondition for social strategies based on a ToM, such as intentional deception demonstrated by chimpanzees but not by rhesus monkeys (Povinelli et al., 1990; Woodruff & Premack, 1979), seems to be absent in 3-year-olds (Perner, 1991; Perner et al., 1987; Wimmer & Perner, 1983), though these children have presumably achieved MSR by the end of their second year (Lewis & Brooks-Gunn, 1979).

There are several proposals to explain why 3-year-olds fail to attribute false beliefs to others (apart from other related conceptual problems they seem to have; see Perner, 1991). Leslie (1988) suggests that the three-year old's problem is not that he cannot represent or report on mental states, as, according to Leslie, such a metarepresentational ability is present already at age 2, as shown by the emergence of pretend play at that age (see also Fodor, 1992). Rather, Leslie believes that the 3-year old child "has yet to understand how beliefs relate *causally* to situations in the world" (p. 35). In contrast, Perner (1991) argues that 3-year-olds lack the ability to metarepresent beliefs, which is necessary to understand that a belief can be false, i.e., that it can misrepresent reality: "The child has to mentally represent that . . . a false statement about reality has an interpretation (described situation) that is at odds with reality (its referent situation)" (p. 89).

The point I wish to make here is that irrespective of whether the 3-year-old's cognitive deficit is related to a lack of metarepresentational ability à la Perner, or to a lack of integrating the ToM with a causal theory of the world à la Leslie, the finding is highly problematic for Gallup's model of MSR, which attributes both of these abilities to children already at $1\frac{1}{2}$–2 years of age (Mitchell, 1993). On one hand, according to Gallup's model, children at this age have the capacity for "being aware of being aware," which certainly qualifies as a metarepresentational ability to mentally represent one's mental state. On the other hand, his proposal that "given a knowledge of my mental states and *their relation to various external events*, I now have a means of gaining inferential access to yours" (Gallup, 1991, p. 123, emphasis added), seems to imply an understanding of the causal relations between beliefs and external situations.

A further difficulty for Gallup's model is posed by the finding that autistic children, who reach the mental age of normal children showing MSR, can also recognize themselves in the mirror (Dawson & McKissick, 1984; Ferrari & Matthews, 1983; Neuman & Hill, 1978). At the same time, irrespective of their mental age, childhood autistics fail on the kind of false-belief attribution tasks developed to test for ToM, as do 3-year-olds (Baron-Cohen et al., 1985; Perner et al., 1989), and show a striking lack of pretend play (Baron-Cohen, 1987), which in normal children appears at $1\frac{1}{2}$–2 years of age simultaneously with MSR. Baron-Cohen et al. (1985) argue that childhood autistics suffer from a specific deficit of lacking the ability to form metarepresentations, which, according to Leslie (1987), is the representational basis for pretend play as well as a precondition for a ToM. Due to their "inability to represent mental states . . . autistic subjects are unable to impute beliefs to others and are thus at a great disadvantage when having to predict the behaviour of other people" (Baron-Cohen et al., 1985, p. 43).

The above results suggest, therefore, that the developmental synchrony prediction based on Gallup's model of MSR does not hold, and that the cognitive capacities involved in MSR are either independent or satisfy only a necessary but not a sufficient condition for a ToM.

Cognitive prerequisites for a theory of mind

MSR in normal infants appears between $1\frac{1}{2}$ and 2 years of age, more or less in synchrony with a set of other qualitatively new behavioral achievements such as pretend play, understanding means–ends relationships, or solving "invisible displacements" tasks (Piaget, 1937/1954). Recently several authors have hypothesized that these behavioral changes correspond to a qualitative maturational change in the representational system that allows the infant to break away from the single, primary model of reality that he was building up until then on the basis of direct perceptual evidence (e.g., Gergely, 1992; Leslie, 1987; Meltzoff, 1990; Meltzoff & Gopnik, 1989; Perner, 1991). These proposals agree that due to the change in the representational system, the infant becomes able to construct multiple representations that can model hypothetical as well as real situations eventually leading to the development of a ToM. There is disagreement, however, concerning the precise nature of this qualitative restructuring of the representational system (see Fodor, 1992; Leslie, 1987; Perner, 1991; Wellman, 1990). One open question is whether the change involves a single, unitary enrichment of the representational system, or whether it can be factored out into a number of independent changes, out of which a ToM is constructed. I shall argue that the developmental asynchronies in the appearance of the qualitatively new skills in question between $1\frac{1}{2}$ and 4 years of age, as well as the pathological dissociation between them in autistics, can be best explained by the latter alternative.

We can identify at least five separable aspects of the qualitative reorganization of the representational system that allows for the development of a ToM:

A1. The ability to modify the primary representation of reality based on information coming from sources other than direct perception, for example, through conceptual inference. (This also seems to be a necessary precondition for success on the Stage 6 "serial invisible displacements" task of Piaget's [1937/1954] object permanence series, in which the child does not directly perceive the last transformation of the position of the object searched for: He has to infer it [see Meltzoff, 1990, p. 23; and Watson, 1989, Chap. 4, on this point]. However, the inferentially modified model is of the same ontological status: It is a model of *reality*.)

A2. The ability to represent reality in multiple models. (This is a further precondition for solving the Stage 6 serial invisible displacements task: The child has to retain the last model of the location of the searched object that was directly perceived ["key in hand"] and compare it with the current one ["key not in hand"]. The mismatch leads to an updating of the current model through inference [see A1]: "Key must have been released under cloth where hand was last seen disappear" → ["key under cloth"]. Note that the multiple models are still of the same ontological status: They are models of reality.)

A3. The ability to create and differentiate between models of hypothetical situations and models of reality. (This is a precondition for pretend play. Note that the multiple models simultaneously involved in pretense are explicitly represented as having different ontological status, e.g., a "real" banana being used as a "hypothetical" telephone [see Leslie, 1987; Meltzoff, 1990; Perner, 1991].)

A4. The ability for metarepresentation, that is, for representing the representational relation itself, which then allows for the differentiation between sense and reference, and the understanding of misrepresentation. (This is a precondition for appreciating the appearance–reality distinction around 4 years [Flavell, Flavell, & Green, 1983] and for the attribution of false beliefs to others also appearing at age 4 (Perner, 1991; Perner et al., 1989; Wimmer & Perner, 1983].)

A5. The ability to understand the causal relations between external events (perceptions, actions) and mental states. (This is a further precondition for false belief attribution [see Leslie, 1988].)

Cognitive prerequisites for mirror self-recognition

Which of the above abilities are necessary for passing Gallup's MSR test, that is, to understand that the mark one perceives on the mirror image of one's not–directly visible body parts (e.g., forehead or ear) is, in fact, on the corresponding part of one's body surface? It seems that minimally the following preconditions need to be fulfilled to pass the test:

1. The organism has to appreciate the duality between the projection of objects in the mirror and their relative position in the environment. In other words, it has to be able to use the spatial and featural correspondence relation

between objects in the mirror and their real world counterpart to locate and identify objects in the world based on their reflection in the mirror.

2. The organism must have the capacity to detect the perfect contingency relation between its body movements and the corresponding movements of its mirror image.

3. It must have constructed a visual feature representation of the typical physical appearance of the not–directly visible parts of its body.

4. Upon detecting the mismatch between this visual representation and the corresponding image in the mirror, the organism must be able to reestablish the mirror–reality-correspondence relation by modifying its self-representation through attributing the mismatching visual features in the mirror image to the representation of its body.

Data on rhesus monkeys (Gallup & Suarez, 1986), pigtailed macaques (Boccia, *SAAH*23), elephants (Povinelli, 1989) and infants less than 15 months of age (Bertenthal & Fisher, 1978; Lewis & Brooks-Gunn, 1979), who do not show MSR, indicate that they nevertheless satisfy conditions 1 and 2 above. These organisms can comprehend the duality implicit in mirrors, shown by the fact that they correctly locate objects (*other* than their selves) in space based on mirror information. Infants as early as three months can detect the contingency relation between their own movements and the corresponding movements in a live video feedback display (Bahrick & Watson, 1985). Povinelli (1989) demonstrated that elephants are capable of mirror-guided search for hidden food items, and Rumbaugh, Richardson, Washburn, Savage-Rumbaugh, and Hopkins (1989) reported that rhesus monkeys using joysticks can guide a spatially displaced contingently moving target on a video monitor.

Nevertheless, these organisms seem unable to construct a representation of the visual appearance of their nonvisible body surface (3) in spite of extensive experience with mirrors, and they treat their own mirror image as a conspecific. In chimpanzees, orangutans (Gallup & Suarez, 1986), and infants (Lewis & Brooks-Gunn, 1979) there is a transitional period before MSR (as indicated by passing the facial mark test) occurs, characterized by the systematic exploration of their not–directly visible body parts in front of the mirror. This can plausibly be interpreted as corresponding to the construction of the visual feature representation (3) of their appearance.

Why is it that the lower organisms examined, who understand the duality of mirrors (1), and can detect the kinesthetic–visual contingency between their movements and the corresponding movements of their image (2), are nevertheless unable to use these informational sources to modify their self-representation (3) and (4)?

I suggest that the nature of the difficulty for these organisms lies in their inability to update their primary model of reality based on information other than direct perception (see ability A1, above). Because, as I shall argue, the modification of their self-representation by the attribution of the visual features of the corresponding mirror image would require a conceptual inference, this kind of change in their model of reality seems beyond their cognitive capacities.

We can assume that the organism placed in front of the mirror has a primary model of reality that contains representations of external objects as well as a representation of the self. However, because this representation of reality is based solely on direct perception, the representation of the nonvisible parts of the self (unlike that of other objects or of the visible parts of the self) is limited to kinesthetic, proprioceptive, and haptic information; as a result it lacks visual feature specification. Thus, the primary representation of the self, often referred to in the literature as the "existential self" (Lewis & Brooks-Gunn, 1979) or the "sense of core self" (Stern, 1985), is formed from the directly perceived invariant properties of the self.[1]

When confronted with a mirror, with experience the organism learns to map the visual information in the mirror onto its primary model of reality, relying on the systematic spatial and featural correspondence relation that it discovers between the two. However, this perfect correspondence relation breaks down when the visual features of the image of the nonvisible parts of the body are mapped onto the representation of the corresponding object in space, that is, the representation of the self. Whereas there is partial match in terms of amodal properties such as shape and movement (cross-modal kinesthetic–visual match), the spatially correlated visual features of the image (such as its color) match nothing in the representation of the self in the reality model.

The systematic correspondence between the primary model of reality and the mirror image can be salvaged only by attributing to the self-representation the mismatching visual features of the corresponding object in the mirror. This, however, can be done only inferentially, through analogy to other external objects: Because the spatial position and visual features of objects in the mirror image correspond perfectly to those of the objects in the primary representation of reality, and because the self is represented as one of these objects in space, therefore, just as other objects at other locations, the self must also be truthfully characterized by the visual features of the corresponding object in the mirror image (see also Mitchell, 1993).

However, organisms that cannot modify their primary representation of reality by inference will not be able to salvage the mirror–reality correspondence in the above manner. Because they also lack the capacity to conceive of misrepresentation (a metarepresentational ability, see A4 above), the only way they can interpret their self-reflection is by abandoning the duality assumption and treating their projection as a "real" conspecific directly perceived "out there."

The claim that non- or preverbal organisms (such as great apes or 15–20-month-old infants) have a cognitive capacity that allows them to go beyond the direct perceptual information given by drawing deductive inferences, is often seen as controversial and is not generally accepted. However, it should be pointed out that in the recent developmental literature several serious proposals have appeared that consider precisely such a possibility in relation to successful performance on such nonverbal tasks as sensorimotor weight conservation (Mounoud & Bower, 1974/1975) and, especially, the Piagetian

Stage 6 invisible displacements task of object permanence (see Bower, 1982, 1989; Meltzoff, 1990, p. 23; and Watson, 1989, chapter 4).

Assuming, therefore, that success on the Stage 6 serial invisible displacements task involves deductive inference (see Watson, 1989, chapter 4, for a thorough discussion of this point), it is clear that the present hypothesis according to which the same kind of inferential capacity also underlies MSR, would be strengthened if we found a developmental correlation between the emergence of success on these two kinds of tasks. In fact, Natale and Antinucci (1989), based on a critical review of the primate literature as well as on further comparative empirical studies, concluded that only apes, but not monkeys, class with humans in their ability to solve the Stage 6 invisible displacements task (see also Parker, 1991, Table 2, p. 441). Further support for the current hypothesis is provided by the finding that childhood autistics of the right mental age, apart from showing MSR (Dawson & McKissick, 1984; Ferrari & Matthews, 1983; Neuman & Hill, 1978), are also able to solve the Stage 6 serial invisible displacements task (e.g., Curcio, 1978; Dawson & Adams, 1984). Finally, that Bertenthal and Fisher (1978) found a high correlation between the development of self-recognition and object permanence (where success on the MSR rouge-mark test strongly correlated with passing the Stage 6 object-permanence task), also supports the present hypothesis that success on these different kinds of tasks is related to the emergence of the same underlying cognitive capacity: the ability for deductive inference.[2]

To sum up: It seems that to pass the MSR facial mark test an organism needs to be able to update its primary model of reality through inference as well as through direct perception, and to represent reality in multiple models, such as reality as reflected in the mirror versus reality as perceived "out there." It seems that only great apes and humans have both of these abilities, while other species such as elephants or rhesus monkeys have the latter but lack the former.

An important feature of this model is that to pass the MSR test, an organism does *not* need to be able to represent hypothetical situations. This further representational precondition must be fulfilled, however, for pretend play in which the organism must be able simultaneously to represent entities as belonging to the primary model of reality and to a hypothetical model in which their normal referential properties are suspended (see Leslie's, 1987, "decoupling" mechanism). This provides an explanation for the autistic data. In childhood autism conditions A1 and A2 are intact, which allows the autistic child to construct a visual representation of himself based on mirror information, thus passing the MSR test. However, autistic children show a striking lack of pretend play, suggesting that their specific impairment relates to condition A3, the ability to represent hypothetical as well as real situations (Leslie, 1987). That autistic children can succeed in simple visual perspective taking tasks (Hobson, 1984) is also in line with the present hypothesis, as such tasks, which necessitate representing reality in multiple models, do not involve representing states that do not correspond to reality.

Finally, for social strategies based on a ToM, which allow for the correct attribution of false beliefs to others, conditions A4 (metarepresentation) and A5 (causal theory of mental states) must also be fulfilled. Three-year-olds lack at least one of these abilities, while 4-year-olds (and, probably, mature chimps) exhibit both.

In closing, I wish to make a final comment on the developmental status of the crucial cognitive change (A1) in the representational system that allows for the modification of the primary model of reality on the basis of inferential rather than direct perceptual evidence. Although this clearly is a less powerful condition than the cognitive changes envisioned in Gallup's model of MSR, it is, nevertheless, developmentally and probably phylogenetically, momentous. This is so because it provides the organism with qualitatively new conditions for constructing and enriching the representation of the "categorical self" through the introjection of inferred properties. This change is clearly cognitive in nature, and therefore I agree with Gallup's criticism (e.g., Gallup, 1991; Gallup & Suarez, 1986) of those alternative approaches that attempt to reduce the development of MSR to an increase in perceptuomotor skills (Bertenthal & Fisher, 1978; Menzel, Savage-Rumbaugh, & Lawson, 1985; Rumbaugh et al., 1989).

Notes

1. Contrary to Gallup, I see no reason why this self-representation could not become the object of the organism's attention in the same way as the representations of other objects can, of course, only in the simple (nonreflexive) sense of being aware of the invariant stimulus properties of the self that are directly perceived (see also Mitchell, 1993).
2. Of course, the competence for deductive inference is claimed here to be only a necessary and not a sufficient condition for MSR. In fact, there is evidence (see Gagnon & Doré, 1992, for a review) indicating success on the Stage 6 serial invisible displacements task in several organisms, such as capuchin monkeys (Mathieu, Bouchard, Granger, & Herscovitch, 1976), dogs (Gagnon & Doré, 1992; Pasnak, Kurkjian, & Triana, 1988), and gorillas (Natale & Antinucci, 1989), in which MSR has not been demonstrated (but see Patterson & Cohn, *SAAH*17, who present evidence for MSR in at least one gorilla, Koko).

References

Astington, J. W., Harris, P. L., & Olson, D. R. (Eds.) (1988). *Developing theories of mind.* Cambridge University Press.

Bahrick, L. R., & Watson, J. S. (1985). Detection of intermodal proprioceptive-visual contingency as a potential basis of self-perception in infancy. *Developmental Psychology, 21*(6), 963–973.

Baron-Cohen, S. (1987). Autism and symbolic play. *British Journal of Developmental Psychology, 5,* 139–148.

Baron-Cohen, S., Leslie, A. M., & Frith, U. (1985). Does the autistic child have a "theory of mind"? *Cognition, 21,* 37–46.

Bertenthal, B., & Fisher, K. (1978). Development of self-recognition in the infant. *Developmental Psychology, 14,* 44–50.

Bower, T. G. R. (1982). *Development in Infancy* (2nd ed.). San Francisco: Freeman.
    (1989). *The Rational Infant.* New York: Freeman.

Curcio, F. (1978). Sensori-motor functioning and communication in mute autistic children. *Journal of Autism and Childhood Schizophrenia, 8,* 281–292.

Dawson, G., & Adams, A. (1984). Imitation and social responsiveness in autistic children. *Journal of Abnormal Child Psychology, 12,* 209–226.

Dawson, G., & McKissick, F. C. (1984). Self-recognition in autistic children. *Journal of Autism and Developmental Disorders, 9,* 247–260.

Ferrari, M., & Matthews, W. (1983). Self-recognition deficits in autism: syndrome-specific or general developmental deficit? *Journal of Autism and Developmental Disorders, 13,* 317–324.

Flavell, J. H., Flavell, E. R., & Green, F. L. (1983). Development of the appearance–reality distinction. *Cognitive Psychology, 15,* 95–120.

Fodor, J. A. (1992). A theory of the child's theory of mind. *Cognition, 44,* 283–296.

Gagnon, S., & Doré, F. Y. (1992). Search behavior in various breeds of adult dogs (*Canis familiaris*): Object permanence and olfactory cues. *Journal of Comparative Psychology, 106,* 58–68.

Gallup, G. G., Jr. (1991). Toward a comparative psychology of self-awareness: Species limitations and cognitive consequences. In G. R. Goethals & J. Strauss (Eds.), *The self: An interdisciplinary approach* (pp. 121–135). New York: Springer-Verlag.

Gallup, G. G., Jr., & Suarez, S. D. (1986). Self-awareness and the emergence of mind in humans and other primates. In J. Suls & A. G. Greenwald (Eds.), *Psychological Perspectives on the Self* (Vol. 3, pp. 3–26). Hillsdale, NJ: Erlbaum.

Gergely, G. (1992). Developmental reconstructions: Infancy from the point of view of psycho-analysis and developmental psychology. *Psychoanalysis and Contemporary Thought, 15*(1), 3–55.

Hobson, R. P. (1984). Early childhood autism and the question of egocentrism. *Journal of Autism and Developmental Disorders, 14,* 85–104.

Leslie, A. M. (1987). Pretense and representation: The origins of "theory of mind." *Psychological Review, 94,* 412–426.

(1988). Some implications of pretense for mechanisms underlying the child's theory of mind. In J. W. Astington, P. L. Harris, & D. R. Olson (Eds.), *Developing theories of mind.* (pp. 19–46). Cambridge University Press.

Lewis, M., & Brooks-Gunn, J. (1979). *Social cognition and the acquisition of self.* New York: Plenum.

Mathieu, M., Bouchard, M. A., Granger, L., & Herscovitch, J. (1976). Piagetian object perman-ence in *Cebus capucinus, Lagothrica flavicauda,* and *Pan troglodytes. Animal Behaviour, 24,* 585–588.

Meltzoff, A. M. (1990). Towards a developmental cognitive science. *Annals of the New York Academy of Sciences, 608,* 1–37.

Meltzoff, A. M., & Gopnik, A. (1989). On linking nonverbal imitation, representation, and language learning in the first two years of life. In G. E. Speidel & K. E. Nelson (Eds.), *The many faces of imitation in language learning* (pp. 23–51). New York: Springer-Verlag.

Menzel, E., Savage-Rumbaugh, E. S., & Lawson, J. (1985). Chimpanzee (*Pan troglodytes*) spatial problem solving with the use of mirrors and televised equivalents of mirrors. *Journal of Comparative Psychology, 99,* 211–217.

Mitchell, R. W. (1993). Mental models of mirror self-recognition: Two theories. *New Ideas in Psychology, 11,* 295–325.

Mounoud, P., & Bower, T. G. R. (1974/1975). Conservation of weight in infants. *Cognition, 3,* 29–40.

Natale, F., & Antinucci, F. (1989). Stage 6 object-concept and representation. In F. Antinucci (Ed.), *Cognitive structure and development in nonhuman primates* (pp. 97–112). Hillsdale, NJ: Erlbaum.

Neuman, C., & Hill, S. (1978). Self-recognition and stimulus preference in autistic children. *Developmental Psychobiology, 11,* 571–578.

Parker, S. (1991). A developmental approach to the origins of self-recognition in great apes. *Human Evolution, 6,* 435–449.

Pasnak, R., Kurkjian, M., & Triana, E. (1988). Assessment of Stage 6 object permanence. *Bulletin of the Psychonomic Society, 26,* 368–370.

Perner, J. (1991). *Understanding the representational mind.* Cambridge, MA: MIT Press.

Perner, J., Frith, U., Leslie, A. M., & Leekam, S. R. (1989). Exploration of the autistic child's theory of mind: Knowledge, belief and communication. *Child Development, 60,* 689–700.

Perner, J., Leekam, S. R., & Wimmer, H. (1987). Three-year-olds' difficulty with false belief: The case for a conceptual deficit. *British Journal of Developmental Psychology, 5,* 125–137.

Piaget, J. (1937/1954). *The Construction of Reality in the Child.* New York: Basic.

Povinelli, D. J. (1989). Failure to find self-recognition in Asian elephants (*Elephas maximus*) in contrast to their use of mirror cues to discover hidden food. *Journal of Comparative Psychology, 103,* 122–131.

Povinelli, D. J., Nelson, K. E., & Boysen, S. T. (1990). Inferences about guessing and knowing by chimpanzees (*Pan troglodytes*). *Journal of Comparative Psychology, 104,* 203–210.

Rumbaugh, D. M., Richardson, W. K., Washburn, D. A., Savage-Rumbaugh, E. S., & Hopkins, W. D. (1989). Rhesus monkeys (*Macaca mulatta*), video tasks, and implications for stimulus–response spatial contiguity. *Journal of Comparative Psychology, 103,* 32–38.

Stern, D. N. (1985). *The interpersonal world of the infant: A view from psychoanalysis and developmental psychology.* New York: Basic.

Watson, J. S. (1989). *The spatial imperatives.* Unpublished manuscript, available from the author.

Wellman, H. M. (1990). *The child's theory of mind.* Cambridge, MA: MIT Press.

Wimmer, H., & Perner, J. (1983). Beliefs about beliefs: Representation and constraining function of wrong beliefs in young children's understanding of deception. *Cognition, 13,* 103–128.

Woodruff, G., & Premack, D. (1979). Intentional communication in the chimpanzee: The development of deception. *Cognition, 7,* 333–362.

5   Mutual awareness in primate communication: A Gricean approach

Juan Carlos Gómez

Introduction

My aim in this paper is to explore the origins of self- and other-awareness in prelinguistic intentional communication. The study of communication (both human and animal) has long been a "messy" area of inquiry – "messy" in the sense of being a mixture of approaches and definitions. For one thing, the term "communication" itself has proved extremely difficult to define. Indeed, it is common to begin textbooks or chapters on the subject with an explicit recognition of the variety of views on communication (e.g., Slater, 1983; M. S. Dawkins, 1986). In this chapter I will explore an aspect of this variety, one concerning the role granted to mutual awareness in the process of communication. I will contrast two radically different views: one, prevalent until recently in the study of animal behavior, that avoids cognitive processes in the study of communication; the other, popular among students of human communication, that apparently relies upon a multiplicity of cognitive processes. I will then refer to recent attempts to apply the latter approach to animal communication, and will suggest what I think is the proper way to do so. Finally, I will explore the implications of this approach to understand the origins of self-awareness and self-consciousness.

In the multiplicity of approaches to communication, an important dimension seems to be the status given to inner cognitive processes. When communication is studied in animals, cognitive components tend to be ignored, whereas students of human communication seem to overemphasize them. Until recently, the traditional ethological approach has prevailed in the study of animals.[1] It is focused upon the identification of signals displayed by organisms and their effects on the behavior of other organisms (see Slater, 1983, for a textbook statement of the approach). The cognitive approach prevalent in human studies, however, tries to understand the psychological processes whereby communication is carried out, that is, the kind of information processing phenomena on which the displays and effects of signals rely. The first approach is reluctant to embrace concepts such as "communicative intentions" or "mutual awareness," whereas for the second those are the very foundation of communication.

61

Communication in animals: The traditional ethological approach

Ethologists like to present their approach to behavior as consisting of four general questions or whys: the evolution, development, function, and causation of behavior. The question of which are the (immediate) mechanisms that perform the functions of communication, however, has long been neglected or even avoided by ethologists. The reason is probably that communication is a phenomenon that inevitably leads to the consideration of highly complex processes such as intentionality and awareness.

For example, in relation to the possibility of defining communication in terms of the sender's intentions, Slater (1983) states that

while these may be a useful criterion for humans, who can be asked about their intentions, it is not easy to establish criteria for intentions in other animals. A definition expressed in terms of the advantage of communicating rather than its causes therefore seems preferable. (pp. 10–11)

According to him,

the essence of animal communication is that one animal influences another in some way. (p. 9)

Richard Dawkins (1987) also begins offering an ample definition of communication in the objective spirit of ethology:

An animal is said to have communicated with another animal when it can be shown to have influenced its behavior. What we cannot discover is the subjective intentions or feelings, if any, of the parties to the communication. We cannot say they have not subjective feelings; very probably they do, but it is impossible for us to know. Our definition of communication, if it is to be useful, has to recognize this impossibility. (p. 79)

But the study of animal communication rarely relies upon such unrestricted definitions as "influencing another animal's behavior." As Dawkins himself points out, such definitions would lead us to consider as communicative the case in which an animal A bites off animal B's foot, thereby affecting its subsequent behavior. To avoid this excessive generality, he adopts the following definition:

Animal A is said to have communicated with B when A's behavior manipulates B's sense organs in such a way that B's behavior is changed. (p. 79)

Although, literally taken, this definition is not entirely satisfactory (because biting out animal B's eye would qualify as communication!), its spirit seems quite appropriate. The basic idea is that A's influence upon B must be informational in nature; that is, it must not be based upon the reception by B of physical forces but upon its reception of information, its ability to extract and process information by means of its senses. To quote Von Glasersfeld (1977):

In communication there is no thermodynamically calculable relationship between the energy change that constitutes the signal and the energy expended as a result of receiving the signal. Instead, in communication there is always an exchange of "information." (p. 58)

Contrary to Slater's too-general definition quoted above, the essence of communication is that one animal influences another in a certain way, and this is informational in nature. When an animal acts with mechanical force upon another causing a change upon it; say, pushing aside or restraining it, most people would agree that this is not communication, but is instead physical manipulation by means of mechanical forces. However, when an animal, for example a chimpanzee, refrains from taking a banana upon being threatened by another, the change of behavior is caused by the information picked up by the senses, not by mechanical causes. Almost everybody would agree that mechanical changes should be left outside the realm of communicative phenomena.

However, the activity of an animal (or its presence), when perceived by another, will almost always influence the behavior of the latter. For example, a subordinate chimpanzee may refrain from taking a banana he has spotted upon seeing a dominant individual sitting close to the banana. The dominant individual affects the other not by means of mechanical forces, but as a consequence of the perception of its mere presence. We have then an informational effect. Many people would agree that this is not a good example of communication; nevertheless, at the very least, everybody would agree that this interaction is different from the situation in which the subordinate chimpanzee leaves the banana upon being threatened by the dominant one.

There is indeed an important difference between these examples: In our everyday language this is best captured by saying that in the first example the effect is intended by the dominant, whereas in the second it is not intended. But, as we have seen, this description is not acceptable for an ethologist. How does, then, the orthodox ethologist capture this important difference? Let me quote once more Richard Dawkins (1987):

An ethologist feels justified in using the word communication, when he thinks that NATURAL SELECTION has acted on the behavior in question to enhance its power to influence the behavior of other animals. . . . All these [communicative] behavior patterns seem to have been adapted to cause some change in the physical environment which is picked up by the senses of another animal, as a consequence of which the behavior of the receiving animal is changed. (p. 79)

In this definition two major features of communication are captured: (1) that signals act upon another organism informationally, and (2) that they are specialized to do so by natural selection. Thus, communication is a way to influence another's behavior by means of signals specifically designed ("ritualized" is the word used by ethologists to denote this process) to take advantage of information transmission between organisms. In the above example the dominant chimpanzee's threat is communicative because it has been selected in evolution for its specific informational effects upon conspecifics. The mechanisms whereby evolution has selected this communicative function do not matter; the important point for the ethologist, according to this view, is to identify behaviors as specifically selected by their informational effects, that is, to identify behaviors as *ritualized signals*. It is only in this meta-phorical and evolutionary sense that the informational effects of signals can be said to be phylogenetically "intended," instead of merely adventitious (cf. Marler, 1977).

In agreement with Slater's (1983) and Dawkins's recommendations, ethologists have usually concentrated upon the description of the signals and their functions and evolutionary origins, leaving aside the thorny issue of causal mechanisms. Ethologists have tended to avoid the details of the information processing ocurring inside the black boxes of the sender and the receiver of signals.[2] However, resorting to the evolutionary history of behavior patterns, they have managed to incorporate an essential distinction between mechanical, informational and communicative effects in animal interactions.

Communication in humans: Intentionality and mutual awareness

In sharp contrast with the cautious approach of traditional ethology is the study of human communication. Psychologists, linguists, and philosophers tend to analyze it in terms of intentions and other psychological processes.

It is possible to describe in humans the same three levels of mutual influence – mechanical, informational, and communicative – I have just discussed in animals. I can push aside a person who is blocking my way, thereby achieving a merely mechanical effect on another.[3] Alternatively, the person in the passage could put himself aside upon seeing me approach him. In this case the effect is informational because it has been provoked by the other person's perception of my approaching body. Finally, I could ask him to move aside, saying "Excuse me!" and adopting a position to show my goal. This would be a genuine communicative act in that its effect is provoked by the perception of a behavior specifically designed to have such informational effect. We would not hesitate to say that in the latter case the behavior is *intended* to have a communicative effect.

However, the student of human communication uses the term "intended" in a different way than that of the ethologist. For the latter it is, if anything, a metaphor: A signal is "intended" to provoke an informational effect in the sense that it has been shaped and selected by that kind of effect in the evolutionary history of the species. In the case of human communication, however, the signal is intended in the sense that, when producing it, the sender is somehow aware of the informational effect it will provoke; indeed this is why he or she uses the signal. This implies some mental or cognitive process, the nature of which we will examine now. Different approaches have been proposed to deal with the problem of human intended communication. In this paper I will refer to the one that, in my view, best captures the complexity of communication. I will refer to it as the *Gricean approach*, that is, one inspired in the work of the philosopher H. P. Grice (1957, 1989). I want to make it clear that what I shall present is my own interpretation of what the Gricean approach implies; this may differ from the interpretations of other people.

The Gricean approach to human communication

The mechanism of human communication seems to involve some explicit representation of the process of information transmission. When I ask someone to allow me to pass through the door, I seem to be aware of the informational nature of the effect I am trying to provoke. Let us examine what this implies by analyzing a simple act of communication (see Figure 5.1). When I say to the waiter in the example "May I have some salt, please," I don't want the waiter just to give me the salt. My representation of the situation is not merely: "If I say, 'May I have some salt, please,' he will give me the salt." I not only understand that my words will provoke a reaction in the waiter, I also understand something about why this will happen. I understand that the waiter will give me the salt because *he will understand* that I want it. When I speak the above sentence, I want the waiter to understand that I want the salt, that is, I am taking into account the mental process in the waiter that will allow him to answer my request. Note that, as shown in Figure 5.1, this implies a remarkable mental structure in which the waiter is represented as achieving a representation of the client's desire (which is, in turn, a representation of something; in this case, the salt). It is frequently assumed that this remarkable series of three embedded representations is the essence of the Gricean account of communication (e.g., Cheney & Seyfarth, 1990; Dennett, 1983), and it has even been termed the *Gricean mechanism* (Bennett, 1976). Nevertheless, according to my own interpretation, the Gricean account involves more than that.

Consider the following situation: I have had a quarrel with the waiter of Figure 5.1, and I promise myself never to talk again to such a nasty man. Then I discover that my salt cellar is empty. To get the salt without breaking my promise, I act as follows: I covertly watch the waiter until I see that he is looking in the direction of my table. Then I take the empty salt cellar and try to get salt out of it acting as if I were discovering its emptiness then. I hope that the waiter, upon seeing my trouble, will bring me some salt. If I succeed, the waiter will bring the salt without my having requested it. Note, however, that if my scheme works, I will have made the waiter to understand that I want the salt. The plan I entertain in my mind can be described as follows: I want the waiter to understand that I want the salt, and this coincides with the alleged Gricean mechanism of Figure 5.1. Now, if an act that is not a request has this structure, a true request must present a different one.

Perhaps someone would want to argue that, though covert, the above action is a request. Nevertheless, everybody would agree that there is an important difference between this action and the overt request in which I turn to the waiter, show him the empty salt cellar, and ask him, "May I have some salt, please?" What is this essential difference? In the covert version, although it is true that I want the waiter to understand that I want the salt, I don't want him to be aware of my purpose, that is, I want him to understand that I want the salt, but I don't want him to understand that I want him to understand that I want the salt. This – the missing element in the above example – is the essence of human communication from a Gricean point of view. In a genuine communicative act, I want the listener to understand that I *intend* him to understand what I want from him (Figure 5.2). Thus, in our example, a true communicative act would involve that I want the waiter to understand that I want him to understand that I want the salt: a series of *five* embedded mental states. This spiral of mutual understandings can be appropriately called a *Gricean loop*, in reference to their circular and entangled nature (see Grice, 1957).

Note that the loop also exists on the receiver's side: In an overt request the waiter will give me the salt because he understands that I want him to understand that I want him to understand that I want the salt. This contrasts with the covert maneuver which, if successful, will lead the waiter to give me the salt only because he becomes aware that I want the salt. The reader can have an idea of the complexity such loops can reach if he considers the case in which the waiter notices the true intention behind my covert maneuver. If the waiter notices my covert watching before enacting my false discovery of the empty salt cellar, he will probably understand something like the following: that I want him to believe that I want some salt, but I don't want him to understand that I want him to know what I want – a remarkably entangled instance of a Gricean loop.

Thus, my interpretation of the Gricean account of communication is that it requires a minimum of five embedded mental states on the part of the speaker, and six on the part of the listener.

Gricean loops, metarepresentation, and awareness

The kind of mental process that seems unavoidable in any Gricean account of communication is *metarepresentation*, or the ability to represent the representation of a thing instead of the thing itself. If you know that I have a red car, you just have a representation of this fact; but if I know that you know that I have a red car, then I need to be able to represent not only facts (that I have a red car) but also mental processes (that you know it). The ability to impute mental states such as knowing or believing to other people requires a metarepresentational capacity (Whiten & Perner, 1991).[4] Organisms capable of attributing mental states are said to possess a theory of mind (Gopnik & Meltzoff, *SAAH*10; Premack & Woodruff, 1978; Whiten, 1991). Metarepresentation is inherently recursive; that is, a representation can refer to another representation that, in turn, refers to another, and so forth. The number of embedded representations involved in particular instance identifies its *order* (Dennett, 1983; Whiten & Perner, 1991). For example, if "I

Figure 5.1. A moderate Gricean interpretation of a simple request for salt. The speaker is assumed to have a third-order intentional representation encompassing his own and his listener's mental states: *A* wants *B* to *understand* that he *wants* some salt. (See text.)

Figure 5.2. A truly Gricean interpretation of a requestive situation. The speaker entertains a fifth-order level of intentionality: *A wants B* to *understand* that he *wants* him to *understand* that he *wants* the salt. (See the text for further explanations.)

want the waiter to *understand* that I want some salt" we have an instance of a third-order representation or, as Dennett (1983) puts it, a third-order level of intentionality.

According to my interpretation of the Gricean approach, the standard communicative act would imply a fifth-order level of metarepresentation by the speaker: When I ask you for the salt, I *want* you to *understand* that I *want* you to *understand* that I *want* the salt. This, despite Grice's claim that he did not intend to multiply mental occurrences (Grice, 1957), requires a fifth-order metarepresentational process. Note that a property of this Gricean account is that, although a fifth-order representation is the minimum required, theoretically nothing prevents people from entertaining higher-order representations. The listener may understand everything that is going on in the head of the speaker (sixth-order level), and nothing prevents the latter from foreseeing that the former will do so (seventh-order), and so on.

This Gricean account of communication involves mutual awareness between the speaker and the listener. This is not an awareness of the physical presence of each other, but an awareness of the mental processes of each person involved in the communicative act, that is, an awareness of the other's awareness. The entangled flavor of the Gricean approach is introduced by the fact that the content of the other's awareness is precisely our awareness of his or her awareness, and so on. The situation has the same mutually embedding structure as two mirrors confronted: Each mirror reflects the other reflecting the other's reflection, ad infinitum. Thus, a Gricean account of communication apparently implies a sophisticated combination of self-awareness – the ability to consider one's own mental processes – and other-awareness, or theory of mind (ToM) – the ability to take into account the mental processes of other people.

The problem with the Gricean account – even in its more usual, moderate, third-order level interpretation – has always been that intuitively we don't feel that in everyday communication we are engaging in such complex and recursive cognitive processes. There is something counterintuitive in this account. Even for human beings, everyday communication seems to be simpler than that. My aim in the remainder of this paper will be to reconcile this intuition with the Gricean view, showing that the reciprocal embeddedness of Gricean loops actually captures a most essential feature of communication, but that it can be achieved without resorting to sophisticated metarepresentational loops. Before this, however, let me say a few words about some recent attempts to explore animal communication in a Gricean mood.

Intentionality and mutual awareness in animal communication

The cautious approach to animal communication, ignoring the causes and mechanisms that make the function possible, has been challenged in the last years by a number of authors. Following Donald Griffin's call "towards a cognitive ethology" (Griffin, 1978) several researchers have dealt with the question of animal (mainly primate) intentionality and awareness in communication. Cheney and Seyfarth (1990), Dennett (1983), Marler, Karakashian, and Gyger (1991), Mitchell (1986), Whiten (1991), Woodruff and Premack (1979), and others have explicitly posed the question of whether primates have intentions to communicate and whether they are aware of them.

For example, when vervet monkeys emit their famous alarm calls, do they take into account the mental processes of the intended recipient, or are they just calling in response to the predator's presence? In a recent review of the problem of intentionality in animal communication, Hauser and Nelson (1991) conclude that current evidence raises "the possibility that call production is mediated by the caller's intentions" (p. 188), and that, although further experiments and observations are needed to settle the question, there is "suggestive evidence that nonhuman animals attribute mental states to one another" (p. 189). The evidence the authors refer to is the existence of audience effects in the calling behavior of vervet monkeys and other animals; that is, that the signaling behavior of an animal is affected by the presence and identity of its potential receivers (see also Marler et al., 1991). In this new "cognitive-ethological" approach there seems to exist the assumption that complex and flexible communicative behaviors such as those involving audience effects must reflect some ability to understand the mental states of other organisms. Indeed, intentional communication seems sometimes to be identified with these metarepresentational capacities. However, we have seen that a metarepresentational view of intentional communication seems to require highly complex cognitive structures. The intention to communicate involves a Gricean loop whose metarepresentational realization, according to the interpretation put forward in the previous section, apparently requires not only intentionality but also a fifth-order level thereof. This possibility is not considered even by authors who explicitly quote Grice (e.g., Cheney & Seyfarth, 1990). Obviously, if wondering whether monkeys are capable of second-order representations (e.g., "Does monkey *A* want *B* to understand that there is a snake?") is seen by many ethologists with suspicion, a truly Gricean question (such as "Does monkey *A* want *B* to understand that he wants him to understand that there is a snake?") could provoke epistemological indignation in them. The implausibility of the recursive structures apparently implied by Grice's approach is underscored when one deals with animals. However, as we have seen, those structures seem to capture some essential feature of intentional communication.

In the next section I will defend the view that the Gricean interpretation of intentional communication is right, but that it does not necessarily require sophisticated metarepresentational abilities, and can be applied to nonhuman primates and prelinguistic human infants. Let us begin by having a look at how intentional communication emerges in human babies and anthropoid infants.

The emergence of intentional communication in gorilla and human infants

There is some variability in what different authors consider to be the onset of true intentional communication in human infants. Most seem to agree, however, that around 9–12 months an important landmark occurs: Human babies first begin to address gestures and vocalizations to people in order to request or show things to them (e.g., pointing to a favorite toy; see, for example, Schaffer, 1984). Most authors[5] think that this kind of behavior marks the onset of intentional communication (e.g., Harding, 1982; Sugarman, 1984). Few studies have tried systematically to address the question of how intentional communication is identified in babies (see Sarriá, 1989, for an exception); but most authors seem to use similar criteria. Consider the following behaviors (see also Figure 5.3):

1.   A baby sitting on the floor tries to reach a toy; he leans forward and stretches as much as he can, totally focused on his goal. An adult sees the scene and gives him the toy (Figure 5.3a).
2.   The same baby, a few months later, after verifying that he cannot reach his goal, turns to an adult, looks at his eyes, points to the toy and waits. The adult gives him the toy (Figure 5.3b).

Example 1 illustrates a case in which the behavior of the infant accidentally transmits information to an adult. Example 2 is typical of what most authors agree to identify as infant *intentional communication*. How does it differ from the first case? Two main changes seem to characterize the behavior of the baby in Example 2: First, he uses an act – pointing – that has no direct mechanical efficiency upon the world; second, he *addresses* the act to a person, looking at his face, in particular at his eyes. Although other criteria are sometimes added (e.g., the persistence and variation of acts until the goal is reached), these two features seem to be the most essential to identify intentional communication versus involuntary information transmission (Gómez, 1992).

Gómez (1990, 1992) has shown that the same criteria could be used to identify the emergence of intentional communication as a strategy to make requests in hand-reared infant gorillas. The first strategies used by the animal he studied involved trying to move the human to the goal (e.g., the latch of a door) much as she would move an object, and then trying to climb upon him and reach the goal. However, a few months later the animal began to show a different approach to the problem: taking the human by the hand and gently leading him toward the goal, alternately looking at his eyes and the goal. This illustrates a transition from noncommunicative, mechanical acts to communicative gestures. The communicative strategies were characterized by the lack of mechanical efficiency of the gorilla's actions and the interpolation of looks at the eyes of the human; that is, by the same criteria that identify intentional communication in human infants.

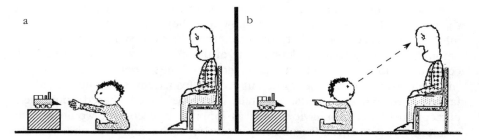

Figure 5.3. The emergence of intentional communication in infancy. (a) A non-communicative strategy to get things: The child strives to get hold of the toy. His father, upon seeing his efforts, will eventually give him the toy. (b) A communicative strategy to get the toy: The child points to the object and looks at his father's eyes.

Why are these criteria interpreted as evidence of communication, moreover of "intentional" communication?

The design features of intentional communication

The first feature – that communicative acts are not designed to reach their goals mechanically but rather through the behavior of the other organism – is related to the criterion of communication used by ethologists. According to them, communicative actions are phylogenetically ritualized; that is, their design is adapted to perform information transmission rather than mechanical functions. The communicative actions of human infants and gorillas are ontogenetically "ritualized." Ontogenetic ritualization implies only some understanding that the effect of the signal – the reaction of the other organism – is not mechanically provoked by the sender. This feature serves only to separate informationally from mechanically effective actions. It does not imply any understanding of the causes that provoke the informational effect, that is, the information processing activity of the receiver.

However, the second feature of communicative actions – their coordination with eye contact behaviors – calls for a different interpretation. Why should the other person's eyes be relevant when the infant or the gorilla uses a communicative action? Drawing on my study of gorilla communicative development (Gómez, 1990, 1991, 1992), I have suggested that looking at the eyes of the receiver is a way of taking into account the information processing that allows gestures to work as communicative actions. The other's eyes are relevant because they are the window through which the receiver perceives the gestures of the person and the things of the world to which the gesture refers. Looking at the eyes of the receiver is evidence that infants understand something of how their gestural behavior is instrumental in provoking their audience's reactions. This behavior can be conceived of as a way to control that the audience is in a special relation to the gesture and its referent: If gestures are to work, the audience must be attending to them.

A function of eye contact would, then, be to check the attentional state of the receiver. Infants would thereby show that they somehow understand the information processing activity on which the effects of signals depend. This amounts to having some understanding of the mental processes that underly communication. This understanding, however, need not yet be a proper (metarepresentational) theory of mind. Whatever human or anthropoid infants understand about other's minds, it can be directly based upon their perception of overt behaviors such as looking. Eye contact would be the result of trying to check the object of attention of the addressee and, as such, would reflect an implicit recognition of visual attention as a "window" to the minds of others (Gómez, 1991). The important point is that, according to this interpretation, human infants and gorillas would be able to understand other people's minds using first-order representations of behaviors that directly reflect mental states such as attention (Gómez, 1991; Gómez et al., 1993). If this analysis is correct, then so-called prelinguistic intentional communication would seem to be far from the elaborate, metarepresentational Gricean loops.

Attention contact: Gricean loops without metarepresentation

The above analysis assumes that infants look at the eyes of other people to check that they are attending to the relevant points of the situation. In a communicative situation such as those depicted in Figures 5.3 and 5.4, the relevant points are the object pointed out and the gesture produced by the child. Thus, when the infant looks at the adult's eyes, he finds out two things: first, that the adult is attending to his gesture; and second, that the adult is attending to the relevant object (Figure 5.4a).

This analysis, however, fails to make an important distinction in attention-checking patterns; actually, it fails to explain the very pattern of eye contact. Proper eye contact consists of looking at the eyes of a person who, in turn, is looking at your eyes ("look or gaze contact" would probably be a more accurate term; see Figure 5.4b). In our above explanation, however, we assumed that when eye contact occurred, the infant was checking whether the adult was attending to his gesture. To do this, however, it is not necessary that the eyes of the sender and the receiver meet at all. Moreover, if the eyes of the child meet the eyes of the adult, it must be because the adult is attending to his eyes, not to his gesture! You can know that someone is attending to you without making eye contact. If someone is looking at your hand or any other part of your body, he or she is attending to you, and in an excellent position to perceive any manual or corporal gesture you carry out. In eye contact, however, the other person is not just attending to any part of your body: For eye contact to occur it is necessary that you attend to the eyes of the other person – checking his or her attention – and that the other person attends to your eyes too, probably checking your attention. This means that he or she is *attending to your attention* (Figure 5.4b).

When we look at the eyes of a person to find out her focus of attention (for

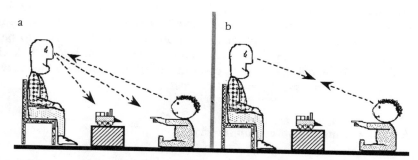

Figure 5.4. Two different components of joint-attention patterns: (a) The child looks at the eyes of the adult when the latter is looking at the toy or the child's gesture; the child can thereby follow the adult's gaze and find out what is the focus of his attention. (b) The child and the adult make *eye contact*, i.e., they look at each other's eyes and find out that the focus of their attention is each other's attention.

example, we attend to her attention) and we make eye contact, what we discover is that the focus of her attention is precisely our own attention, that she is doing in relation to us what we are doing in relation to her. Eye contact implies mutual attention to each other's signs of attention, and this I will call *attention* contact. Attention contact consists of attending to the attention of a person who, in turn, is attending to our attention.

Now, if one thinks about what this implies, it becomes clear that eye contact involves an attentional loop similar to the Gricean loop discussed earlier. "*A* is attending to *B* attending to *A* attending to *B*," and so on. Mutual attention has that endless entangled structure of two confronted mirrors that defines a Gricean loop; but this is so only if one thinks about it. In order to engage in an actual situation of attention contact one need not think about its Gricean structure. This is created by the situation itself. Organisms endowed with metarepresentational abilities are able to think about the cognitive implications of looking at each other's eyes and translate this into analytical chains of embedded thoughts. My contention, however, is that neither infant apes nor human infants possess the cognitive tools to do this. Attention contact in them is not the result of complex metarepresentational abilities. It is primary, in the sense that it is first established; then, if one has the necessary cognitive tools, one can construct a metarepresentation of it and its implications. However, the special mutuality of intentional communication characteristic of the Gricean loop can exist without metarepresentations.

It is important to bear in mind that although eye contact is a powerful way of marking and showing attention contact, the latter cannot be reduced to the former. Attention contact can exist without eye contact. Indeed, in sighted persons, eye contact is a privileged way to establish and mark attention contact; but it is not the only one. There is, so to speak, more in attention than meets the eye.[6]

In conclusion, the pattern of looking at the eyes of others, identified as a criterion of intentional communication in studies of human and anthropoid infants, seems to reflect two things. First, it is an early indicator of *awareness* of the role played by the other's attention in allowing gestures to be functional; second, it is a way of establishing the *mutual awareness* link characteristic of Gricean loops. In both cases the awareness can be achieved without recourse to metarepresentational processes.

Attention contact and communication

What are the implications of this analysis for the issues we have been considering in this paper? Concerning human communication, my suggestion is that the concept of attention contact provides a solution to the problems posed by the Gricean analysis of communication (see Avramides, 1989, and Sperber & Wilson, 1986, for a review of these problems). It provides a direct way of establishing something that can be called mutual awareness without recourse to metarepresentational powers. Remember our example of the waiter. When I tried to let him know my need of salt without actually making a request, I performed a little pantomime without looking at him after having covertly verified that he was watching in the right direction. All I should do to transform this covert maneuver into a proper request is to look at the eyes of the waiter while performing the pantomime. Simply by doing this I would confer to the situation the Gricean structure it lacked before. Even in human linguistic interchanges the intention to communicate is frequently marked with nonverbal cues such as eye contact. Perhaps human adults first engage in intentional communication using attention contact and then, if necessary, use their inferential powers to modulate their interaction, always from the secure base of attentional loops. The concept of "attention contact" provides a solid basis upon which to raise the cognitive buildings of embedded mental states that develop the implications implicit in mutual attention.

This solution to the "Gricean problem" is similar in many respects to that proposed by Sperber and Wilson (1986). These authors propose to substitute the cognitively weaker concept of "mutually manifest" for the cognitively more complex concept of "mutually known," purportedly necessary to engage in Gricean interchanges. The concept of attention contact identifies a particular mechanism capable of making things mutually manifest and provides a connection with an important body of empirical research in development and comparative psychology (Gómez, 1992).

Intentional communication gains in subtlety and strategy when our understanding of attention and intentions goes beyond their external manifestations. Nevertheless, a metarepresentational understanding is not the condition for communication to occur: It is, rather, a consequence of the possibilities opened up by attention contact – a consequence that is anything but trivial, because it dramatically affects the complexity of human communication.

The concept of attention contact also allows us to understand how organisms presumably devoid of metarepresentational abilities, such as human infants and anthropoids, can nevertheless engage in intentional communication. They do not need to understand their partners' ideas and intentions by means of second- or higher-order representations. All they need is to represent overt phenomena such as gaze direction, eye contact, facial expressions, etc. (Gómez, 1991; Gómez et al., 1993).

Concerning animal communication, we have seen that some recent attempts at overcoming the traditional ethological neglect of communicative mechanisms have led to a consideration of mental and intentional representations in animals. Phenomena such as audience effects call for complex interpretations of animal communication; but models from human studies tend to emphasize sophisticated processes such as theory of mind and metarepresentation. I propose that the Gricean structure of attention contact and, in general, the ability to handle the overt signs of attention, may provide a way to analyze intentional communication in animals without metarepresentational abilities.

Thus, animals unable to attribute mental states to others might approach communication in an intelligent, "nonreflexive" way (to use the terminology of Marler et al., 1991). They could not only pursue gross behavioral goals (e.g., making B run away), but also attentional ones (e.g., making B look at a particular place). They could be able to understand that without showing the behavior pattern of attending, an animal will not benefit from a signal unless the signal itself evokes such an attentional state, as happens with vocal signals. They could understand that the display of attention patterns by the receiver is one of the consequences of the "speaker's" calling behavior. They would understand that their calls not only cause others to run away, but also to redirect their attention. In the well-known case of vervet alarm calls, A's computations need not include "B will believe that there is a leopard" but rather something like "B will look for a leopard and then run away" or, more adaptively, "run away and then or simultaneously look for a leopard." Indeed, Cheney and Seyfarth's analysis of the semanticity of vervet monkeys' alarm calls is largely based upon the different attentional reactions they evoke. Vervets look at the sky when they hear an eagle alarm call, but they inspect the ground upon hearing the leopard alarm call (Cheney & Seyfarth, 1990).

Attentional reactions are the visible part of the information processing that mediates between the stimuli in the environment and the behavior of the organism. It can be understood as the link that connects stimuli and behavior – a link that actually hides a whole chain: the mental processes that mediate the production of responses (Gómez, 1991). But the mental chain behind the attentional link cannot be perceived; it can only be imagined or inferred. It is only accessible to organisms capable of "imagining" things that cannot be perceived, such as the mental representations of other animals. That is, it is only accessible to organisms capable of entertaining metarepresentions.

The Gricean origins of self-awareness

What happens with the phenomenon of mutual awareness in our account of intentional communication? Does it disappear in our attentional version of the Gricean loop? On the contrary, I would like to suggest that mutual awareness begins as mutual attention to the signs of attention, as attention contact. When two organisms are confronting each other, they can be said to be aware of each other's physical presence; but if they also are in *attention contact*, they are aware of something else. Their perception is focused on each other's signs of attention, and insofar as we admit that the signs of attention are signs of awareness, they can be said to be aware of each other's awareness. An organism can handle another's awareness at least in two ways: by imagining the other's mental experiences (i.e., using a metarepresentational theory of mind[7]) or by perceiving the external signs of this awareness. The mutual awareness involved in attention contact is more perceptual than intellectual: Organisms in attention contact don't think of each other thinking of each other; they perceive each other attending to each other.

The best way to understand this is to think in terms of the ability to follow the line of regard. One of the consequences of being able to follow another organism's line of regard is that one can attend to the same object to which the other organism is attending. For example, chimpanzees can discover hidden food by monitoring the attention of a leader who knows where the food is (Menzel, 1974). One's attention may become alerted toward particular objects through the attention of others. When the attention of another person is focused upon you – that is, on your body – you can get oriented to yourself as a physical entity. However, when the focus of another person's attention is your own attention, this may act as a clue toward your own attentional activity. The other's attention *points to* your own attention and, as a result, you are led to your own attention as a focus of attention. If we consider the signs of attention as signs of awareness, the structure of attention contact seems to lead to a first version of self-awareness (both as a physical and as an "aware" or "attending" entity).

G. Gallup (1983) proposes that "an organism is self-aware to the extent that it can be shown capable of becoming the object of his own attention" (p. 474), and justifies the power of the mirror situation to test self-awareness because mirrors "enable organisms to see themselves as they are seen by others" (p. 475). If mirrors are the best way to test self-awareness, attention contact is perhaps *the* way to develop it. This would explain why chimpanzees without social experience fail to pass the mirror test (Gallup et al., 1971). They would have been deprived, among other things, of attention-contact experiences.

The hypothesis I am putting forward is that mutual awareness (with its two components – self- and other-awareness) first appears as a perceptual and attentional phenomenon with a peculiar, Gricean structure. By its own nature this phenomenon has a number of mental implications that may be computed *only after* an organism possesses particular cognitive abilities. However, in its original nonintellectual version, attention contact, due to its Gricean structure, might constitute the core situation from which more sophisticated versions of other-awareness and self-awareness appear.

This hypothesis implies that self-awareness is achieved through the mediation of other organisms – a proposal quite in the line of G. H. Mead's (1934) and Vygotsky's (1934/1986) ideas. The ability to understand the other's attention or awareness would lead to the ability to understand one's own awareness. Thus, in the dispute about which comes first, understanding others through self-knowledge or understanding oneself through the knowledge of others (see Gopnik and Meltzoff, *SAAH* 10), my position would seem to favor the latter option. However, my view is that in the development of such a complex ability as self-awareness different steps must be involved. Perhaps understanding one's attention from its reflection in others is a step; but the next step could involve projecting onto others what one learns about the inner side of awareness once one is alerted to oneself as an aware entity. Perhaps one reaches a metarepresentational understanding of others (i.e., a theory of mind, or ToM) partly through this projection. This would confer to the genesis of self-awareness and ToM a curious spiral structure, the development of which would occur in different moments. When we analyze the implications of attention contact, it seems that many important later developments are embedded in its structure: self-conciousness, theory of mind, and complex intentional communication, for example. In what sense do these developments grow out of this germinal situation? It is very important to avoid a developmental fallacy, namely, that once you identify a situation whose structure somehow implies later developments, you have explained these developments. Theory of mind and self-awareness are implied in attention contact, but their realization does not consist of merely unwrapping something that is already there. We know that they are implied because we posses analytical tools that allow us to develop the implications, but these analytical tools (mainly made of metarepresentation) are the very thing we want to explain. My proposal, then, is a suggestion that the structure of a simpler situation – attention contact – provides a basis for the development of more complex abilities if adequate developmental mechanisms are at work, but I don't address here the question of which these mechanisms are. In Gómez et al. (1993) a similar approach to ToM and its precursors is outlined with a tentative mention of possible developmental mechanisms.

In conclusion, when approaching the problem of awareness and consciousness in nonhuman animals, it seems important to bear in mind that, before this awareness is explicit or metarepresentational, a simpler version of it can be achieved by taking into account the external manifestations of such mental activities as intending and attending. There is specially a situation – attention contact – that can be accounted for without metarepresentation, and that seems to be the key to the Gricean structure of intentional communication. Moreover, attention contact seems to be a seminal situation for the development of self-consciousness and ToM, although the mechanisms of this development are still to be identified. This approach can contribute to bridging the gap between ethological and psychological approaches to communication.

Acknowledgments

Research mentioned in this paper was partly funded by a grant from the CIDE (MEC, 1988–1990). I want to thank the Zoo of Madrid for their support in the work with gorillas mentioned in this paper. The three editors of this book made valuable suggestions; I wish, however, to single out Robert Mitchell's thorough comments and incisive suggestions under the drive of our common "afición" for Paul Grice's thinking. E. Sarriá, S. Baron-Cohen, J. Tamarit, A. Brioso and E. León participated in discussions of some of the ideas presented in the paper.

Notes

1. Sometimes, something called "the classical ethological approach to the study of communication" has been contrasted to the "games theory or behavioral ecology approach" (Hinde, 1981). For the purposes of this chapter both approaches are considered to reflect the traditional noncognitive approach of ethology to communication.
2. These caveats against intentions are strongly reminiscent of behaviorism's reasons to abandon mental processes as a scientific subject of study because they are unobservable entities. See Gómez (1992) for a discussion of the double epistemology that leads some ethologists to reject the study of cognition because it is an unobservable process, whereas they accept the study of past evolutionary processes that are equally unobservable.
3. Of course, carrying out a mechanical action upon someone such as pushing him aside will usually have some informational consequences apart from its physical results. The victim of my push can interpret it as a signal of aggression and adjust his or her subsequent behavior accordingly; but the very effect of displacing the other person is merely mechanical.
4. There is some disagreement about the concept of metarepresentation. The definition I am adopting here is closer to those of Perner (1991) and Dennett (1983). Leslie (1987), however, uses the term in a slightly different sense.
5. Some authors disagree with this view because they think intentional communication is a much later achievement. For example, Shatz (1983) argues on Gricean grounds that the ability to attribute mental states cannot be present in infants so young. Other authors, however, argue that the intention to communicate is an early development discernible in 6-month-olds or even younger babies (e.g., Tronick, 1981).
6. Blind people are a good example of how attention contact can be established without eye contact. In communicative situations where visual contact is precluded, such as a telephone interchange, attention contact is also marked by nonvisual means. Attention is something that goes beyond the eyes.
7. See Note 4.

References

Avramides, A. (1989). *Meaning and mind: An examination of a Gricean account of language.* Cambridge, MA: MIT Press.
Bennett, J. (1976). *Linguistic behaviour.* Cambridge University Press.
Bruner, J. S. (1974). From communication to language – a psychological perspective. *Cognition,* 3, 255–287.

Cheney, D. L., & Seyfarth, R. M. (1990). *How monkeys see the world.* Chicago: University of Chicago Press.

Dawkins, M. S. (1986). *Unravelling animal behaviour.* London: Longman.

Dawkins, R. (1987). Communication. In D. McFarland (Ed.), *The Oxford companion to animal behaviour* (pp. 78–91). Oxford University Press.

Dennett, D. C. (1983). Intentional systems in cognitive ethology: The panglossian paradigm defended. *Behavioral and Brain Sciences, 6,* 343–390.

Gallup, G. G. (1983). Towards a comparative psychology of mind. In R. L. Mellgren (Ed.), *Animal cognition and behavior* (pp. 473–510). Amsterdam: North Holland.

Gallup, G. G., McClure, M. K., Hill, S. D., & Bundy, R. A. (1971). Capacity for self–recognition in differentially reared chimpanzees. *Psychological Record, 21,* 69–74.

Gómez, J. C. (1990). The emergence of intentional communication as a problem-solving strategy in the gorilla. In S. T. Parker & K. R. Gibson (Eds.), *"Language" and intelligence in monkeys and apes: Comparative developmental perspectives* (pp. 333–355). Cambridge University Press.

 (1991). Visual behavior as a window for reading the minds of others in primates. In A. Whiten (Ed.), *Natural theories of mind: Evolution, development and simulation of everyday mindreading* (pp. 195–207). Oxford: Blackwell Publisher.

 (1992). *El Desarrollo de la comunicación intencional en el gorila.* Unpublished doctoral dissertation, Universidad Autónoma de Madrid.

Gómez, J. C., Sarriá, E., & Tamarit, J. (1993). The comparative study of early communication and theories of mind: Ontogeny, phylogeny and pathology. In S. Baron-Cohen, H. Tager-Flusberg, & D. Cohen (Eds.), *Understanding other minds: Perspectives from autism* (pp. 397–426). Oxford University Press.

Grice, H. P. (1957). Meaning. *Philosophical Review, 66,* 377–388.

 (1989). *Studies in the way of words.* Cambridge, MA: Harvard University Press.

Griffin, D. R. (1978). Prospects for a cognitive ethology. *Behavioral and Brain Sciences, 1*(4), 527–538.

Harding, C. G. (1982). Development of the intention to communicate. *Human Development, 25,* 140–151.

Hauser, M. D., & Nelson, D. A. (1991). "Intentional" signaling in animal communication. *Trends in Ecology and Evolution, 6*(6), 186–189.

Hinde, R. A. (1981). Animal signals: Ethological and games theory approaches are not incompatible. *Animal Behavior, 29,* 535–542.

Leslie, A. (1987). Pretense and representation: The origins of "theory of mind." *Psychological Review, 94,* 412–426.

Marler, P. (1977). The evolution of communication. In T. A. Sebeok (Ed.), *How animals communicate* (pp. 45–70). Bloomington: Indiana University Press.

Marler, P., Karakashian, S., & Gyger, M. (1991). Do animals have the option of withholding signals when communication is inappropriate? The audience effect. In C. Ristau (Ed.), *Cognitive ethology: The minds of other animals* (pp. 187–208). Hillsdale, NJ: Erlbaum.

Mead, G. H. (1934). *Mind, self and society from the standpoint of a social behaviorist.* Chicago: University of Chicago Press.

Menzel, E. (1974). A group of young chimpanzees in a one-acre field. In M. Schrier & F. Stolnitz (Eds.), *Behavior of nonhuman primates* (Vol. 5, pp. 85–153). New York: Academic Press.

Mitchell, R. W. (1986). A framework for discussing deception. In R. Mitchell & N. S. Thompson (Eds.), *Deception: Perspectives on human and nonhuman deceit* (pp. 3–40). Albany, NY: SUNY Press.

Perner, J. (1991). *Understanding the representational mind.* Cambridge, MA: MIT Press.

Premack, D., & Woodruff, G. (1978). Does the chimpanzee have a theory of mind? *Behavioral and Brain Sciences, 1,* 515–526.

Sarriá, E. (1989). *La intención comunicativa preverbal: Observación y aspectos explicativos.* Unpublished doctoral Dissertation, UNED.

Schaffer, R. H. (1984). *The child's entry into a social world.* London: Academic Press.

Shatz, M. (1983). Communication. In J. H. Flavell & E. M. Markman (Eds.), *Handbook of child psychology*, 4th ed., *Vol. 3: Cognitive Development* (pp. 841–889). New York: Wiley.

Slater, P. (1983). The study of communication. In T. R. Halliday & P. J. B. Slater (Eds.), *Animal behavior, Vol. 2: Communication* (pp. 9–42). Oxford: Blackwell Scientific Publications.

Sperber, D., & Wilson, D. (1986). *Relevance: Communication and cognition*. Cambridge, MA: Harvard University Press.

Sugarman, S. (1984). The development of preverbal communication: Its contribution and limits in promoting the development of language. In R. L. Schiefelbusch & J. Pickar (Eds.), *The acquisition of communicative competence* (pp. 23–67). Baltimore: University Park Press.

Tronick, E. (1981). Infant communicative intent: The infant's reference to social interaction. In R. E. Stark (Ed.), *Language behavior in infancy and early childhood* (pp. 5–16). New York: Elsevier.

von Glasersfeld, E. (1977). Linguistic communication: Theory and definition. In D. M. Rumbaugh (Ed.), *Language learning by a chimpanzee* (pp. 55–71). New York: Academic Press.

Vygotsky, L. S. (1934/1986). *Thought and language*. Cambridge, MA: MIT Press.

Whiten, A. (Ed.) (1991). *Natural theories of mind*. Oxford: Blackwell Publisher.

Whiten, A., & Perner, J. (1991). Fundamental issues in the multidisciplinary study of mindreading. In A. Whiten (Ed.), *Natural theories of mind* (pp. 1–17). Oxford: Blackwell Publisher.

Woodruff, G., & Premack, D. (1979). Intentional communication in the chimpanzee: The development of deception. *Cognition, 7*, 333–362.

6   Multiplicities of self

Robert W. Mitchell

As Gregor Samsa awoke one morning from
uneasy dreams he found himself transformed in
his bed into a gigantic insect. . . . What has happened to me? he thought.
                                    – Kafka (1912 / 1979), p. 1

Introduction

An examination of Western philosophical and psychological perspectives on
the self reveals that the self can be viewed as a soul, an underlying substrate,
an activity, an explanatory hypothesis, a cognitive structure, a verbal activity,
an experience, a process, or a normative attainment (Levin, 1992, pp. 204–205).
These various perspectives on "the" self present a problem for a unitary view
of the self, but in fact the variety is even greater than this. According to
some, multiple selves are simultaneously present in dissociative phenomena
such as hypnosis or multiple personality disorder (E. Hilgard, 1986/1991), as
well as in the play and exploration of young children (Wolf, 1990). According
to others, the notion of multiple selves should not be taken literally, either
because it stops us from recognizing that a single "person" is in control
(Elster, 1986) or because it reifies what are actually only "parts" of a mythical
unit (Orne & Bauer-Manley, 1991). Although many agree that there is a
distinct number of types or aspects of self, they disagree as to the exact
number: Self is viewed as single (Gallup, 1985), double (Lewis & Brooks-
Gunn, 1979), triple (Freud, 1923/1962; Lichtenberg, 1975; Schafer, 1968),
quadruple (Stern, 1985), quintuple (Neisser, 1988), septuple (Stern, 1985,
p. 7), or octuple (Allport, 1961). According to others there can be no *number*
of selves because the diverse processes that constitute the self are too integrated and thereby indivisible (Nagel, 1979, pp. 147–164).

I hit myself; I hate myself; I'm self-conscious; I'm self-sufficient; I feel like my old self;
I'm selfish; my humiliation was self-inflicted; and I couldn't contain myself. Self does
not mean exactly the same thing from one of these sentences to the next. It means my
body, my personality, my actions, my competence, my continuity, my needs, my agency,
and my subjective space. Self is thus a diffuse, multipurpose word; like the pronouns
"I" and "me," of which it is after all a variant, self is a way of pointing. (Schafer, 1976,
pp. 189–190)

81

For some, however, there is nothing to point to: The self is observationally nonexistent or a fiction (Hume, 1738/1971; Humphrey & Dennett, 1989; see discussion in Mackie, 1980; Glover, 1988), or merely a heuristic which has "caught on": "Self and identity are not facts about people; they are technical ways of thinking about people; and they have become ways in which people think about themselves" (Schafer, 1976, pp. 188–189).

Even focusing on one aspect of self leaves one with a diversity of perspectives. For example, the self-concept has been viewed as singular (Gallup, 1985), as having three parts (cf. Rosenberg, 1979; Trafimow, Triandis, & Goto, 1991), as having five structures, each with two substructures with up to six categories, all of which develop through six stages (L'Ecuyer, 1981), or as having an unlimited number of hierarchically related levels (Loevinger & Blasi, 1991, p. 164). Even the body image, a seemingly simpler aspect of the self, is itself not a single, unified schema, but rather one that dynamically incorporates knowledge of spatial, categorical, perceptual, experiential, imaginal, and evaluative aspects of one's body (Fisher, 1970, 1990; Schilder, 1950; Schontz, 1969; Werner, 1965).

Must we *assume* that all these "selves" are one and the same? That assumption is hard to justify: a unitary "self" has philosophical charms, but can hardly embrace all the modes of reflexive thought and conduct that we refer to by "self-" and "-self" terms. All that such compound terms have in common is reflexivity: instead of having someone or something else as a target, the thoughts and actions in question, *in some way or other*, turn back on the agent. The specific target of this "turning back" is not the same in all cases: it may be directed at any feature of an agent's habits or feelings, personality, body or skills. (The terms "self-esteem," "self-awareness" and "self-abuse" point to quite different features of the individual concerned!) (Toulmin, 1986, p. 52)

Nevertheless, for all its variety, the self continues to serve as a framework for human social and developmental psychology (Baumeister, 1986; Cash & Pruzinsky, 1990; Cicchetti & Beeghly, 1990; Hales, 1985; Harter, 1983; Heelas & Lock, 1981; Lynch, Norem-Hebeisen, & Gergen, 1981; Mischel, 1977; Rosenberg & Kaplan, 1982; Young-Eisendrath & Hall, 1987).

For the comparative psychologist, in contrast, the application of "self" terms to nonhumans can be problematic (Mitchell, 1992; 1993a,b). How is one meaningfully to compare cephalopods that are "visibly aware of their appearance" in communication and thus "self-aware" (Moynihan, 1985, p. 104), the chimpanzee and $1\frac{1}{2}$-year-old human infant who recognize themselves in a mirror and are thus "self-aware" (Gallup, 1985, p. 632), and the adolescent human who is "aware of the self as a self" (Loevinger & Blasi, 1991, p. 155)? The problem is not just that there are so many dimensions of self, but that any given instance of self-knowledge in humans involves so many different dimensions simultaneously, whereas the same need not be true in nonhumans. For example, many (but not all) people recognize their own voice within 5 sec of hearing it unexpectedly from a tape recording (Holzman & Rousey, 1966). The person experiences "a sudden confrontation with aspects of one's self mirrored in the voice and conveyed in the nonlexical qualities of speech"

(Holzman, Berger, & Rousey, 1967, p. 428). Following recognition, people's verbal responses suggest the following sequence within the first 2 min: They experience the discrepancy between their expectation and perception, respond affectively to it, describe their own voice's sound qualities, evaluate their voice, and accept the voice as their own (Holzman, Berger, & Rousey, 1967).

As I speak, I do not sound to myself as I sound to others (or to myself when I hear my voice from a tape recorder). This difference results from "physical differences between bone- and air-conducted sounds" (Holzman, Berger, & Rousey, 1967, p. 428). Often the first clue that the voice is my own is not the quality of my voice, but the knowledge that I said the things I am hearing from the tape recorder. In addition, the reason I believe that I sound like I do on a tape recorder is that I *was told by other people that my voice sounds like that to them.* Thus, own-voice recognition in humans entails self-evaluative, affective, semantic, and intersubjective components. If I say that an animal or very young human infant recognizes its own voice from a tape recording (or from a simultaneous video recording), does that mean that this animal shows all the same processes present in adult human beings' own voice recognition? Clearly not. Animals such as pygmy marmosets act differently toward their own calls than toward those of others (Snowdon & Cleveland, 1980), but this difference need not indicate that they identify the calls as their own; They could classify the calls as from a strange marmoset.

Such potential problems should not lead one to avoid any extrapolation from human to nonhuman animals. Rather, the psychology of the self in humans should be used to devise methods for studying the self in nonhuman animals (Mitchell, 1992; Parker & Milbrath, *SAAH*7) on the grounds that "anthropomorphism is the null hypothesis" for many investigations of animal behavior (Parker, in press-a). The similarities and differences between human and nonhuman behaviors and their explanations must be examined critically. Conceptual problems are likely to arise about the applicability of any given term to both nonhumans and humans, as different processes and mechanisms may produce similar results (see, e.g., Mitchell, 1986, 1987, 1990, 1992, 1993a,d).

A case in point is mirror self-recognition (MSR). This common human awareness was examined in nonhumans, with the initially surprising result that apes achieve what human children do (Gallup, 1970). There are, however, differences between apes' and humans' uses of mirrors. For example, humans use mirrors for self-adornment to make themselves more attractive to themselves and others, whereas nonhumans rarely if ever do such self-creation (Mitchell, 1993a, in press; cf. Miles, *SAAH*16). We still don't know how much nonhumans understand about mirrors, not even chimpanzees. Nor do we know how flexible their understanding is. For example, would a chimpanzee with two mirrors use both together to look at parts of itself not visible with only one?

Other problems of comparison between the self-knowledge of humans and nonhumans are salient in studies of mirror self-recognition. For example, monkeys have been described as having no sense of self because they fail to recognize themselves in a mirror (e.g., Gallup, 1985). According to this view, it is "a mistaken view of self-awareness" to think that species incapable of MSR "could evidence self-recognition in some other modality" because

[t]his would imply the existence of several selves (e.g., a visual self, olfactory self, etc.). One's sense of self transcends modality specific information and as such requires (or presupposes) the capacity for intermodal equivalence. (Gallup, 1983, p. 484)

The self-awareness present in mirror self-recognition is, in this view, an awareness of the only self there is. However, if one grants the belief that the self "transcends modality specific information," it is surprising that many adult humans fail to recognize their own voice from a recording (Holzman, Rousey, & Snyder, 1966; Huntley, 1940). Rather than presuming a transcendent self, it is more reasonable to assume that different modalities contribute to recognition of the self in the mirror and the self in a tape recording.

Given the many different senses of self in human beings, it is unfortunate that the research on the self in nonhuman animals has become so focused on the study of MSR (Gallup, *SAAH*3). Gallup's comparative approach has helped to maintain interest in psychological activities of animals such as deception, pretense, and self-knowledge; and the search for other forms of self-knowledge, such as shadow recognition, has been initiated by Gallup (Boysen, Bryan, & Shreyer, *SAAH*13; Cameron & Gallup, 1988). Nevertheless, we need to go further and examine forms of self unrelated to MSR. To assist in the search for nonhuman selves, I describe different aspects or forms of self in humans and offer any evidence of these different aspects of self in nonhumans. Rather than offer a global typology of self, I instead offer, as much as possible, particular aspects of self so that the comparisons between humans and animals are relatively focused. Whether these aspects of self are identical in humans and nonhumans, even where I might assume they are, is still open to inquiry. The headings provided are not intended to indicate that the aspects of self presented within them derive from identical processes; development is possible within each type of self, and the different headings present aspects of self that are often overlapping.

The perceiving self and the self perceived

If you ask anybody about the localization and boundaries of the self, . . . confusions are immediately displayed. . . . consider a blind man with a stick. Where does the blind man's self begin? At the tip of the stick? At the handle of the stick? Or at some point halfway up the stick?
                                        – Bateson (1972), p. 318

Most perceptions (olfactory, auditory, visual, and somatosensory) are of things outside the body, such that these things are perceived more accurately or intensely when the body is closer to them. These perceptions thus localize the body in relation to these outside things (Butterworth, 1992a), and provide the perceiver with a particular point of view (Nagel, 1979, pp. 165–180). This point of view is, however, not open to observation, as Hume (1738/1971, p. 257) noted: "For my part, when I enter most intimately into what I call myself, . . . I never can catch *myself* at any time." The perceiver is inherently present, observing but unobservable.

Neonatal imitation of facial gestures suggests that some matching *from* visual *to* kinesthetic experience is present at birth (Gopnik & Meltzoff, *SAAH*10), although this imitation may be represented amodally (Stern, 1985, pp. 47–53). Such early imitation is different from later imitations (Butterworth, 1992b, p. 135; Gibson, 1982; Mitchell, 1987; Parker, in press-b) and seems to result from a program that instructs the infant to "replicate the model upon or after perceiving it" (Mitchell, 1987, p. 203). Proprioception is innately tied to vision in humans, such that infants react to stabilize their body in relation to apparent visual movement outside their body and consequently lose their balance (Butterworth, 1992a, p. 105). Loss of balance in response to visual perception of only apparently moving rooms ceases at around 7 months in human infants, when these previously competing systems come under self-control (Butterworth, 1992a, p. 106). Visual–proprioceptive contingency is also recognized as early as 3 months of age (Watson, *SAAH*8), and play with self-contingency when watching videotapes of oneself increases from 30% of 9–12-month-old human infants to 90% of 21–24-month-olds (Lewis & Brooks-Gunn, 1979, p. 109). Self-contingency testing toward the mirror image is also observed in apes (Custance & Bard, *SAAH*12; Hyatt & Hopkins, *SAAH*15; Parker, 1991), and similar (though perhaps not identical) behavior is reported for pygmy marmosets and macaques (Eglash & Snowdon, 1983; Boccia, *SAAH*23).

Kinesthesis, somasthesis, and some pains, in contrast to other perceptions, are usually located inside or on the body; they give the body feeling and inform one of the spatial extent of the body. If a part of the body is amputated, people often feel as if that part of the body is still present kinesthetically and somasthetically, even when visually the body part is obviously not present; similarly, amputees often feel pain emanating from the absent area (Melzack, 1989, 1992; von Fieandt, 1966, pp. 327–332). These "phantom limbs" include not only missing arms and legs, hands and feet, and fingers and toes, but also breasts, penises, bladders, and anuses. Surprisingly, people born without appendages also experience phantom limbs, suggesting that the kinesthetically and somatically felt *experiences* of the body are the result of genetically determined neural networks. The same neural networks appear to be present in monkeys as well (Melzack, 1989).

Phantom limbs can be modified by experience. A bunion may be felt on the phantom of an amputated foot, or a wedding ring on the phantom of an amputated finger; the phantom may be experienced as taking on the form of a prosthesis or, if the stump is divided, the phantom may itself feel divided (Melzack, 1989); and after a second amputation of the same region, the phantom limb is experienced as the shape of the first amputation (von Fieandt, 1966).

If the neural networks that define the perceived self are disturbed, people may deny that visibly attached parts of their body are their own, and may find themselves on the floor after trying to push their own "alien" leg out of bed (Melzack, 1989). Even without neural damage, at times we do not recognize our body as our own when we are directly perceiving it. For example,

a psychologist [during] basic training ... was told to run, so he ... began running. After going on for as long as he could he finally stopped, puffing, "I can't go on any more." At this point he was ordered to continue running until he was told to stop or until he fell on his face ... so he began running again. ... After a short period he lost all voluntary control. He looked down and saw somebody's legs pumping away under him and was mildly surprised to notice that they were attached to his own body. (Miller, Galanter, & Pribram, 1960, p. 110)

Apparently the self begins with kinesthesis/somasthesis, and the visual–kinesthetic match that develops between kinesthetic and visual experiences as the result of their proximity (Koffka, 1935/1963, pp. 328–329) can be derailed under stress.

The self extended

In its widest possible sense ... a man's Self is the
sum total of all that he CAN *call his,* not only his
body and his psychic powers, but his clothes and
his house, his wife and children, his ancestors
and friends, his reputation and works, his lands
and horses, and yacht and bank account.
                                  – James (1890), p. 291

Material possessions are, as James suggests, often experienced as extensions of the self. On a less material and more symbolic level, self-extensions include gender, class, and regional and national identity. Hairstyles, clothing, jewelry, and other adornments are among the most personal and expressive extensions of self. As such, when people are placed in asylums or enter the military, they are (or have been) stripped of these self-extensions to be given a new sense of self (Goffman, 1961; Mares, 1971). How important other people are as self-extensions in people's experiences probably determines the extent to which they later think of themselves as in relation to others (Markus & Kitayama, 1991). However, the private, public and collective selves, or independent and interdependent selves, more likely point to differing foci of an individual's self-cognitions that are accessed in different situations (Trafimow et al., 1991) than to distinctly different types of self created from culture (Markus & Kitayama, 1991).

Sometimes during their first year human infants incorporate an external object into their personal activities such that this object is treated by the infant as part of itself, yet is also recognized (at some level) as "not-me." This "transitional object" is cuddled, loved, and mutilated; the infant "assumes rights over the object" and is disturbed if the object is taken away or if its relation with the child is in any way threatened (Winnicott, 1951/1958, p. 5). These possessions are extensions of the self (James, 1890, p. 291; Lancaster & Foddy, 1988; Rosenberg, 1979), and can become symbolically codified as "mine" in humans and apes (Itakura, *SAAH*14; Lewis, *SAAH*2; Miles, *SAAH*16; Patterson & Cohn, *SAAH*17). However, humans are the only organisms to create external representations of and/or for the self or another (Mitchell, in press; Reynolds, 1981, p. 164).

Other people can also be experienced as self-extensions, as revealed in jealousy. Early experiences of jealousy are based upon perception of another's interaction with the loved one or anticipation of being the object of the loved one's attention (Stern, 1914/1924; Valentine, 1942/1950, pp. 327–329).

[The jealous child's] claim of monopoly has reference, practically always, only to individual demonstrations of affection, whether it be one present or hoped for later. Thus, he pushes away his little brother who, in his desire for love, has secured the place nearest his mother. Thus, too, he is pleased when his father goes away, because he can then be the sole object of his mother's attention. But that he should harbour this desire of sole possession when away from her and constantly feed his imagination on such an idea is entirely improbable. (Stern, 1914/1924, p. 516)

In contrast, mature forms of jealousy depend upon imagination as well as taking the perspective of the other: "The bare thought that the object of affection can either give or receive love in any other direction is simply unbearable for the monopolising attitude of the lover" (Stern, 1914/1924, p. 516). The lover is "overcome . . . when he sees the loved being's interest attracted or distracted by persons, objects, or occupations which in [the lover's] eyes function as so many secondary rivals" (Barthes, 1977/1990, p. 110). The lover imagines or perceives the loved person with the other, and feels the loss of the other's attention to the self, which he or she craves.

Behaviors suggestive of self-extensions are observed in nonhumans and might be explained as forms of attachment, though the processes that result in these behaviors are unexamined. For example, one pet dog I studied wandered everywhere with a ball in his mouth, and was very upset when his owner took it from him to play. However, the dog's ball was unlike the transitional object of the human infant because the dog accepted substitute balls as replacements. Dogs and other mammals, including humans, also pretend that an object is their own and develop object-keepaway games to entice others to try to obtain the object (Aldis, 1975; Mitchell & Thompson, 1991). The jealousy of nonhumans appears more akin to that of human infants than of adults, though more research is needed here to parse out the affective and cognitive components of such behavior. Dog owners frequently tell me that their dog is jealous of any new relationship they develop, and describe instances in which their dog has placed itself between the owner and the new person. Primate mothers maintain control over their infant's contact with others, and other primates show apparent jealousy toward those interacting with their mother or with a friend (Breuggeman, 1978, pp. 186–187;

Goodall, 1986; Mitchell, 1989; Smuts, 1985, 224–225). In baboons, "vocalizations often reflect a vicarious involvement in the interactions of others" (Smuts, 1985, p. 225). Grief at the loss of a mother or friend may be present in baboons and chimpanzees (Goodall, 1986; Smuts, 1985, pp. 229–231), and the loss often leads to an earlier-than-expected death in chimps (Goodall, 1986). Whether any of these activities are based on processes identical to those in humans deserves further scrutiny.

The self identified

I like to look closely in the mirror and I see only
one eye in my forehead.
                      – Nijinsky (1919/1964), p. 424

Connections between vision and kinesthesis/proprioception apparently do not begin to be *bidirectional* matches until around 14 months of age in humans. At this time children are not only imitating actions, but using this matching ability to recognize that they are *being* imitated when another is imitating their actions (Gopnik & Meltzoff, *SAAH*10). Once an organism has such kinesthetic–visual matching (KVM), it knows what it looks like when it acts, such that it can recognize the identity between its kinesthetically experienced actions and the visual appearance of these actions (Mitchell, 1993a). Mimetic enactment of another's simultaneous activities, such that one identifies with the other's current actions so much as to react as if in the same situation, is observed in dolphins, chimpanzees, orangutans, and humans, as is delayed matching of other's activities (Allport, 1937, 1961, 1968; Köhler, 1925/1969; Russon & Galdikas, 1993; Taylor & Saayman, 1973; see Mitchell, in press). Whether nonhuman organisms can recognize another's imitation of their own activities is unknown.

   Kinesthetic–visual matching allows an organism to recognize itself in a mirror (Mitchell, 1993a; see also Parker, 1991). If an organism is looking in a mirror and it has KVM, it recognizes that the visual display in the mirror is the same as (matches) its kinesthetic experience, and can infer (given that it understands mirrors somewhat) that the visual image in the mirror is its own. Knowledge that mirrors reflect one's own body is found in most humans and some apes, and perhaps in dolphins (Carpenter, 1975; Gallup, 1985; Hyatt & Hopkins, *SAAH*15; Lewis & Brooks-Gunn, 1979; Marino, Reiss, & Gallup, *SAAH*25; Marten & Psarakos, *SAAH*24; Patterson & Cohn, *SAAH*17). Human children recognize their own image in mirrors and concurrent video images starting at 15–18 months of age (Amsterdam, 1972; Lewis & Brooks-Gunn, 1979), apes some time later (Lin, Bard, & Anderson, 1992; see also Custance & Bard, *SAAH*12; Miles, *SAAH*16; Patterson & Cohn, *SAAH*17). Some children at 9–12 months of age looking at themselves in a mirror express embarrassment (smiling at their image and then turning away), and at 15–24 months of age these signs of embarrassment are predictive of MSR: 78% of those who showed embarrassment also showed self-recognition; 22%

showed only embarrassment (Lewis, Sullivan, Stanger, & Wiess, 1989; see also Amsterdam, 1972). Before the end of the second year of life, children also gain "an appreciation of standards of proper behavior," noticing deviations from normality in objects and their own actions (Kagan, 1989, p. 236) that may influence their interest in the mark on their own face during the mark test devised by Gallup (1970) and Amsterdam (1972).

To what degree a young child is influenced by the appearance of caregivers and playmates or other close associates in formulating its own body image is unclear. Some infants, when observing a dot on their mother's face or on the face of another infant, touch their own face in the same place to determine if they too have a dot there (Johnson, 1983; Lewis & Brooks-Gunn, 1979). In the semi-autobiographical novel *Their Eyes Were Watching God* (Hurston, 1937/1978), a black girl raised with white children fails to recognize that the only black child in a photograph is herself because she thought she was white. Further, Viki, a human-reared chimp, classified a picture of herself, but not those of other chimps, as belonging with photos of humans (Hayes & Nissen, 1971). Although recognition that one has two eyes and a nose just like other people seems a rather obvious conclusion for a human child (Epstein, 1973, p. 413), in fact this recognition is not immediately salient (Piaget, 1945/1962, pp. 55–60) and seems to result from imitative games involving KVM in which children are asked by their parents – verbally or nonverbally – to locate their eyes, nose, forehead, or ears (see, e.g., Bretherton, McNew, & Beeghley-Smith, 1981, pp. 342–343; Guillaume, 1926/1971, p. 140; Piaget, 1945/1962, p. 55). Surprisingly, chimpanzees and 4-year-old children have difficulty creating an accurate composite image of a face out of two-dimensional images of pieces of a face (Premack & Premack, 1983).

The cross-modal perceptual processes that define one's experience of one's body are not precisely encoded in memory. Exact knowledge of one's visual body is not known from proprioception in adult humans, who under- and over estimate the spatial extent of their bodily dimensions in estimating the length of their limbs and in judging the accuracy of their image in distorting mirrors (Fisher, 1970; Fisher & Fisher, 1964; Orbach, Traub, & Olson, 1966; Wapner & Werner, 1965). When trying to create an accurate image of themselves in a distorting mirror, people "frequently state that they do not remember precisely what they look like" and show "a wide range of acceptability" toward distorted body images (Orbach et al., 1966, pp. 44, 45).

Although monkeys (as well as very young humans and apes) can use the image of objects in mirrors and videotapes to locate these objects (Anderson, 1986; Menzel, Savage-Rumbaugh, & Lawson 1985), they treat their own mirror image as another organism (Anderson 1984, *SAAH*21). The strong realistic image of a monkey beyond the plane of a normal mirror probably contributes to these monkeys' responses. However, it is interesting that the image is experienced as another monkey even when the nonreality of that monkey is made salient, as Jim Anderson and I discovered when we gave a long-tailed macaque named Rodrique an upside-down virtual image of himself. Rodrique was given a concave (bowl-shaped) mirror, which inverted his image and made this inverted image appear to be a 3-D monkey moving inside the bowl of the mirror. Rodrique groomed and licked the inverted monkey head as if it were actually present, in mid air, inside the bowl-shaped mirror. He was not noticeably disturbed by the lack of tactile or olfactory confirmation of this apparent monkey in the bowl. He failed to pass a mark test, and also failed to learn to imitate a human's scratching when rewarded for scratching in the same place as the human (Mitchell & Anderson, 1993), suggesting a difficulty with KVM.

Sometimes mirror self-recognition is viewed as indicative of the whole array of the diverse types of self-awareness found in humans – imagining ourselves, recognizing others' perspectives on us, and evaluating ourselves according to their perspectives (e.g., Gallup, 1985; Merleau-Ponty, 1960/1982). It is more parsimonious to suggest, however, that the kinesthetic–visual matching likely responsible for MSR creates a self-representation of a very specific and more limited sort – one that can be used to imagine and recognize ourselves (Mitchell, 1993a). Once I recognize myself in a mirror, "I leave the reality of my lived *me* in order to refer myself constantly to the ideal, fictitious, or imaginary *me*, of which the specular image is the first outline" (Merleau-Ponty, 1960/1982, p. 136).

The self imagined

Through KVM an organism can imagine how it looks, recognize itself in mirrors, imagine itself engaging in future activities, and imagine acting as another acts. The mirror self-recognizing organism can refer to its specular image – the visual image of its own body – because it can imagine at least the outline of that image in its fantasies: "When the ... [human] subject is given the task of imagining his own body while closing his eyes, the *somatic periphery*, that is, the peripheral body surface, clearly directs the formation of this mental image" (von Fieandt, 1966, p. 327). Kinesthetic–visual matching apparently works such that an organism can match what it does kinesthetically to a visual representation, and translate from a visual image to a kinesthetic recreation of that image. Thus, the organism can plan because it can imagine itself in its fantasies and can go from those visuallike images to produce a kinesthetically experienced action identical, in outline, to that visual image (Mitchell, 1993a, in press). Kinesthetic–visual matching spatializes the self and makes information about one's external appearance accessible to internal imagery (conscious or unconscious): Organisms with KVM not only have visuallike mental images, as other organisms do (Griffin, 1976), but can imagine themselves within these images (see Mitchell, 1993a, in press). Such an organism can plan its actions beyond simple plans such as "do X to get Y" in that it now has a means of imagining itself as a body acting in various scenarios. By the end of the 2nd year, children have the "capacity to hold cognitive representations on the stage of active memory" (Kagan, 1989, p. 237).

Pretenders in play are "self-consciously imitative" and self-aware in that they recognize their simulations as such (Mitchell, 1987, p. 207; 1990, p. 215). Young humans are highly proficient at bodily imitation and pretense of others. By 12 months of age children fleetingly pretend to use prototypical objects in pretense (e.g., using a spoon to "eat") based on imitation of others' actions (these children were fed with bottles), as well as elaborating on these imitations by using atypical objects (e.g., using a cup to "eat") (Fein & Apfel, 1979). They are less likely to imitate themselves than to imitate others at this point in pretense. Dolphins, chimpanzees, orangutans, gorillas, and perhaps (though rarely) macaques also pretend to enact other's actions (Miles, *SAAH*16; Mitchell, 1987, 1990, 1991, 1993a,d; Parker & Milbrath, *SAAH*7; Patterson & Cohn, *SAAH*17).

Children's pretenses are "self-building" (Fein, 1984): They become more elaborate as the self matures, such that "there is a marked increase in the number of roles and the order and coherence of actions reproduced in make-believe, accompanied by a decreasing reliance on veridical props" (Bretherton, 1984, p. 8). Older children and adults extend their activities and knowledge in imagination with role-playing, imaginary companions, daydreams, sexual fantasies, and representational arts (Bretherton, 1984; Parker & Milbrath, *SAAH*7; Partington & Grant, 1984; Singer & Singer, 1990; Walton, 1990; Watkins, 1986). One man "In his imagination . . . was constantly making away with people so as to show his heartfelt sympathy for their bereaved relatives" (Freud, 1909/1960, p. 370). Apes also apparently exhibit fantasies, but these are less elaborate than those of adult humans: The home-reared chimp Viki repeatedly pretended to have an imaginary pulltoy (Hayes, 1951, pp. 80–84), and sign-trained apes – a chimpanzee, a gorilla, and an orangutan – all appeared to attribute psychological qualities to dolls (Gardner & Gardner, 1969; Miles, 1990; Patterson & Linden, 1981). In pretense, fantasies, and daydreams, as well as in artistic representations, the self is present in imagination either as an observer or as an agent within the fantasy:

[V]irtually all our imaginings are partly about ourselves. Even when we are not central characters, *heroes* of our dreams, daydreams, and games of make-believe, we usually have some role in them – at least that of an observer of other goings-on. (Walton, 1990, p. 28)

In sexual fantasies sometimes the fantasizer is fictionally observing an encounter, sometimes he or she is fictionally acting within an encounter (see, e.g., Friday, 1975). These sexual fantasies are often plans for future sexual activities. For example, some men in sexual fantasy imagined themselves with an amputation and people's response to it, which greatly increased their excitement; these men later sought real amputations (Money, Jobaris, & Furth, 1977). These and other plans may include verbal instructions to oneself (Mischel & Mischel, 1977; Morin & Everett, 1990; Vygotsky, 1935/1978), a form of planning also used by sign-using apes when signing to themselves (Miles, *SAAH*16).

The self objectified and intersubjective

Kinesthetic–visual matching paves the way for understanding others' perspectives because in imitating the other we can pretend to be another and, perhaps, experience the world from a perspective that we recognize as different from our own.

> By imitating the other we come to know how he feels and thinks simply because we know how we ourselves thought and felt when we made similar movements or held similar postures in the past. Empathy [in the sense of objective motor mimicry] would seem to be simply a case of "kinesthetic inference." (Allport, 1961, p. 536)

However, kinesthetic–visual matching can only go so far. I can pretend facially and bodily to be sad without the accompanying feeling of sadness. Similarly, having empathy for another requires not only feeling what another feels, but also trying to communicate this shared feeling – for example, 10–12-month-old infants became visibly distressed by another's distress, but only at around 18 months were they able to offer comfort (Harris, 1989, pp. 30–31). Our empathic responses are apparently a communicative portrayal of the other's expressed feelings, kinesthetic and/or emotional (Bavelas, Black, Lemery, & Mullett, 1987). For example, "Watch the facial expressions of a sympathetic audience – the chances are that they show strains, smiles, and changes like those of the speaker" (Allport, 1961, p. 534). Along with empathy, other self-conscious emotions of embarrassment and envy appear in human children "before or around the second birthday" (Lewis et al., 1989, p. 148), and also occur in apes (Miles, *SAAH*16; Patterson & Cohn, *SAAH*17).

Before the end of the 2nd year of life, human children show affective self-consciousness (embarrassment, shyness, or coyness) from the look of another (Amsterdam & Greenberg, 1977). By 20 months of age, "being the focus of another's direct admiring attention is more likely to evoke self-conscious feeling than being the focus of one's own attention" (Amsterdam & Levitt, 1980, p. 72; see also Lewis et al., 1989). Recognition that one is an object of attention for another, then, is present early in human development, developing after mirror self-recognition (Lewis, 1992; Mitchell, 1993a). Developments in introspection, language, imagination, perspective taking, and self-evaluation further transform self-conscious emotions. For example, 5–8-year-old children expect to be embarrassed only when an audience ridicules their performance, but at 11–13 years they expect to be embarrassed when performing before *any* audience, because they expect to be evaluated (Bennett, 1989). Children at about 8 years of age begin spontaneously to imagine how they might feel in another's situation and thereby feel empathy, whereas children as young as 5 require coaching to so imagine (Hughes, Tingle, & Sawin, 1981). This late development of a spontaneous inference from one's own experience to another's in a similar situation argues against Gallup's (1983, 1985) idea that children and other organisms should begin to make such inferences once they recognize themselves in mirrors (see also Mitchell, 1993b).

In human children 2-years of age and older, the ability to take another's point of view concurs with the correct use of pronouns such as "I," "you," "me," and "mine" (Loveland, 1984). Autistics show a deficit in both of these abilities (Ricks & Wing, 1975). Verbal self-recognition and perspective taking are present in sign-using apes, who use their own names as well as the sign ME to refer to themselves (Miles, *SAAH*16; Patterson & Cohn, *SAAH*17). Nonlinguistic but verbal self-recognition was exhibited by one human male with a severed corpus callosum, who also showed a dual perspective in a joke: When photographs of familiar and unfamiliar faces were presented only to his right brain, he recognized some faces and responded with a thumbs-up gesture for those he liked and the opposite for those he did not. When shown his own photograph, he gave a thumbs-down gesture "accompanied by a broad grin indicating not only self-recognition in the minor hemisphere but also a subtle sense of humor and self-conscious perspective" (Sperry, Zaidel, & Zaidel, 1979, p. 163). The perspective taking of children becomes more complete at around 4 years of age, when they recognize that another may have a false belief (Sodian, 1991). At the same time, children become aware that they and others have private mental experiences (Neisser, 1988). The theory of mind developed by apes appears more rudimentary than that of 4-year-old children (Povinelli, Nelson, & Boysen, 1992; Premack, 1988).

The self presented and evaluated

Style means the presentation of the self as a three-dimensional art object, to be wondered at and handled.
                                    – Carter (1967/1982), p. 86

To influence others and to feel comfortable with him- or herself, the human subject comports his or her body and actions to present "an image of self delineated in terms of approved social attributes" (Goffman, 1967, p. 5). The subject uses impression management – accentuating body position and voice tone, manipulating facial muscles to produce expressions, and wearing clothes that support the presentation of self intended – to influence others to believe in his or her presentation of self and to instantiate his or her own ideal. Peoples' presentations of self indicate that they recognize that others are perceiving and evaluating them in relation to social norms, and to some degree monitor themselves to fit those social norms (Duval & Wicklund, 1972; Goffman, 1959, 1967; Lewis et al., 1989; Snyder, 1987). Chimpanzees also appear to recognize that others are perceiving and evaluating them, and monitor their behaviors to avoid repercussions (de Waal, 1982, 1986); but they and other apes appear to be without social norms (Mitchell, 1993c, in press; de Waal, 1991).

People control their actions in relation to standards derived from other people, thereby exhibiting a self-awareness that reflects these standards (and that can reflect on them). This "reflective self-awareness" is different from that evident in mirror self-recognition (Mitchell, 1993a), and is present in self-evaluative emotions such as mature embarrassment, pride, shame, and guilt, which begin to develop as early as the 3rd year (Lewis, 1992; Lewis et al., 1989). The concerns of more mature humans with how they are perceived by others, and with evaluating themselves in terms of others' perspectives are, I think, possible only with language or some other form of external symbolization with which to present and sustain standards (see Mitchell, 1993a, in press). The beginnings of such reflective self-awareness are present in some of the sign-using apes when they evaluate their actions as GOOD or BAD (for example, Miles, 1990, 1993, SAAH16; Patterson & Linden, 1981), but the reflective self-awareness of most mature humans seems more pervasive than that of sign-using apes, and suggests a mature form of self-understanding and self-knowledge unknown among our phylogenetically closest relatives. As mentioned, unlike apes, humans use mirrors to make themselves more attractive to others and to themselves; indeed, humans are the only organisms to so transform the look of their own body (see Mitchell, 1993a, in press).

Because human beings "evaluate their own behavior in light of internalized standards" (Rosenberg, 1988, p. 554; see also Duval & Wicklund, 1972, p. 3), they are also influenced by an *imagined* audience based on particular significant others (Cooley, 1902/1964, p. 184; Rosenberg, 1979, pp. 83–84). For example, people's self-evaluations change when asked to visualize faces of family members (Baldwin & Holmes, 1987), and the overwhelming experience of being negatively evaluated may stimulate some people to become binge eaters in order to divert their attention from this self-evaluative awareness (Heatherton & Baumeister, 1991). The presentation of the body is also influenced by an internal audience. When asked to transform the image in a distorting mirror to accurately reflect their own body, people tended to select images in conformity with societal standards (Lerner & Javanovic, 1990). For males, acceptable images "were disproportionately wide and 'husky-looking'"; for females, more acceptable images "were too long and 'slender-looking'" (Schneiderman, 1956, p. 98). The presence of an internal audience interferes with self-deceptive attempts to avoid unfavorable criticism (Baumeister & Cairns, 1992), influences how we express our emotions (Fridlund et al., 1990), and may cause embarrassment upon recognition by another that one falls short of some standard of self-identity (Goffman, 1967, p. 105). For some people, embarrassment is experienced even when another is or should be embarrassed (Miller, 1987). Before they commit suicide, some people imagine how their deaths might affect others, and some hope for revenge by enacting their death so that another will suffer as a result (Rosenberg, 1988, p. 560).

Social interaction is clearly necessary for the development of self (see Baldwin, 1894/1903; Gopnik & Meltzoff, SAAH10; Hart & Fegley, SAAH9; Mead, 1934/1974; Parker & Milbrath, SAAH7). Although some argue that the self is determined by social interaction (Cooley, 1902/1964; Gergen, 1977; Harré, 1984), living up to social expectations does not always lead one to experience one's actions as consonant with one's self. On the contrary, such expectations may assist one in creating a persona consistent with these expectations rather than with reality.

My father was a Jew. This did not seem to him a good idea, and so it was his notion to disassemble his history, begin at zero, and re-create himself.... He would not make peace with his actualities, and so he was the author of his own circumstances, and indifferent to the consequences of this nervy program.... I wish he hadn't selected from among the world's possible disguises the costume and credentials of a yacht club commodore. Beginning at scratch, he might have reached further, tried something a bit more bold and odd, a bit less inexorably conventional, a bit less calculated to please. (Wolff, 1979, p. 9)

We can move through social interactions with a great deal of ease by living according to the roles and rules expected (Rosenberg, 1988), but we may still recognize a discrepancy between ourselves and these roles and rules.

This discrepancy results from our evaluations, which provide us with desires for things we wish we did not desire – that is, with second-order desires (Frankfurt, 1971). Sometimes we learn what we want or feel from our dreams or daydreams, even when we do not want to want or feel what we fantasize about (Freud, 1900/1963). For example, a common belief for a young gay man in America is that "he [is] really heterosexual despite the fact that his fantasies and activities [had been] exclusively homosexual for years" (Friedman, 1989, p. 227).

Many of these young men go on to develop a stable sense of homosexual identity later in life. Many, however, internalize negative attitudes and values from the general culture and are consumed by self-hate because of their sexual orientation.... (Friedman, 1989, pp. 227–228)

Clearly cultural expectations of sexuality are in part creative of homosexual identity and negative self-evaluations, in that in other cultures these aspects of self do not usually develop from homosexual behavior and ideation during adolescence (Stoller & Herdt, 1985). The conflict between cultural expectations and one's experience of one's own desires creates a divided self.

Dissociation of the self

Although temporary attentional dissociation is present when driving on a well-known route and automatically attending to traffic signs while thinking about something else, other forms of dissociation suggest a more dynamic separation of parts of the self. Dissociation operates as a defense against unbearable pain following or during traumas, such that feelings of "bodily anesthesia" and "invisibility" occur (Terr, 1991). Dissociation of this kind also occurs in self-deception and denial, in which awareness of a version of reality in which one wants to believe is kept separate from awareness of both this version's fictionality and a more accurate version (see Mitchell, 1993d). It also occurs in multiple personality disorder, in which several selves having different memories and divergent knowledge of each other are present in the same individual (E. Hilgard, 1986/1991; Janet, 1920). For example, one man had four personalities. The primary one lacked awareness of the other three, although these three were somewhat aware of him and of each other. Each of the three subsidiary personalities developed in response to particular problems; they mediated violent family squabbles, dealt with sexual confusions, and protected the man from violence (E. Hilgard, 1986/1991).

Surprisingly, the self created through successful integration of multiple personalities can leave the individual with a loss of a sense of self. In one case, the previous, seemingly fragmented self in fact comprised these distinct entities, all of which were experienced as the individual's self (see Sizemore & Pittillo, 1977). Although any multiplication of selves suggests that the self or selves in people with multiple personality disorder – as well as in normal individuals – can reasonably be viewed as fictional in the sense of not being "real" (Humphrey & Dennett, 1989; Orne & Bauer-Manley, 1991), I would suggest instead that we come to know, conceive, and experience ourselves in part as a result of our fictional experiences in make-believe: We come to be what we are as a result of our imagination (Baldwin, 1894/1903, pp. 360–365; Mead, 1934/1974; Parker & Milbrath, *SAAH*7; Rosenberg, 1979, pp. 44–45).

Imagination is involved in the dissociation of self and the temporary incorporation of another following hypnosis (J. Hilgard, 1970). Hypnotized subjects internalize the hypnotist's voice, either letting it take control or saying the hypnotist's words to themselves (J. Hilgard, 1970, pp. 253–255; Miller et al., 1960). The consciousness of the hypnotized subject dissociates into two separate consciousnesses, one controlled by the hypnotist, the other by the subject (E. Hilgard, 1986/1991). For example, when a woman was told under hypnosis that she would feel no pain, part of her felt no pain when her hand was maintained in freezing cold water, while another part of her recorded increasing pain. When a blind man was told under hypnosis that he would become deaf unless touched on the shoulder by the hypnotist, he became unresponsive to sound and bored from lack of stimulation, and attended to a statistics problem in his imagination. Nevertheless, when asked to raise his finger if a part of him could hear, he did so. Simultaneously, the part of him that had been attending to the statistics problem became disturbed when he felt his hand moving upward. This part had no awareness of the question and no understanding of why he was raising his finger.

The self evaluated by the self

I hope you realize that I am remembering all this now. I am looking back and realizing what the truth was. The motives back of the action. I don't think it was all as cold-blooded as it sounds. I hope not. But the truth has to be that I was a terrible pig. My aim was ME ME ME.
                                          – Hepburn (1991), p. 158

Self-reflective self-awareness is the form of self-awareness in which the self examines itself from its own perspective. Once people have the ability to evaluate and imagine themselves, they can to some degree create their own audience and evaluate themselves according to their own precepts; they "can ponder past and future and weigh alternative courses of action in the light of some vision of a whole life well lived" (Arnhart, 1990, p. 528). The idea of the self as one who reflects upon his or her own actions, thoughts, and beliefs, as well as their consequences, is prominent in Western culture, implicit in the Old Testament, and explicit in writings of Plato, Aristotle, and other early Greeks (see Arnhart, 1990; Ey, 1963/1978, p. 229; Lichtenberg, 1975, p. 468; Mischel & Mischel, 1977; Morin & Everett, 1990). For example, Antigone, the title character in Sophocles' fifth-century-B.C. play, is motivated by self-reflection to defy a decree by burying her brother. Self-reflective self-awareness is derived from the reflective self-awareness using internalized audience, yet differs by allowing for self-examination and self-evaluation by the self rather than through others' prescriptions (see Babcock, 1988), such that one can strive to achieve a consistent and coherent image of oneself independent of others' images.

Inner speech is often directed toward examining the self, and as such serves as a mediator of self-knowledge (Morin & Everett, 1990), although in excess this self-examination may lead to fragmentation of the self (Sass, 1987). That we can examine our beliefs in the privacy of our own mind suggests some separation from the internalized audiences with which social psychologists people our head, but the separation is likely incomplete. Although people can act as their own audience, such that they make themselves more consistent with their desired self-presentation upon viewing themselves in a mirror, previous evaluations from others can produce the same self-consistency (Duval & Wicklund, 1972; Wicklund, 1982). Moreover, even if our self-evaluations occur privately, the results of these evaluations are often subject to others' evaluations and may need to be defensible morally in argumentation (Arnhart, 1990).

In autobiography, one can examine one's life against what one hoped to accomplish or create, and find one's own understanding. Nevertheless, the self we recognize as our own may not always be consistent with that self as seen from other people's perspective.

What makes autobiography valuable ... is not its fidelity to fact but its revelations – to the writer as much as to the reader – of self. Rather than blaming our autobiographers for discrepancies between their stories and supposedly verifiable facts, we should realize, on the one hand, that memory's deceptions are not always conscious and, on the other, that the duplicity of memory affords us one of the most powerful avenues of entry into the self-identity of the writer. ... what we remember – which is a mysterious amalgam of what we choose, what we really want, what actually happened, and what we are forced to remember – once turned into language and written down, becomes our personal truth without much consideration for its literal accuracy. (Adams, 1990, pp. 170–171)

The autobiographer writes from a feeling of discovery of the unifying (or discontinuous) factors of his or her experience, and in doing so recognizes

Table 6.1. *Aspects of self in humans, and the nonhuman animals that so far provide some evidence of at least* similar *aspects of self*

Human forms	Similar nonhuman forms in:
The perceiving self and the self perceived	
perception and point of view	most animals
visual proprioception	?
neonatal imitation	?
visual–proprioceptive contingency	many primates
kinesthetic somasthetic body image	monkeys
contingency testing toward mirror image	apes; perhaps pygmy marmosets, macaques
The self extended	
transitional objects and possessions	many mammals
object keepaway	many mammals
external representations	?
grief	monkeys, apes
jealousy	dogs, monkeys, apes
mature jealousy	?
The self identified	
kinesthetic–visual matching	apes, dolphins; perhaps macaques (rare)
recognize that self is being imitated	?
mirror self-recognition	apes, dolphins; perhaps macaques (rare)
own-voice recognition	?
expectation that one looks like others	human-reared chimp
imperfect kinesthetic–visual inference	?
The self imagined	
pretend play	apes, dolphins; perhaps macaques (rare)
planning	chimpanzees, sign-using apes
internal imagination	sign-using apes
sexual fantasy	?
inner speech, self-signing	sign-using apes
The self objectified and intersubjective	
taking perspective of the other	chimpanzees, sign-using apes
self-conscious emotions	sign-using apes
pronouns of self	sign-using apes
theory of mind	apes (rudimentary)
The self presented and evaluated	
self-presentation	chimpanzees, sign-using apes
self-monitoring	chimpanzees, sign-using apes
evaluative terms	sign-using apes
self-evaluative terms	sign-using apes
socially approved image of self	?
transform body for social approval	?
self-evaluative emotions	?

Table 6.1. (*Cont.*)

Human forms	Similar nonhuman forms in:
invented persona	?
imagined audiences	?
second-order desires	?
Dissociation of the self	
self-deception	perhaps chimpanzees
dissociation in hypnosis	?
multiple personality disorder	?
The self evaluated by the self	
self-reflective action	?
self-reflective self-awareness	?
evaluation of one's life	?
autobiography	?
moral argumentation	?

and evaluates the identity between his or her past and current selves. An autobiographer's self-evaluation however, may be just as biased as anyone else's evaluation of him or her.

Although humans through their imagination may provide an autobiography for a speaking chimpanzee (Kafka, 1919/1979), a diary of a dog (Gilman, 1921), or even the thoughts and experiences during the final days of a human recently transformed into a dung beetle (Kafka, 1912/1979), nonhumans themselves are unable to take the initiative. Humans also use their imagination to create images of a self that lives on after death, though usually this self is not examined too closely (Updike, 1989).

Concluding comments

The wealth and variety of aspects of self present in human beings suggest new directions for research in nonhuman beings. Three broad levels appear in the development of the self, although they do not by any means characterize all the different aspects of the self discussed in the chapter (see Table 6.1):

1. the self as largely implicit, a point of view that experiences, acts and, at least in the case of mammals and birds, has emotions and feelings;
2. the self as built on kinesthetic–visual matching, leading to MSR, imitation, pretense, planning, self-conscious emotions, and imaginative experiences of fantasy and daydreams; and contributing to perspective taking and the beginnings of theory of mind; and
3. the self as built on symbols, language and artifacts, which provide external support for shared cultural beliefs, social norms, inner speech, dissociation, and evaluation by others as well as self-evaluation (see also Mitchell, 1993a, in press).

Within adult humans all three levels of self are present. In fact adult humans are distinguished from members of other species by the multiplicity of their selves. Animals such as insects may show only one aspect – that of having perception and therefore a point of view – of the first level, whereas mammals show many more. Apes, especially sign-using apes, appear to provide extensive evidence of the first two levels of self and some evidence of the third level not dependent upon complex linguistic skills.

Note, however, that because my examination of aspects of self has been focused on humans, it may be hopelessly anthropocentric. Apes and other organisms may have self-understandings never dreamed of by humans, or they may have more or less the same understandings we have within the limits of their cognitive capacities. Our attempts to understand them may distort their sense of self conceptually, much as Gregor Samsa's transmogrification into a gigantic insect distorted him physically. Such conceptual translation problems are also present in cross-cultural comparisons (Rosaldo, 1984). Still, unless we start somewhere, we will never find a path we can follow to see where it leads.

Acknowledgments

I greatly appreciate the assistance and support of Sue Parker and Lyn Miles in thinking through the ideas in this chapter. Their comments helped me to find where I wanted to go.

References

Adams, T. D. (1990). *Telling lies in modern American autobiography*. Chapel Hill: University of North Carolina Press.
Aldis, O. (1975). *Play fighting*. New York: Academic Press.
Allport, G. W. (1937). *Personality: A psychological interpretation*. New York: Holt.
　(1961). *Pattern and growth in personality*. New York: Holt, Rinehart & Winston.
　(1968). The historical background of social psychology. In G. Lindzey & E. Aronson (Eds.), *Handbook of social psychology, Vol. 1: Theory and method* (3rd ed., pp. 1–46). New York: Random House.
Amsterdam, B. K. (1972). Mirror self-image reactions before age two. *Developmental Psychobiology*, 5, 297–305.
Amsterdam, B. K., & Greenberg, L. G. (1977). Self-conscious behavior of infants. *Developmental Psychobiology*, 10, 1–6.
Amsterdam, B. K., & Levitt, M. (1980). Consciousness of self and painful self-consciousness. *Psychoanalytic Study of the Child*, 35, 67–83.
Anderson, J. R. (1984). Monkeys with mirrors: Some questions for primate psychology. *International Journal of Primatology*, 5, 81–98.
　(1986). Mirror-mediated finding of hidden food by monkeys (*Macaca tonkeana* and *M. fascicularis*). *Journal of Comparative Psychology*, 100, 237–242.
Arnhart, L. (1990). Aristotle, chimpanzees, and other political animals. *Social Science Information*, 29, 477–557.
Babcock, M. K. (1988). Embarrassment: A window on the self. *Journal for the Theory of Social Behaviour*, 18, 459–483.
Baldwin, J. M. (1894/1903). *Mental development in the child and the race*. New York: Macmillan.

Baldwin, M. W., & Holmes, J. G. (1987). Salient private audiences and awareness of the self. *Journal of Personality and Social Psychology, 52*, 1087–1098.

Barthes, R. (1977/1990). *A lover's discourse*. New York: Noonday.

Bateson, G. (1972). The cybernetics of "self": A theory of alcoholism. In G. Bateson, *Steps to an ecology of mind* (pp. 309–337). New York: Ballantine.

Baumeister, R. F. (Ed.) (1986). *Public self and private self*. New York: Springer-Verlag.

Baumeister, R. F., & Cairns, K. J. (1992). Repression and self-presentation: When audiences interfere with self-deceptive strategies. *Journal of Personality and Social Psychology, 62*, 851–862.

Bavelas, J. B., Black, A., Lemery, C. R., & Mullett, J. (1987). Motor mimicry as primitive empathy. In N. Eisenberg & J. Strayer (Eds.), *Empathy and its development* (pp. 317–338). Cambridge University Press.

Bennett, M. (1989). Children's self-attribution of embarrassment. *British Journal of Developmental Psychology, 7*, 207–217.

Bretherton, I. (1984). Representing the social world in symbolic play: Reality and fantasy. In I. Bretherton (Ed.), *Symbolic play: The development of social understanding* (pp. 3–41). Orlando, FL: Academic Press.

Bretherton, I., McNew, S., & Beeghly-Smith, M. (1981). Early person knowledge as expressed in gestural and verbal communication: When do infants acquire a "theory of mind"? In M. E. Lamb & L. R. Sherrod (Eds.), *Infant social cognition: Empirical and theoretical considerations* (pp. 333–373). Hillsdale, NJ: Erlbaum.

Breuggeman, J. A. (1978). The function of adult play in free-ranging *Macaca mulatta*. In E. O. Smith (Ed.), *Social play in primates* (pp. 169–191). New York: Academic Press.

Butterworth, G. E. (1992a). Origins of self-perception in infancy. *Psychological Inquiry, 3*, 103–111.

(1992b). Self-perception as a foundation for self-knowledge. *Psychological Inquiry, 3*, 134–136.

Cameron, P. A., & Gallup, G. G., Jr. (1988). Shadow self-recognition in human infants. *Infant Behavior and Development, 11*, 465–471.

Carpenter, E. (1975). The tribal terror of self-awareness. In P. Hockings (Ed.), *Principles of visual anthropology* (pp. 451–461). Paris: Mouton.

Carter, A. (1967/1982). Notes for a theory of sixties style. In *Nothing sacred: Selected writings* (pp. 85–90). London: Virago.

Cash, T. F., & Pruzinsky, T. (Eds.) (1990). *Body images: Development, deviance, and change*. New York: Guilford.

Cicchetti, D., & Beeghly, M. (1990). *The self in transition: Infancy to childhood*. Chicago: University of Chicago Press.

Cooley, C. H. (1902/1964). *Human nature and the social order*. New Brunswick, NJ: Transaction.

Duval, S., & Wicklund, R. A. (1972). *A theory of objective self awareness*. New York: Academic Press.

Eglash, A. R., & Snowdon, C. T. (1983). Mirror-image responses in pygmy marmosets (*Cebuella pygmaea*). *American Journal of Primatology, 5*, 211–219.

Elster, J. (1986). Introduction. In J. Elster (Ed.), *The multiple self* (pp. 1–34). Cambridge University Press.

Epstein, S. (1973). The self-concept revisited, or a theory of a theory. *American Psychologist, 28*, 404–416.

Ey, H. (1963/1978). *Consciousness: A phenomenological study of being conscious and becoming conscious*. Bloomington: Indiana University Press.

Fein, G. G. (1984). The self-building potential of pretend play, or "I got a fish, all by myself." In T. D. Yawkey & A. D. Pellegrini (Eds.), *Child's play: Developmental and applied* (pp. 125–141). Hillsdale, NJ: Erlbaum.

Fein, G. G., & Apfel, N. (1979). Some preliminary observations on knowing and pretending. In N. R. Smith & M. B. Franklin (Eds.), *Symbolic functioning in childhood* (pp. 87–100). Hillsdale, NJ: Erlbaum.

Fisher, S. (1970). *Body experience in fantasy and behavior*. New York: Appleton-Century-Crofts.

(1990). The evolution of psychological concepts about the body. In T. F. Cash & T. Pruzinsky (Eds.), *Body images: Development, deviance, and change* (pp. 3–20). New York: Guilford.

Fisher, S., & Fisher, R. (1964). Body image boundaries and patterns of body perception. *Journal of Abnormal and Social Psychology, 68*, 255–262.

Frankfurt, H. (1971). Freedom of the will and the concept of a person. *Journal of Philosophy, 67*, 5–20.

Freud, S. (1900/1963). *The interpretation of dreams*. New York: Wiley.

(1909/1960). Notes upon a case of obsessional neurosis. In S. Freud, *Collected papers* (pp. 293–383). New York: Basic.

(1923/1962). *The ego and the id*. New York: Norton.

Friday, N. (1975). *Forbidden flowers: More women's sexual fantasies*. New York: Pocket Books.

Fridlund, A. J., Sabini, J. P., Hedlund, L. E., Schaut, J. A., Shenker, J. I., & Knauer, M. J. (1990). Audience effects on solitary faces during imagery: Displaying to the people in your head. *Journal of Nonverbal Behavior, 14*, 113–137.

Friedman, R. C. (1989). Denial in the development of homosexual men. In E. L. Edelstein, D. L. Nathanson, & A. M. Stone (Eds.), *Denial: A clarification of concepts and research* (pp. 219–230). New York: Plenum.

Gallup, G. G., Jr. (1970). Chimpanzees: Self-recognition. *Science, 167*, 86–87.

(1983). Toward a comparative psychology of mind. In R. L. Mellgren (Ed.), *Animal cognition and behavior* (pp. 473–510). Amsterdam: North Holland.

(1985). Do minds exist in species other than our own? *Neurosciences and Biobehavioral Review, 9*, 631–641.

Gardner, R. A., & Gardner, B. T. (1969). Teaching sign language to a chimpanzee. *Science, 165*, 664–672.

Gergen, K. J. (1977). The social construction of self-knowledge. In T. Mischel (Ed.), *The self: Psychological and philosophical issues* (pp. 139–169). Totowa, NJ: Rowman & Littlefield.

Gibson, K. (1982). Comparative neuro-ontogeny: Its implications for the development of human intelligence. In G. Butterworth (Ed.), *Infancy and epistemology: An evaluation of Piaget's theory* (pp. 52–82). New York: St. Martin's.

Gilman, E. (1921). A dog's diary. *Journal of Comparative Psychology, 1*, 309–315.

Glover, J. (1988). *I: The philosophy and psychology of personal identity*. London: Allen Lane.

Goffman, E. (1959). *The presentation of self in everyday life*. Garden City, NY: Doubleday.

(1961). *Asylums: Essays on the social situation of mental patients and other inmates*. Chicago: Aldine.

(1967). *Interaction ritual*. Garden City, NY: Doubleday.

Goodall, J. (1986). *The chimpanzees of Gombe: Patterns of behavior*. Cambridge, MA: Harvard University Press.

Griffin, D. (1976). *The question of animal awareness*. New York: Rockefeller University Press.

Guillaume, P. (1926/1971). *Imitation in children*. Chicago: University of Chicago Press.

Hales, S. (Ed.) (1985). Special Issue: The rediscovery of self in social psychology: Theoretical and methodological implications. *Journal for the Theory of Soical Behaviour, 15*(3)–*16*(1).

Harré, R. (1984). *Personal being*. Cambridge, MA: Harvard University Press.

Harris, P. L. (1989). *Children and emotion: The development of psychological understanding*. Oxford: Blackwell Publisher.

Harter, S. (1983). Developmental perspectives on the self-system. In P. Mussen (series Ed.) and E. Hetherington (volume Ed.), *Handbook of child psychology, 4th ed., Vol. 4: Socialization, personality, and social development* (pp. 275–385). New York: Wiley.

Hayes, C. (1951). *The ape in our house*. New York: Harper Bros.

Hayes, K. J., & Nissen, C. H. (1971). Higher mental functions in a home-raised chimpanzee. In A. M. Schrier & F. Stollnitz (Eds.), *Behavior of nonhuman primates, Vol. 4* (pp. 59–115). New York: Academic Press.

Heatherton, T. F., & Baumeister, R. F. (1991). Binge eating as escape from self-awareness. *Psychological Bulletin, 110*, 86–108.

Heelas, P., & Lock, A. (Ed.) (1981). *Indigenous psychologies: The anthropology of the self*. New York: Academic Press.

Hepburn, K. (1991). *Me: Stories of my life.* New York: Ballantine.

Hilgard, E. (1986/1991). Dissociative phenomena and the hidden observer. In D. Kolak & R. Martin (Eds.), *Self and identity: Contemporary philosophical issues* (pp. 89–114). New York: Macmillan.

Hilgard, J. (1970). *Personality and hypnosis: A study of imaginative involvement.* Chicago: University of Chicago Press.

Holzman, P. S., Berger, A., & Rousey, C. (1967). Voice confrontation: A bilingual study. *Journal of Personality and Social Psychology, 7,* 423–428.

Holzman, P. S., & Rousey, C. (1966). The voice as a percept. *Journal of Personality and Social Psychology, 4,* 79–86.

Holzman, P. S., Rousey, C., & Snyder, C. (1966). On listening to one's own voice: Effects on psychophysiological responses and free associations. *Journal of Personality and Social Psychology, 4,* 432–441.

Hughes, R., Jr., Tingle, B. A., & Sawin, D. B. (1981). Development of empathic understanding in children. *Child Development, 52,* 122–128.

Hume, D. (1738/1971). *An enquiry concerning human understanding.* LaSalle, IL: Open Court.

Humphrey, N., & Dennett, D. C. (1989). Speaking for ourselves: An assessment of multiple personality disorder. *Raritan, 9,* 68–98.

Huntley, C. W. (1940). Judgments of self based upon records of expressive behavior. *Journal of Abnormal and Social Psychology, 35,* 398–427.

Hurston, Z. N. (1937/1978). *Their eyes were watching God.* Urbana: University of Illinois Press.

James, W. (1890). *The principles of psychology* (Vol. 1). New York: Holt.

Janet, P. (1920). *The major symptoms of hysteria* (2nd ed.). New York: Macmillan.

Johnson, D. B. (1983). Self-recognition in infants. *Infant Behavior and Development, 6,* 211–222.

Kafka, F. (1912/1979). The metamorphosis. In *The basic Kafka* (pp. 1–54). New York: Pocket Books.

(1919/1979). A report to an academy. In *The basic Kafka* (pp. 245–255). New York: Pocket Books.

Kagan, J. (1989). *Unstable ideas: Temperament, cognition, and self.* Cambridge, MA: Harvard University Press.

Koffka, K. (1935/1963). *Principles of Gestalt psychology.* New York: Harcourt Brace & World.

Köhler, W. (1925/1969). *The mentality of apes.* New York: Liveright.

Lancaster, S., & Foddy, M. (1988). Self-extensions: A conceptualization. *Journal for the Theory of Social Behaviour, 18,* 77–94.

L'Ecuyer, R. (1981). The development of the self-concept through the life span. In M. D. Lynch, A. A. Norem-Hebeisen, & K. J. Gergen (Eds.), *Self-concept: Advances in theory and research* (pp. 203–218). Cambridge, MA: Ballinger.

Lerner, R. M., & Javanovic, J. (1990). The role of body image in psychosocial development across the life span: A developmental contextual perspective. In T. F. Cash & T. Pruzinsky (Eds.), *Body images: Development, deviance, and change* (pp. 110–127). New York: Guilford.

Levin, J. D. (1992). *Theories of the self.* Washington, DC: Hemisphere.

Lewis, M. (1992). *Shame: The exposed self.* New York: Free Press.

Lewis, M., & Brooks-Gunn, J. (1979). *Social cognition and the acquisition of self.* New York: Plenum.

Lewis, M., Sullivan, M., Stanger, C., & Weiss, M. (1989). Self development and self-conscious emotions. *Child Development, 60,* 146–156.

Lichtenberg, J. D. (1975). The development of the sense of self. *Journal of the American Psychoanalytic Association, 23,* 453–484.

Lin, A. C., Bard, K. A., & Anderson, J. R. (1992). Development of self-recognition in chimpanzees (*Pan troglodytes*). *Journal of Comparative Psychology, 106,* 120–127.

Loevinger, J., & Blasi, A. (1991). Development of the self as subject. In J. Strauss & G. R. Goethals (Eds.), *The self: Interdisciplinary approaches* (pp. 150–167). New York: Springer-Verlag.

Loveland, K. A. (1984). Learning about points of view: Spatial perspective and the acquisition of "I / you." *Journal of Child Language, 11,* 535–556.

Lynch, M. D., Norem-Hebeisen, A. A., & Gergen, K. J., Eds. (1981). *Self-concept: Advances in theory and research*. Cambridge, MA: Ballinger.

Mackie, J. L. (1980). The transcendental "I." In Z. Van Straaten (Ed.), *Philosophical subjects* (pp. 48–61). Oxford: Oxford University Press.

Mares, W. (1971). *The marine machine*. Garden City, NY: Doubleday.

Markus, H. R., & Kitayama, S. (1991). Culture and the self: Implications for cognition, emotion, and motivation. *Psychological Review, 98*, 224–253.

Mead, G. H. (1934/1974). *Mind, self and society*. Chicago: University of Chicago Press.

Melzack, R. (1989). Phantom limbs, the self, and the brain. *Canadian Psychology, 30*, 1–16.

—— (1992). Phantom limbs. *Scientific American, 266*, 120–126.

Menzel, E. W., Jr., Savage-Rumbaugh, E. S., & Lawson, J. (1985). Chimpanzee (*Pan troglodytes*) spatial problem solving with the use of mirrors and televised equivalents of mirrors. *Journal of Comparative Psychology, 99*, 211–217.

Merleau-Ponty, M. (1960/1982). The child's relations with others. In J. M. Edie (Ed.), *The primacy of perception* (pp. 96–155). Evanston, IL: Northwestern University Press.

Miles, H. L. (1990). The cognitive foundations for reference in a signing orangutan. In S. T. Parker & K. Gibson (Eds.), *"Language" and intelligence in monkeys and apes: Comparative developmental perspectives* (pp. 511–539). Cambridge University Press.

—— (1993). Language and the orangutan: The old "person" of the forest. In P. Singer & P. Cavalieri (Eds.), *A new equality: The great ape project*. London: Fourth Estate.

Miller, G. A., Galanter, E., & Pribram, K. H. (1960). *Plans and the structure of behavior*. New York: Holt.

Miller, R. S. (1987). Empathic embarrassment: Situational and personal determinants of reactions to the embarrassment of another. *Journal of Personality and Social Psychology, 53*, 1061–1069.

Mischel, T. (Ed.) (1977). *The self: Psychological and philosophical issues*. Totowa, NJ: Rowman & Littlefield.

Mischel, W., & Mischel, H. N. (1977). Self-control and the self. In T. Mischel (Ed.) *The self: Psychological and philosophical issues* (pp. 31–64). Totowa, NJ: Rowman & Littlefield.

Mitchell, R. W. (1986). A framework for discussing deception. In R. W. Mitchell & N. S. Thompson (Eds.), *Deception: Perspectives on human and nonhuman deceit* (pp. 3–40). Albany, NY: SUNY Press.

—— (1987). A comparative-developmental approach to understanding imitation. In P. P. G. Bateson & P. H. Klopfer (Eds.), *Perspectives in ethology* (Vol. 7, pp. 183–215). New York: Plenum.

—— (1989). Functions and social consequences of infant–adult male interaction in a captive group of lowland gorillas (*Gorilla gorilla gorilla*). *Zoo Biology, 8*, 125–137.

—— (1990). A theory of play. In M. Bekoff & D. Jamieson (Eds.), *Interpretation and explanation in the study of animal behavior, Vol. 1: Interpretation, intentionality, and communication* (pp. 197–227). Boulder, CO: Westview.

—— (1991). Bateson's concept of "metacommunication" in play. *New Ideas in Psychology, 9*, 73–87.

—— (1992). Developing concepts in infancy: Animals, self-perception, and two theories of mirror-self-recognition. *Psychological Inquiry, 3*, 127–130.

—— (1993a). Mental models of mirror self-recognition: Two theories. *New Ideas in Psychology, 11*, 295–325.

—— (1993b). Recognizing one's self in a mirror? A reply to Gallup and Povinelli, Byrne, Anderson, and de Lannoy. *New Ideas in Psychology, 11*, 351–377.

—— (1993c). Humans, nonhumans, and personhood. In P. Singer & P. Cavalieri (Eds.) *A new equality: The great ape project* (pp. 237–247). London: Fourth Estate.

—— (1993d). Animals as liars: The human face of nonhuman duplicity. In M. Lewis & C. Saarni (Eds.), *Lying and deception in everyday life* (pp. 59–89). New York: Guilford.

—— (in press). The evolution of primate cognition: Simulation, self-knowledge, and knowledge of other minds. In D. Quiatt & J. Itani (Eds.), *Hominid culture in primate perspective*. Denver: University Press of Colorado.

Mitchell, R. W., & Anderson, J. (1993). Discrimination learning of scratching, but failure to obtain imitation and self-recognition in a long-tailed macaque. *Primates, 34,* 301–309.

Mitchell, R. W., & Thompson, N. S. (1991). Projects, routines, and enticements in dog–human play. In P. P. G. Bateson & P. H. Klopfer (Eds.), *Perspectives in ethology* (Vol. 9, pp. 189–216). New York: Plenum.

Money, J., Jobaris, R., & Furth, G. (1977). Ampotemnophilia: Two cases of self-demand amputation as a paraphilia. *Journal of Sex Research, 13,* 115–125.

Morin, A., & Everett, J. (1990). Inner speech as a mediator of self-awareness, self-consciousness, and self-knowledge: An hypothesis. *New Ideas in Psychology, 8,* 337–356.

Moynihan, M. (1985). *Communication and noncommunication by cephalopods.* Bloomington: Indiana University Press.

Nagel, T. (1979). *Mortal questions.* Cambridge University Press.

Neisser, U. (1988). Five kinds of self-knowledge. *Philosophical Psychology, 1,* 35–59.

Nijinsky, V. (1919/1964). The doctors do not understand my illness. In B. Kaplan (Ed.), *The inner world of mental illness* (pp. 422–430). New York: Harper & Row.

Orbach, J., Traub, A. C., & Olson, R. (1966). Psychophysiological studies of body-image: II. Normative data on the adjustable body-distorting mirror. *Archives of General Psychiatry, 14,* 41–47.

Orne, M. T., & Bauer-Manley, N. K. (1991). Disorders of self: Myths, metaphors, and the demand characteristics of treatment. In J. Strauss & G. R. Goethals (Eds.), *The self: Interdisciplinary approaches* (pp. 93–106). New York: Springer-Verlag.

Parker, S. T. (1991). A developmental approach to the origins of self-recognition in great apes. *Human Evolution, 6,* 435–449.

(1993). Imitation and circular reactions as evolved mechanisms for cognitive construction. *Human Development, 36,* 309–323.

(in press-a). Anthropomorphism is the null hypothesis in comparative developmental evolutionary studies. In R. W. Mitchell, N. S. Thompson, & H. L. Miles (Eds.), *Anthropomorphism, anecdotes, and animals: The emperor's new clothes?* Lincoln: University of Nebraska Press.

Partington, J. T., & Grant, C. (1984). Imaginary playmates and other useful fantasies. In P. K. Smith (Ed.), *Play in animals and humans* (pp. 217–240). Oxford: Blackwell Publisher.

Patterson, F., & Linden, E. (1981). *The education of Koko.* New York: Holt, Rinehart & Winston.

Piaget, J. (1945/1962). *Play, dreams, and imitation in childhood.* New York: Norton.

Povinelli, D. J., Nelson, K. E., & Boysen, S. T. (1992). Comprehension of role reversal in chimpanzees: Evidence of empathy? *Animal Behaviour, 43,* 633–640.

Premack, D. (1988). "Does the chimpanzee have a theory of mind?" revisited. In R. Byrne & A. Whiten (Eds.), *Machiavellian intelligence: Social expertise and the evolution of intelligence in monkeys, apes, and humans* (pp. 160–179). Oxford University Press (Clarendon Press).

Premack, D., & Premack, A. J. (1983). *The mind of an ape.* New York: Norton.

Reynolds, P. C. (1981). *On the evolution of human behavior: The argument from animals to man.* Berkeley and Los Angeles: University of California Press.

Ricks, D., & Wing, L. (1975). Language, communication and the use of symbols in normal and autistic children. *Journal of Autism and Childhood Schizophrenia, 5,* 191–221.

Rosaldo, M. Z. (1984). Toward an anthropology of self and feeling. In R. A. Shweder & R. A. LeVine (Eds.), *Culture theory: Essays on mind, self, and emotion* (pp. 137–157). Cambridge University Press.

Rosenberg, M. (1979). *Conceiving the self.* New York: Basic.

(1988). Self-objectification: Relevance for the species and society. *Sociological Forum, 3,* 548–565.

Rosenberg, M., & Kaplan, H. B. (Eds.) (1982). *Social psychology of the self-concept.* Arlington Heights, IL: Davidson.

Russon, A. E., & Galdikas, B. M. F. (1993). Imitation in free-ranging rehabilitant orangutans (*Pongo pygmaeus*). *Journal of Comparative Psychology, 107,* 147–161.

Sass, L. A. (1987). Introspection, schizophrenia, and the fragmentation of self. *Representations, 19,* 1–34.

Schafer, R. (1968). *Aspects of internalization.* New York: International Universities Press.

(1976). *A new language for psychoanalysis.* New Haven, CT: Yale University Press.

Schilder, P. (1950). *The image and appearance of the human body.* New York: International Universities Press.

Schneiderman, L. (1956). The estimation of one's own bodily traits. *Journal of Social Psychology, 44,* 89–99.

Schontz, F. C. (1969). *Perceptual and cognitive aspects of body experience.* New York: Academic Press.

Singer, D. G., & Singer, J. L. (1990). *The house of make-believe: Play and the developing imagination.* Cambridge, MA: Harvard University Press.

Sizemore, C., & Pittillo, E. (1977). *I'm Eve.* New York: Doubleday.

Smuts, B. (1985). *Sex and friendship in baboons.* New York: Aldine.

Snowdon, C. T., & Cleveland, J. (1980). Individual recognition of contact calls by pygmy marmosets. *Animal Behaviour, 28,* 717–728.

Snyder, M. (1987). *Public appearances / private realities: The psychology of self-monitoring.* New York: Freeman.

Sodian, B. (1991). The development of deception in young children. *British Journal of Developmental Psychology, 9,* 173–188.

Sperry, R. W., Zaidel, E., & Zaidel, D. (1979). Self-recognition and social awareness in the deconnected minor hemisphere. *Neuropsychologia, 17,* 153–166.

Stern, D. N. (1985). *The interpersonal world of the infant: A view from psychoanalysis and developmental psychology.* New York: Basic.

Stern, W. (1914/1924). *Psychology of early childhood up to the sixth year of age.* New York: Holt.

Stoller, R. J., & Herdt, G. H. (1985). Theories of origins of male homosexuality: A cross-cultural look. In R. J. Stoller, *Observing the erotic imagination* (pp. 104–134). New Haven, CT: Yale University Press.

Taylor, C. K., & Saayman, G. S. (1973). Imitative behaviour by Indian Ocean bottlenose dolphins (*Tursiops aduncus*) in captivity. *Behaviour, 44,* 286–298.

Terr, L. C. (1991). Childhood traumas: An outline and overview. *American Journal of Psychiatry, 148,* 10–20.

Toulmin, S. (1986). The ambiguities of self-understanding. *Journal for the Theory of Social Behaviour, 16,* 41–55.

Trafimow, D., Triandis, H. C., & Goto, S. G. (1991). Some tests of the distinction between the private self and the collective self. *Journal of Personality and Social Psychology, 60,* 649–655.

Updike, J. (1989). On being a self forever. In *Self-consciousness* (pp. 212–257). New York: Knopf.

Valentine, C. W. (1942/1950). *The psychology of early childhood: A study of mental development in the first years of life.* London: Methuen.

von Fieandt, K. (1966). *The world of perception.* Homewood, IL: Dorsey.

Vygotsky, L. S. (1935/1978). The prehistory of written language. In M. Cole, V. John-Steiner, S. Scribner, & E. Souberman (Eds.), *Mind in society: The development of higher psychological processes* (pp. 105–119). Cambridge, MA: Harvard University Press.

de Waal, F. B. M. (1982). *Chimpanzee politics: Power and sex among apes.* New York: Harper & Row.

(1986). Deception in the natural communication of chimpanzees. In R. W. Mitchell & N. S. Thompson (Eds.) *Deception: Perspectives on human and nonhuman deceit* (pp. 221–244). Albany, NY: SUNY Press.

(1991). The chimpanzee's sense of social regularity and its relation to the human sense of justice. *American Behavioral Scientist, 34,* 334–349.

Walton, K. (1990). *Mimesis as make-believe: On the foundations of the representational arts.* Cambridge, MA: Harvard University Press.

Wapner, S., & Werner, H. (1965). An experimental approach to body perception from the organismic-development point of view. In S. Wapner & H. Werner (Eds.), *The body percept* (pp. 9–25). New York: Random House.

Watkins, M. (1986). *Invisible guests: The development of imaginal dialogues.* Hillsdale, NJ: Analytic Press.

Werner, H. (1965). Introduction. In S. Wapner & H. Werner (Eds.), *The body percept* (pp. 3–8). New York: Random House.

Wicklund, R. A. (1982). How society uses self-awareness. In J. Suls (Ed.), *Psychological perspectives on the self* (Vol. 1, pp. 209–230). Hillsdale, NJ: Erlbaum.

Winnicott, D. W. (1951/1958). Transitional objects and transitional phenomena. In *Play and reality* (pp. 1–25). London: Tavistock.

Wolf, D. P. (1990). Being of several minds: Voices and versions of self in early childhood. In D. Cicchetti & M. Beeghly (Eds.), *The self in transition: Infancy to childhood* (pp. 183–212). Chicago: University of Chicago Press.

Wolff, G. (1979). *The Duke of deception: Memories of my father.* New York: Random House.

Young-Eisendrath, P., & Hall, J. A. (Eds.) (1987). *The book of the self: Person, pretext, and process.* New York: New York University Press.

7 Contributions of imitation and role-playing games to the construction of self in primates

Sue Taylor Parker and Constance Milbrath

Introduction

All mammals exhibit some form of awareness: Attention, proprioception, the capacity for pain, as well as a sense of permanence, agency, and continuity constitute "the machinery of the self" (Lewis, *SAAH*2), which underlies simple awareness. At least some aspects of the machinery of the self are probably universal among animals because they are necessary for goal-directed behavior by any kind of entity (Fehling, personal communication). Self-awareness, in contrast, goes beyond the machinery of the self to the idea of me (Lewis, *SAAH*2). As the phrase "the idea of me" implies, self-awareness involves cognitive and affective components that exceed those necessary for simple awareness. In this paper we are concerned first with self-awareness and how it develops in human children. Second, we are concerned with the implications that developmental models of self-awareness may have for understanding self-awareness in our closest relatives, great apes, lesser apes, and Old World monkeys.

Drawing on William James's (1892/1961) famous distinction between the subjective and objective self, we understand subjective self-awareness in human adults to be an individual's awareness of his own bodily, emotional, mental, and social characteristics, his own goals, plans, and intentions, and of the strategies he might employ in the service of those goals. Conversely, we understand objective self-awareness to be an individual's judgments regarding his bodily, emotional, mental and social characteristics, his goals, his success at achieving his goals, his skill as a strategist, the acceptability of the strategies he has employed in the service of those goals, and his feelings about himself consequent on those judgments (e.g., Damon & Hart, 1988). These two aspects of self-awareness develop in tandem with each another (e.g., Damon & Hart, 1988) and with other-awareness (see Gopnik & Meltzoff, *SAAH*10) from infancy to adulthood. Following Baldwin (1897) and Mead (1934/1970), as well as others, we understand that self- and other-awareness are constructed through social interactions, and particularly through imitation and role-playing in social games. As Fein (1984) notes, Mead's ideas about the importance of pretend play in the development of the self have been

108

largely ignored by developmental psychologists, perhaps because they need to be operationalized before they will be useful for empirical studies.

In the following discussion of the construction of self-awareness through social games we rely on a Piagetian stage model to explore Mead's and Baldwin's theories of self construction. This model is useful for this purpose because it explicitly addresses the development of play and imitation in relation to other cognitive achievements. In particular it addresses

1. the relationship between the so called circular reactions in play and imitation in the sensorimotor period of infancy (Piaget, 1952);
2. the relationship between play and imitation in the symbolic period of early childhood (Piaget, 1962); and
3. the relationship between games-with-rules and role-playing in later childhood (Piaget, 1965).

Likewise, just as Piagetian stages provide a chronological framework for elaborating Baldwin's and Mead's ideas about the development of the self through play, the Piagetian concept of "construction" of new schemes and concepts through coordination of old (Case, 1985; Piaget, 1985) provides a vehicle for describing the processes of development of self in human children.[1]

Mother–infant play: Contingency games of infancy

The earliest manifestations of social play occur during the 4th month in the third stage of the sensorimotor period. They occur coincident with the onset of the "secondary circular reactions" in which infants repeat actions on objects that create "interesting spectacles" contingent on their previous actions (Piaget, 1952).

These contingent spectacles can involve animate and/or inanimate objects:

1. In "object contingency games" infants repeatedly shake, hit, or swing objects through various bodily movements, displaying smiles and other evidence of pleasure in the outcomes contingent on their efforts.
2. In "social contingency games" infants engage in repeated bouts of facial and bodily movements and/or vocalizations in response to recognizing that the facial and vocal displays of their caretakers are contingent on their own prior actions (Watson, 1972).

The earliest social contingency games play a significant part in the development of the subjective self, which precedes and gives rise to the objective self (Damon & Hart, 1988). These games provide structured opportunities for infants to discover their own agency and volition (Stern, 1985) by providing them with the critical contrast between the "perfect contingency" of the response of their own bodies to their actions and the "imperfect contingency" of social objects' responses to their actions (Bahrick & Watson, 1985; Watson, 1984). The emotional contexts of social contingency play, the experience of pleasure and frustration in a social context, and the repetitive, highly stereotyped demonstration of emotional expressions by mothers (Stern,

1977) provide infants with consistent affective experiences that form the basis for a sense of personal history and continuity (Stern, 1985).

Social contingency games become more differentiated at 9 or 10 months of age, during the 4th sensorimotor stage, when infants engage in games involving voluntary facial imitation (Piaget, 1952), and in imitation of simple novel object manipulation schemes (Meltzoff, 1990). Caretakers not only imitate their infants, but engage in "affect attunement" by matching some amodal aspect of the infant's behavior, such as rhythm or intensity, that reflects the infant's feeling state without matching the behavior itself, for example, tapping the baby's rump at the same rhythm he is moving (Stern, 1985).

Beginning in the 5th stage of the sensorimotor period, at about 12–15 months of age, infants extend their sense of agency by engaging in more complex games incorporating systematic variations in contingent responses in the "tertiary circular reactions," for example, dropping objects from various heights, submerging various objects, or stacking or inserting them (Piaget, 1952). At this stage they also begin to imitate more complex novel schemes, and begin to incorporate objects of joint attention (Trevarthan & Hubley, 1978) in such games as peekaboo (Bruner, 1983). Sometime during the 5th or 6th stage, when infants begin to use single words, parents engage in social contingency games involving verbal labeling of body parts, for example, "Where's your nose? Where's my nose?" Such interactions seem to provide the scaffold for "contingency self-recognition" of the face in mirrors, which occurs at some time between 15 and 21 months (Lewis & Brooks-Gunn, 1979; Mitchell, 1993).

As Baldwin (1897) long ago suggested, imitation seems to play a key role in the development of the self by allowing the infant to take others into himself and then project himself back onto them. Exactly how imitative games might aid infants in self-recognition is suggested by Meltzoff's (1988) recent research on imitation in 14-month-old infants (whose sensorimotor period stages were undiagnosed).

Meltzoff (1990) interprets his results as indicating that parental imitation of the infant acts as a "social mirror" by externalizing the infant's own actions for him to assimilate through observation. The infant's ability to use the social mirror depends upon his capacity for visual–kinesthetic matching of another person's actions with his own, and vice versa (Mitchell, 1993). These mirrored actions have greater appeal to the infant because he recognizes them as being just like his own. Conversely, through the infant's own immediate imitation of the model's schemes, he consolidates the new schemes in his memory, and relabels them as a self schema. This process contributes to the infant's sense of distinctiveness and continuity as well as adding to his sense of agency. In later childhood, parents may further nurture the sense of continuity and distinctiveness by coaching their children in remembering and recounting personal narratives (Snow, 1990).

An important aspect of imitation that has received little attention is its selectivity: Infants and children select from among an infinity of possible models and actions only particular ones (Russon & Galdikas, 1992). Among young orangutans, models were selected on the basis of familiarity, affective ties, and dominance; actions were selected on the basis of functional value and contextual relevance as well as on receptive familiarity (Russon & Galdikas, 1992). As among humans, model selection in infants focused on the mother, and in juveniles, on peers and salient adults.

Peer play: Symbolic play and pretend games of early childhood

Play in early childhood (i.e., in children 2–5 years old) is primarily pretend play: "Play in this sense, especially the stage which precedes the organized games, is a play at something. A child plays at being a mother, at being a teacher, at being a policeman, that is, it is taking different roles, as we say" (Mead, 1934/1970; p. 150). Pretend play with peers emerges just before 3 years of age and seems to be based on a coordination of social play and solitary pretend play (Howes, Unger, & Beizer Seider, 1989). It also grows out of role play with mothers and older siblings who provide considerable scaffolding (Dunn & Dale, 1984; Miller & Garvey, 1984).

According to Piaget, symbolic play or pretense is a developmentally more advanced form of imitation that emerges with the capacity for symbolic representation at about 2 years of age (Piaget, 1962). Pretense goes beyond simple imitation of a model to actual impersonation of a model, that is, to the implicit or explicit adoption of some aspect of the model's identity. In pretending, as contrasted with imitating, the child is putting herself in someone else's shoes temporarily by assuming a role in make-believe games. In this way she develops culturally approved scripts for her emerging psychological and social self at the same time she is increasing her sense of agency.[2]

Pretend play is a highly ritualized form of "event representation" (Nelson & Gruendel, 1986; Nelson & Seidman, 1984) involving actions and objects. It develops through a series of 11 increasingly complex stages (Bretherton, 1984) each of which seems to embody a different level of complexity of reflection of the self in relation to others: In Stage 1 the child represents her own behavior, in effect quoting herself by repeating a particular behavior for an audience of others, thereby acknowledging her role as a social agent. In Stage 2 the child represents another's behavior, quoting another person, or consciously repeating a particular behavior of another person, projecting herself into that person's agency. In Stage 3 the child represents others as passive recipients of her own actions, highlighting her own agency through actions on the recipient. By Stage 4 the child represents herself and others in parallel roles, beginning to simultaneously represent two agents, herself and another. By the penultimate stage, Stage 10, she uses replicas in several interacting roles, constructing several independent agents and their willful actions, and mutually accommodating them. Finally, in Stage 11 she coordinates several interacting roles of self and others, and practices one of several possible roles.

Pretend play incorporates the burgeoning universe of symbolically represented roles, feelings, perceptions, and so on. By highlighting and repeating
simple scenarios based on everyday events, it allows children to practice pro-
jections of self-agency and volition through various stereotyped roles, to
experience the reciprocal role-playing reactions of their playmates, and hence
to extend and develop their repertoire of objectified identities through role
expansion and role reversal. In her Meadean analysis of play, Fein (1984)
distinguishes four levels of role playing; her 4th level, like Bretherton's 4th
stage, involves role reversals, as when a child switches back and forth between
playing the role of the mother and the child.

In peer play, the choice of models and actions becomes a social phenomena
shaped by older peers and by negotiation among peers. Increasingly complex
forms of role reversal feed back into affect about the self and other by
providing children with a new basis for experiencing empathetic responses to
the experiences of others (Hoffman, 1983). Role reversals also aid in the
internalization of standards of conduct, which leads to the self-evaluative
emotions of shame and pride (Lewis et al., 1989).

Repetitive role-playing in pretend play provides the young child with the
opportunity to experience a few highly stereotyped roles, to observe the few
reciprocal stereotyped roles of her playmates, and to discover and practice
the stereotyped lines and scenes associated with those roles. It also provides
a context for coordinating interchanges through complex social contingency
interactions (Mueller & Lucas, 1975). The stereotypy and repetition of these
make-believe games are reminiscent of those of the early conversation games
between young infants and their caretakers described by Stern (1977). It is
tempting to conclude that in this case, as in the case within infant games, the
stereotypy is designed to isolate and highlight elements to facilitate learning
within the information processing limitations of the young organisms while
simultaneously pushing children into their "zone of proximal development"
(Vygotsky, 1962).

Peer play: Games-with-rules of later childhood

"Games-with-rules" (Piaget, 1965) are also highly ritualized forms of event
representation, but the events they represent are more complex than those
represented in pretend play, and the social interchanges much more complex
and variable. Games-with-rules have precursors in such infant–caretaker games
as peekaboo and pat-a-cake, but they only develop fully in children between
the ages of 6 and 12. Of course, in such forms as iconic (board and card)
games and field games, they may continue to engage the energy of older
children and even adults.

Games-with-rules are important to the continuing development of the
objectified self because, like pretend games, they provide highly ritualized,
repetitive frames of interaction in which children assume and practice a variety
of interacting social roles. Unlike the roles in pretend games, roles in these

games involve long-range strategies for achieving various subgoals within a larger prescribed goal. Unlike pretend play, games-with-rules often represent roles in extradomestic, extrawork settings with territorial/political significance, as, for example, chess and go represent battles over land. Perhaps for this reason, games-with-rules tend to be more sex segregated than pretend games (e.g., Lever, 1976; Parker, 1984).

Games-with-rules depend on and practice social skills of perspective taking and mind reading that promote the continuing objectification of the physical, psychological, and social self. We suggest that children construct more complex objectified notions of their own physical, behavioral, social, and psychological selves relative to those of their teammates by repeatedly assuming various complementary interactive roles associated with achieving prescribed goals of the game (e.g., Sachs, 1980; Stevens, 1977).

In other words, games-with-rules provide scaffolds for elaborating a concept of the objective self in various social roles. In contrast to earlier games, games-with-rules repeat and ritualize relatively complex strategic interactions among several individuals in various theaters. These games elicit awareness of new elements of self and other through repeated ritualized enactments requiring anticipation of teammates' tactics and strategies and anticipation of reactions by teammates to one's own tactics and strategies through mental role reversal. They culminate in a hypotheticodeductive model of rules and principled morality (Piaget, 1965). Mental role reversal may also contribute to the development of an autobiographical self seen retrospectively as the actor of various roles.

In summary, we propose that self-construction through social play proceeds stage by stage parallel to cognitive development from infancy through adolescence, just as play itself proceeds stage by stage parallel to cognitive development (Piaget, 1962, 1965). By late adolescence children have constructed a multidimensional objective self characterized by systematic beliefs and plans based on concepts of the physical, social, and psychological self, as well as a subjective self characterized by an autobiographical sense of continuity and distinctiveness of past, present, and future selves. These elements seem to develop roughly in parallel through childhood, though some elements are more closely related than others; for example, continuity and distinctiveness are more closely related to the objective self than is agency (Damon & Hart, 1988).

Self-schemata in monkeys and great apes

Following Gallup's (1970) publication of his landmark paper describing the mark test for mirror self-recognition (MSR) in monkeys and apes, animal psychologists have shown an increased interest in the self. Gallup's new technique has stimulated replication experiments on many primate species. These experiments (e.g., Anderson, *SAAH*21; Gallup, 1977) have revealed the ability for self-recognition in great apes (or at least in the chimpanzees and orangutans), and the absence of this ability in monkeys (but see Boccia, *SAAH*23; Thompson & Boatright-Horowits, *SAAH*22). Beyond revealing a disability of monkeys as compared to great apes, however, the mark test has done little to elucidate the nature of self-schemata in these primate species. It could be argued that monkeys have awareness without self-awareness, but given evidence that seems to suggest that they anticipate the outcome of some behaviors, we think it is more likely that they have some rudiments of self-awareness. Such anticipation could of course be mediated by non–self-conscious means. The problem is that we need a broader taxonomy to compare and contrast self-awareness among a broader array of species.

In line with the previous discussion of self-construction though social play in humans, the following sections of this paper are devoted to a discussion of what the play patterns of monkeys and apes imply about self-awareness of these taxa. We should point out in this context that great apes (chimpanzees, gorillas, and orangutans) are the closest living relatives of humans while Old World (OW) monkeys (macaques, baboons, mangabeys, guenons, and leaf monkeys) are slightly more distant relatives: The relative closeness of great apes is reflected in their long infancy and juvenility, which extend until they are 10–13 years old, while the relative distance of OW monkeys is reflected in their shorter infancy and juvenility, which extend until they are only 4–6 years old.

Overview of social play in Old World monkeys

The kinds of behaviors enacted in social play of monkeys are limited in number and scope compared to those enacted by human and chimpanzee children. They fall almost exclusively into the following categories:

1. agonistic behaviors (fighting, fleeing, or submission);
2. locomotoric behaviors (climbing, jumping, or bouncing);
3. sexual behaviors; and
4. parenting behaviors (retrieving, carrying, or protecting).

With the exception of the play face and play vocalizations, behaviors are exclusively whole-body, tactile, and mouthing schemata. They do not evoke, highlight, or repeat facial expressions, vocalizations, or actions on objects, as happens in human play. The absence of these behaviors is consistent with other evidence that monkeys lack the capacity for trial-and-error imitation of novel behaviors and the level of representation that this implies (e.g., Parker, 1977; Russon, 1990; Visalberghi & Fragaszy, 1990).

The kinds of play roles enacted in monkey play are also highly limited and stereotyped in nature:

1. agonistic roles of attacker, fleer, submitter, ally, and perhaps, provocateur;
2. the sexual roles of mounter and mountee; and
3. the parental role of mother or protector.

These roles are enacted primarily in relation to bodily and tactile schemes rather than facial, vocal, or object manipulation schemes. Moreover, the role reversals they entail are concomitantly limited and stereotyped. Finally, they seem to be limited to tripartite interactions (Kummer, 1967/1988). Episodes, in the sense of a series of related events culminating in an end point, are virtually nonexistent.

Finally, the structure of the feedback entailed in social play is largely reactive rather than imitative or anticipatory:

1. Alternation of roles in play is limited to performance of bodily schemes lacking the intersubjectivity characteristic of turn taking in human play.
2. Monkeys respond to but do not imitate the behaviors of their partners; for instance, (a) mothers respond to but do not imitate their infant's behaviors and (b) juveniles attend to but rarely imitate behaviors of adults or peers, nor do they engage in pretend play.

Play in Old World monkeys

Mother–infant play

Unlike both human and great ape mothers, monkey mothers rarely engage their infants in play. In the few cases that have been reported, maternal play seems to be limited to wrestling with their infants to distract them from the nipple (Breuggeman, 1978). Mothers, however, do show such other affectionate behaviors as grooming, licking, gazing at the face, and cuddling. They also tolerate considerable clambering on, off, and around their bodies when their infants are just beginning to climb (e.g., Parker, 1977).

Peer play

Like human playmates, monkey infants and juveniles frequently engage in rough-and-tumble play: play fighting, play chasing, and play fleeing. Also like human playmates, monkey infants in species such as macaques, baboons, vervets, and langurs, who often live in relatively large social groups, tend to segregate themselves by age and sex (Dolhinow & Bishop, 1970; Owens, 1975; Symons, 1978).

The play fighting of juvenile male monkeys seems to function as practice for adult male fighting and dominance competition. Play fighting in monkeys involves alternation between play attack and play-flight roles: chasing, wrestling, sparring, mauling, and mock biting (Owens, 1975). Practicing fighting skills can have a significant payoff in reproductive success insofar as high dominance rank can be achieved, and insofar as it correlates with access to females (Symons, 1978). Both of these conditions appear to obtain among rhesus macaques, and baboons, and vervets, who generally leave their natal groups when they are subadults, seeking better status and mating opportunities in other groups.

Although they do not engage in pretend play, young monkeys sometimes play agonistic exercise games that resemble the simplest games of preschool children, for example, playing keep-away and king-of-the-mountain on some high place, and among patas monkeys, "freeze and jump on the first one who moves" (Dolhinow & Bishop, 1970; Parker, 1977; Symons, 1978). They also engage in alternating and parallel climbing, sliding, jumping, bouncing, and somersaulting. These vestibular-stimulating (Aldis, 1975) activities practice locomotor skills important in evading predators and competitors (Dolhinow & Bishop, 1970).

One of the most significant but little-emphasized aspects of peer play in monkeys is the ubiquitous role of mothers and other relatives: Relatives monitor the intensity of play and, when their status allows them to, intervene on behalf of protesting kin (e.g., Fagen 1981; Symons, 1978). Awareness of the presence of monitoring adults undoubtedly contributes to the restraint that older, larger animals display toward younger, weaker playmates, and to their understanding of the roles of protector and protectee in tripartite relationships.

Play parenting

Although monkeys do not engage in pretend play, they do engage in one class of activities that has some interesting parallels with a theme common in the make-believe play of young human children: play parenting or play mothering (Lancaster, 1972). Unlike make-believe play, however, play mothering in monkeys involves practicing a single role, and does not seem to involve pretense. Young female vervets, baboons, and macaques spend a great deal of time around mothers with young infants, watching them, grooming them, and when possible carrying and playing with their infants (Lancaster, 1972). Whereas human children act as infant caretakers in many societies (Weisner & Gallimore, 1977), they also engage in pretend play with dolls. Young monkeys do not pretend that objects are other objects, but they often carry and play with real infants for extended periods.

Although play mothering has not been extensively described, it seems to be rare compared to other forms of social play (Owens, 1975), and to focus on a few motor patterns involved in picking up, placing, repositioning, and carrying infants on the venter. Although play mothering entails responding to the movements and vocalizations of the infant, it does not involve any clear alternation of roles, except perhaps in cases when two or three juvenile females compete for possession of an infant. In these cases, described in baboons, one or two females may chase the female who is carrying the infant, and grab and wrestle with her; although the infant may cling to the carrier, the other females may grab it during the rough-and-tumble play (Owens, 1975).

As in human children who act as infant caretakers, in monkeys, play mothering seems to function as practice for mothering. The potential importance of practice for mothering is indicated by the relatively poor mothering skills of primiparous as compared to multiparous monkeys (e.g., Altmann, 1970; Hooley, 1983). As Lancaster (1972) points out, practice in play fighting is less important for female vervets, macaques and baboons because, in contrast to males, females in these species remain in their natal groups and assume their mother's dominance rank and therefore need not compete to increase their rank.

Overview of social play in great apes

The kinds of behaviors enacted in social play of great apes fall under the same rubrics as monkey play. The numbers of behaviors, and the differentiation and recombination of motor patterns in each of the foregoing categories, however, is much greater in great apes than in monkeys (Russon, 1990). This is especially true of locomotor, facial, gestural, and object manipulation behaviors: for example, brachiating and knuckle-walking behaviors, more numerous and differentiated facial expressions and gestures (e.g., van Lawick-Goodall, 1968), tickling and poking as elements of play, and various tool-using behaviors.

The kinds of roles enacted in great ape play are also richer, more complex, and more variable in content than those in monkey play. Role reversals are concomitantly richer and more flexible. Some forms of play might be called protogames because, although they lack the explicit rules of human games, they are like human games in practicing the more complex roles and multipartite relations that occur among adults: for example, social coordination and stealth in hunting and border patrolling in male chimpanzees, and in infanticidal attacks by females (Goodall, 1986); in coordinating and planning, for gathering tools and nuts at an anvil site for nut cracking, also in chimpanzees (Boesch & Boesch, 1984); and systematically separating potential allies from rivals (de Waal, 1982).

Finally, in contrast to that of monkeys, the structure of the feedback entailed in social play in great apes has some reflective or circular elements similar to those of human play. Like human infants (Trevarthan & Logotheti, 1987), great apes also "show off" by displaying in front of other animals, and even by making odd facial movements (de Waal, 1989; S. T. Parker, personal observ.). Like human children and unlike monkey children, great ape children engage in gamelike activities involving (1) complex object manipulation schemata, and (2) some elements of imitation of novel schemes. These include, for example, nest building, bipedal branch waving, and tool-using play in wild chimpanzees (van Lawick-Goodall, 1976c; McGrew, 1977), bipedal branch waving, chest beating, and other forms of "showing off" in gorillas (Fossey, 1983; S. T. Parker, personal observ.); imitative object manipulation in ex-captive orangutans (Russon & Galdikas, 1993); and funny faces in captive bonobos (de Waal, 1989). These might be called proto–pretend games.

On the other hand, only a few cases of pretend play have been reported and these only in cross-fostered great apes: for example, pulling an invisible string on an invisible toy (Hayes, 1952), playing tea party and makeup (Patterson, 1978); hiding and eating pretend objects, and pretending to sleep (Savage-Rumbaugh & MacDonald, 1988). None of these games involve the elaboration of roles characteristic of pretend play in human children (Mignault, 1985). At their most ambitious, these proto–pretend games involve only the first three stages of pretend play described by Bretherton (1984) in human children (see earlier in this section).[3]

Great ape children, like monkeys, frequently engage in such agonistic exercise games as keep-away, tug-of-war, and king-of-the-mountain, which are structured somewhat like simple games with rules played by preschool children. Unlike monkeys, however, gorillas and chimpanzees also frequently engage in hide-and-seek and peekaboo, evading both their playmates and their mothers and other group members, and then surprising them by suddenly reappearing (Parker, 1993). Immature chimpanzees frequently annoy adults (especially adult females) with teasing games involving such actions as throwing sand, splashing water, poking with sticks, pulling and twisting limbs, and running and "bluffing." These games apparently prepare young males for the process of climbing the dominance hierarchy (de Waal & Hoekstra, 1980; Goodall, 1986). Immature gorilla males annoy adults (especially adult males) with similar teasing games, which presumably prepare them for adult male competition (Parker, 1993). These games are another form of contingency games, in this case involving tertiary circular reactions rather than secondary circular reactions (Parker, 1993). In these and other playful interactions, great apes display awareness that they can mislead, surprise and provoke other animals (Mitchell, 1989). This capacity seems to be based on their awareness of the locations and attentional foci of all the members of the group, and their ability to adjust their tactics accordingly (Menzel, 1971; Savage-Rumbaugh & MacDonald, 1988). This capacity is also implicit in teaching behavior of wild chimpanzee mothers (Boesch, 1991).

Social play in great apes

Mother–infant play

Chimpanzee mothers begin to play with their infants when they are about 3 months old and continue until they are about 12 months old. Mothers often lift infants onto their feet or hands with limbs extended, bouncing the infants up and down as they dangle. They also tickle, poke, and mock-bite their young infants while the infants display the play face, and control access of older playmates to their infants by either playing with the playmate themselves or joining in play with their infant and the playmate. Chimpanzee mothers use play to distract their infants from leaving them to explore the environment (van Lawick-Goodall, 1976a). Similar patterns of mother–infant play occur in orangutans (MacKinnon, 1974), and gorillas (Fossey, 1978; Mitchell, 1989; Parker, 1991).

Like human mothers, great ape mothers engage their motorically helpless infants in vestibular play (Aldis, 1975), dangling and swinging them in the air, tickling and poking them, and pushing and pulling them. Like human infants, great ape infants respond with the play face (relaxed, open-mouth face) and panting vocalizations, and expectant watching of their mothers (van Lawick-Goodall, 1976a; Parker, 1993). Unlike humans, however, great ape mothers and infants do not imitate their infant's facial expressions and vocalizations, or engage in the face-to-face version of circular play that Watson (1972) calls "the game." Mothers, however, do engage in some sort of contingent responses that seem to reflect their infant's tempo and affect as they play (Parker, in press).

Peer play and play parenting

Peer play among chimpanzee children less than 3 years old is primarily sibling play owing to the frequent fissioning of chimpanzee groups into smaller foraging subunits. Older siblings begin trying to play with infants by touching and then patting and tickling when the infants are about 3 months old. Infants of this age reciprocate by reaching, touching, pouting, and hooing; mothers control access to infants by playing with older siblings. When their infants are 4 or 5 months old, mothers sometimes let older siblings carry them and play with them. Peer play thereby functions to some degree as play parenting. By 5 or 6 months of age, infants are moving about independently and playing for increasing periods with their older siblings, who handicap themselves in order to continue play bouts without parental intervention. As infants grow older, the length of their play bouts increases from about 5 to about 40 min, but tapers off after puberty (van Lawick-Goodall, 1976b).

Among young chimpanzees peer play involves "chasing, wrestling, sparring, play biting, thumping and kicking, butting with the head and a variety of tickling and poking movements . . . during play sessions, chimpanzees . . . climbed, jumped, swung and dangled from branches of trees, chased round tree trunks, broke off and waved or carried branches, leaves, or fruit clusters, grappled with each other for an assortment of small objects, dragged and hit each other with branches, and so on" (van Lawick-Goodall, 1976b, p. 303). These and other peer-play sequences involve some "joint circular reactions" (Russon, 1990). Gorillas show very similar patterns of peer play (Fossey, 1983; Parker, 1993).

Like play in monkeys, peer play in great apes is intrinsically tripartite. Great apes, like monkeys, learn self-handicapping and the roles involved in protected threat as a result of the strategic intervention of mothers and others who monitor their play. Unlike monkeys, great ape youngsters also learn to calibrate their dominance displays by engaging in teasing, provoking, and imitation games aimed at adults (e.g., de Waal & Hoekstra, 1980). These games generate new tripartite relationships among peers, their relatives, and other animals.

Cognition and construction of self-schemata in Old World monkeys and great apes

This brief comparative sketch of the social play of OW monkeys and great apes reveals differing levels of complexity in play that are consistent with the differing levels of cognitive abilities of the two taxa (Russon, 1990): Most OW monkeys fail to achieve fifth- and sixth-stage sensorimotor intelligence stages in most series, while great apes achieve and surpass these stages (e.g., Antinucci at al., 1982; Chevalier-Skolnikoff, 1977; Mathieu & Bergeron, 1983; Parker, 1977; Redshaw, 1978; Russon & Galdikas, in press). Under training great apes display symbolic capacities similar to those of human children in the early preoperations period (e.g., Gardner & Gardner, 1989; Miles, 1990; Patterson, 1980; Premack, 1988). (See Table 7.1.)

The common failure of OW monkeys to pass the mark test, and the success of great apes in doing so (e.g., Anderson, *SAAH*21; Gallup, 1970; 1977; Miles, *SAAH*16; Patterson & Cohn, *SAAH*17) seem to correlate specifically with the monkeys' inability to engage in fifth- and sixth-stage imitation in the facial and gestural modalities, whereas the success of great apes at mirror self-recognition correlates with their abilities to engage in fifth- and sixth-stage imitation (Mitchell, 1992, 1993; Parker, 1991). Taken together these contrasting cognitive patterns in great apes versus Old World monkeys invite two related questions about the abilities of monkeys and apes relative to those of human children:

1. What kinds of selves do monkeys and apes display?
2. To what extent are these selves constructed through play?

The voluntary, purposeful nature of their play suggests that monkeys and apes have some sense of agency. Likewise the social nature of play suggests that monkeys and apes have rudimentary physical, active, and social selves. As a starting point for analyzing these manifestations, we suggest that the nature of the self is shaped by the species-typical repertoire of schemata, that is, by such things as the quantity, modifiability, modalities, and nature of the elicitors and reinforcers of the schemes. We further argue that the nature of the schemes determines their developmental potential, that is, their susceptibility to construction as opposed to modification through learning. By construction we refer to behavioral reorganization on a new level of capacity, while by learning we refer to behavioral reorganization at a new level of efficiency, but, at the same level of capacity. Whereas learning is characteristic of all animals, construction is associated with intelligence and its development (Piaget, 1985).

In the case of OW monkeys, social play is necessary to learn and practice both their agency and their social and active selves: Such play roles as chaser/chasee, biter/bitee, and mounter/mountee and their relative success vis-à-vis various playmates can only by realized through social play. The role reversals entailed in this kind of play are limited to stereotyped behaviors. Social play is primarily an arena for exercising innate schemes and discovering their

Table 7.1. *Classification of games and self-awareness in humans, great apes, and Old World monkeys*

Age range	Primate taxa	Social relationship	Play category	Piagetian intellectual development period
Infancy	Humans	Mother–infant	Contingency games (imitation & circular reactions in facial & vocal modalities)	Sensorimotor (to age 2 yr)
	Great apes	Mother–infant	Contingency games (circular reactions in gestural & manual modalities)	Sensorimotor (to age 4 yr)
	OW monkeys	Mother–infant	No contingency games	Truncated sensorimotor (to old age)
Early childhood[a]	Humans	Peer–peer	Pretend games (symbolic & verbal role-playing with symbolic-level role reversals)	Preoperations (symbolic subperiod, age 2–4 yr)
	Great apes	Peer–peer	Proto–pretend games (imitation in facial, gestural, & manual modalities)	Preoperations (symbolic subperiod, age 4 yr to adulthood)
	OW monkeys	Peer–peer	No pretend games	No preoperations
Middle & late childhood[a]	Humans	Peer–peer	Games-with-rules (mental role reversal with theory of mind)	Concrete & formal operations (age 6–15 yr)
	Great apes	Peer–peer	Proto–games-with-rules (symbolic-level role reversal)	Preoperations (symbolic subperiod continues); no concrete & formal
	OW monkeys	Peer–peer	No games-with-rules	No concrete & formal

[a] All three primate groups engage in agonistic exercise games (play fighting, chasing, etc.) and species-typical stereotyped role reversals throughout childhood.

efficacy. It is not an arena for systematically transforming and reorganizing old schemes into new more complex schemes; in other words, it is not an arena for constructing new self-schemes. Given the vital part that imitation plays in self construction in human children, we suspect one reason for the virtual absence of self construction is the lack of imitative capacity in monkeys.

The sense of physical agency in OW monkeys seems to be limited primarily to the direct mechanical effects of such voluntary bodily and manual schemes as locomotion, and to such simple object manipulation schemes as shaking, pulling, pushing, opening, twisting, and rubbing. Their sense of agency in the social realm seems to go beyond a representation of the direct mechanical effects of their voluntary schemes to include the indirect effects of a small repertoire of largely involuntary social signals on other animals. In the objective realm, the physical and active selves of monkeys apparently entail representation of their own bodies and their spatial orientations, but not of their own facial features.

The social and active selves of monkeys entail knowledge of their own direct bodily agency and that of other animals, the likely effect of specific voluntary and involuntary schemes on the behavior of second and third parties (e.g., Cheney & Seyfarth, 1990), and the knowledge that they must be in the line of sight (or within earshot) of these animals to elicit these effects. Their social strategies are limited to more or less direct action on first and second parties, and to monitoring the action of these parties. Their means of control of themselves and others is primarily limited to applying a few friendly or agonistic schemes such as grooming, huddling, threatening, or appeasing, to gain access to desired objects or to protect themselves, to increasing and decreasing their spatial proximity to and position relative to second and third parties (e.g., Kummer, 1967/1988), or to manipulating their visibility to other animals, hence managing their own levels of arousal and the broadcast and receipt of signals.

In the case of great apes, social play is also necessary to learn and practice their agency and social and active selves, but in their case, social play also entails some self construction. The imitative capacity of great apes allows them to import new schemes into their own repertoire; but more important, it allows them vicariously to experience the subjective state of other animals at least at the level of enacting their behaviors. This capacity is also reflected in playfully reversing a learned role, for example, by playing teacher: Examples include Koko teaching signs to her doll (Patterson & Cohn, *SAAH*17), Chantek asking his teacher to make a sign (Miles, Mitchell, & Harper 1992), and Washoe washing her doll (Gardner & Gardner, 1969). These abilities may allow great apes to anticipate the reactions of other animals, to plan strategies (e.g., Premack & Woodruff, 1978; de Waal, 1982), to deceive them (e.g., Mitchell, 1986; Whiten & Byrne, 1988) or to adopt complementary roles (Miles et al., 1992; Povinelli, Nelson, & Boysen, 1992). Concomitant with the level of imitation and role reversal, the level of self construction through play is minimal in great apes, in contrast to human children.

In accord with their greater imitative abilities, great apes seem to have more elaborated representations of elements of the subjective and objective self than monkeys do. In the subjective realm, their sense of physical agency includes some understanding of object–object interactions such as the transmission of forces through intermediate objects, say, the use of instruments. Their sense of social agency includes knowledge of a greater number of voluntary social schemes, and how to experiment in using them to affect other animals instrumentally. Their capacity for imitation of novel manual, gestural, and facial schemes opens the door to acquiring new schemes and hence to expanding their sense of physical and social agency. Their capacity for role reversal allows them to anticipate the effects of their actions on others.

In the objective realm, the physical and active selves of great apes apparently entail representations of their facial features and bodies, as well as of a broader range of activities involving these body parts. Their social and active selves are also richer in accord with their broader sense of agency and their ability to experimentally modify and recombine their social schemes to distract or deceive other animals, in order to get resources or form strategic alliances. Unlike monkeys, they seem to display rudimentary psychological selves that encompass some representation of their own attention, emotional states, and intentionality, as well as those of other animals (Gómez, *SAAH5*).

In conclusion, Baldwinian, Meadean, and Piagetian ideas about the development of self-awareness through imitation and role-playing games, and their modern sequels, provide some useful frameworks for diagnosing the relative complexity of the self in monkeys, apes, and humans. When applied to comparative data on monkeys and apes, the social-psychological model we have drawn from these sources implies that the level of self-awareness in each taxon seems to be determined by its capacity for imitation, pretense, role-playing, and role reversal.

Notes

1. Piaget argues that play and imitation develop independently in infancy and remain separate until the emergence of symbolic representation at the end of the second year. He describes "circular reactions" as occurring in both infant play and in infant imitation (Piaget, 1952, 1954, 1962), but argues that they have different roles in the two activities: In infant play, circular reactions are primarily assimilative, while in infant imitation, circular reactions are primarily accommodative. In early childhood play, in contrast, he argues that symbolic play combines assimilation and accommodation (Piaget, 1962). This distinction between the assimilative and accommodative roles of circular reactions in infant imitation and play is not significant for our analysis.
2. Some psychologists interested in the child's developing "theory of mind" (ToM) i.e., the understanding that other people have minds, see pretense as the earliest manifestation of metarepresentation (e.g., Leslie, 1988), while others argue that metarepresentation develops a few years after pretense (e.g., Flavell, 1988; Perner, 1988). This controversy projects onto controversies over the presence or absence of "ToM" in chimpanzees (e.g., Premack, 1990; Woodruff & Premack, 1988). It also projects onto the child's development of a theory of his own mind.

3. Enormous controversy swirls around the question of imitation in nonhuman primates. Some investigators disavow the relevance of observations of spontaneous behaviors, while others argue that internal controls can render such observations valid (Russon & Galdikas, 1993). Some investigators define imitation in such a rigorous fashion as to preclude all but exact matching of novel behaviors (e.g., Tomasello, 1990; Visalberghi & Fragaszy, 1990), others include partial matching of novel behaviors and even matching of customary behaviors; still others distinguish levels of imitation (Mitchell, 1987, 1992). We favor a developmental approach derived from Piaget's stages (Parker, 1993b), which distinguishes various levels of imitation.

References

Aldis, O. (1975). *Play Fighting*. New York: Academic Press.

Altmann, J. (1970). *Baboon mothers and infants*. Cambridge, MA: Harvard University Press.

Antinucci, F., Spinozzi, G., Visalberghi, E., & Volterra, V. (1982). Cognitive development in a Japanese macaque (*Macaca fuscata*). *Annali dell' Istituto Superiore di Santa'*, *18*(2), 177–184.

Baldwin, J. M. (1897). *Social and ethical interpretations in mental development*. New York: Macmillan.

Bahrick, L., & Watson, J. S. (1985). Detection of intermodal proprioceptive–visual contingency as a potential basis of self-perception in infancy. *Developmental Psychology*, *21*, 963–973.

Boesch, C. (1991). Teaching among wild chimpanzees. *Animal Behaviour*, *41*, 530–533.

Boesch, C., & Boesch, H. (1984). Mental maps in wild chimpanzees: An analysis of hammer transports for nut cracking. *Primates*, *25*, 160–170.

Bretherton, I. (1984). Introduction: Piaget and event representation in symbolic play. In I. Bretherton (Ed.). *Symbolic play* (pp. 3–41). New York: Academic Press.

Breuggeman, J. (1978). The function of adult play in free-ranging *Macaca mulatta*. In E. O. Smith (Ed.), *Social play in primates*, pp. 169–191. New York: Academic Press.

Bruner, J. (1983). *Child's talk: Learning to use language*. New York: Norton.

Case, R. (1985). *Intellectual development, birth to adulthood*. New York: Academic Press.

Cheney, D., & Seyfarth, R. (1990). *How monkeys see the world*. Chicago: University of Chicago Press.

Chevalier-Skolnikoff, S. (1977). A Piagetian model for describing and comparing socialization in monkey, ape, and human infants. In S. Chevalier–Skolnikoff & F. Poirier (Eds.), *Primate biosocial development* (pp. 159–188). New York: Garland.

Damon, W., & Hart, D. (1988). *Self-understanding in childhood and adolescence*. Cambridge University Press.

Dolhinow, P., & Bishop, N. (1970). The development of motor skills and social relationships among primates through play. In J. P. Hill (Ed.), *Minnesota symposium on child psychology* (Vol. 4, pp. 141–198). Minneapolis: University of Minnesota Press.

Dunn, J., & Dale, N. (1984). I a daddy: 2-year-olds' collaboration in joint pretend play with sibling and with mother. In I. Bretherton (Ed.), *Symbolic play* (pp. 131–158). New York: Academic Press.

Fagen, R. (1981). *Animal play behavior*. New York: Oxford University Press.

Fein, G. G. (1984). The self-building potential of pretend play, or "I got a fish all by myself." In T. D. Yawkey & A. D. Pelligrini (Eds.), *Child's play: Developmental and applied* (pp. 125–141). Hillsdale, NJ: Erlbaum.

Flavell, J. (1988). The development of children's knowledge about the mind: From cognitive connections to mental representation. In J. Astington, P. Harris, & D. Olson (Eds.), *Developing theories of mind* (pp. 244–269). Cambridge University Press.

Fossey, D. (1978). Development of the mountain gorilla (*Gorilla gorilla beringei*): The first thirty-six months. In D. Hamburg & E. McCown (Eds.), *The great apes* (pp. 139–185). Menlo Park, CA: Benjamin/Cummings.

(1983). *Gorillas in the mist*. New York: Houghton Mifflin.

Gallup, G. G., Jr. (1970). Chimpanzees: Self-recognition. *Science, 167*, 86–87.

(1977). Self-recognition in primates. *American Psychologist, 32*, 329–338.

Gardner, R. A., & Gardner, B. T. (1969). Teaching sign language to a chimpanzee. *Science, 165*, 664–672.

(1989). Expression of person, place, and instrument in ASL utterances in children and chimpanzees. In R. A. Gardner, B. T. Gardner, & T. E. van Cantfort (Eds.), *Teaching sign language to chimpanzees* (pp. 240–267). Albany, NY: SUNY Press.

Goodall, J. (1986). *Chimpanzees of the Gombe.* Cambridge, MA: Harvard University Press. [*See also* Lawick-Goodall.]

Hayes, C. (1952). *Ape in our house.* New York: Harper Bros.

Hoffman, M. (1983). Empathy, guilt and social cognition. In W. F. Overton (Ed.), *The relationship between social and cognitive development* (pp. 1–51). Hillsdale, NJ: Erlbaum.

Hooley, J. (1983). Primiparous and multiparous infants and their mothers. In R. Hinde (Ed.), *Primate social relationships* (pp. 142–145). New York: Sinauer.

Howes, C., Unger, O., & Beizer Seider, L. (1989). Social pretend play in toddlers: Parallels with social play and with solitary play. *Child Development, 60*, 77–84.

James, W. (1892/1961). *Psychology: The briefer course.* New York: Harper & Row.

Kummer, H. (1967/1988). Tripartite relations in hamadryas baboons. In S. A. Altmann (Ed.), *Social communication among primates.* Chicago: University of Chicago Press.

Lancaster, J. (1972). Play mothering: The relations between juvenile females and young infants among free-ranging vervets. In F. Poirier (Ed.), *Primate socialization* (pp. 83–104). New York: Random House.

Lawick-Goodall, J. van (1968). A preliminary report on expressive movements and communication in the Gombe Stream chimpanzees. In P. C. Jay (Ed.), *Primates: Studies in adaptation and variability* (pp. 313–374). New York: Holt, Rinehart & Winston.

(1976a). Mother chimpanzees play with their infants. In J. Bruner, A. Jolly, & K. Sylva (Eds.), *Play* (pp. 262–266). New York: Basic.

(1976b). Sibling relations and play among wild chimpanzees. In J. Bruner, A. Jolly, & K. Sylva (Eds.), *Play* (pp. 300–311). New York: Basic.

(1976c). Early tool use in chimpanzees. In J. Bruner, A. Jolly, & K. Sylva (Eds.), *Play* (pp. 222–225). New York: Basic.

Leslie, A. M. (1988). Some implications of pretense for mechanisms underlying the child's theory of mind. In J. Astington, P. Harris, & D. Olson (Eds.), *Developing theories of mind* (pp. 19–46). Cambridge University Press.

Lever, J. (1976). Sex differences in the games children play. *Social Problems, 23*, 478–487.

Lewis, M., & Brooks-Gunn, J. (1979). *Social cognition and the acquisition of self.* New York: Plenum.

Lewis, M., Sullivan, M., Stanger, C., & Weiss, M. (1989). Self development and self-conscious emotions. *Child Development, 60*, 148–156.

McGrew, W. C. (1977). Socialization and object manipulation of wild chimpanzees. In S. Chevalier-Skolnikoff and F. Poirier (Eds.), *Primate bio-social behavior* (pp. 261–288). New York: Garland.

MacKinnon, J. (1974). The behaviour and ecology of wild orang-utans. *Animal Behaviour, 22*, 3–75.

Mathieu, M., & Bergeron, G. (1983). Piagetian assessment on cognitive development in chimpanzee (*Pan troglodytes*). In A. B. Chiarelli & R. S. Corruccini (Eds.), *Primate behavior and sociobiology* (pp. 143–147). Berlin: Springer-Verlag.

Mead, G. H. (1934/1970). *Mind, self and society.* Chicago: University of Chicago Press.

Meltzoff, A. N. (1988). The human infant as Homo imitans. In T. R. Zentall & B. G. Galef, Jr. (Eds.), *Social learning: Psychological and biological perspectives* (pp. 319–341). Hillsdale, NJ: Erlbaum.

(1990). Foundations for developing a concept of self: The role of imitation in relating self to other and the value of social mirroring, social modeling, and self practice in infancy. In

D. Cicchetti & M. Beeghly (Eds.), *The self in transition: Infancy to childhood* (pp. 139–164). Chicago: University of Chicago Press.

Menzel, E. (1971). Communication about the environment in a group of young chimpanzees. *Folia Primatologica, 15,* 220–222.

Mignault, C. (1985). Transition between sensorimotor and symbolic activies in nursery-reared chimpanzees (*Pan troglodytes*). *Journal of Human Evolution, 14,* 747–758.

Miles, H. L. (1990). The cognitive foundations for reference in a signing orangutan. In S. T. Parker & K. R. Gibson (Eds.), *"Language" and intelligence in monkeys and apes* (pp. 511–539). Cambridge University Press.

Miles, H. L., Mitchell, R., & Harper, S. (1992). *Imitation and self-awareness in a signing orangutan.* Paper delivered at the 14th congress of the International Primatological Society, Strasbourg, France.

Miller, P., & Garvey, C. (1984). Mother–baby role play: Its origins in social support. In I. Bretherton (Ed.), *Symbolic play* (pp. 101–130). New York: Academic Press.

Mitchell, R. W. (1986). A framework for discussing deception. In R. M. Mitchell & N. S. Thompson (Eds.), *Deception: Perspectives on human and nonhuman deceit* (pp. 3–39). Albany, NY: SUNY Press.

(1987). A comparative-developmental approach to understanding imitation. In P. Klopfer & P. Bateson (Eds.), *Perspectives in ethology* (Vol. 7, pp. 183–215). New York: Plenum.

(1989). Functions and social consequences of infant–adult male interaction in a captive group of lowland gorillas (*Gorilla gorilla gorilla*). *Zoo Biology, 8,* 125–137.

(1992). Developing concepts in infancy: Animals, self-perception, and two theories of mirror self-recognition. *Psychological Inquiry, 3,* 127–130.

(1993). Mental models of mirror self-recognition: Two theories. *New Ideas in Psychology, 11,* 295–325.

Mueller, E., & Lucas, T. (1975). A developmental analysis of peer interaction among toddlers. In M. Lewis & L. Rosenblum (Eds.), *Friendship and peer relations* (pp. 223–257). New York: Wiley.

Nelson, K., & Gruendel, J. (1986). Children's scripts. In K. Nelson (Ed.), *Event knowledge* (pp. 21–45). Englewood Cliffs, NJ: Prentice-Hall.

Nelson, K., & Seidman, S. (1984). Playing with scripts. In I. Bretherton (Ed.), *Symbolic play* (pp. 45–72). New York: Academic Press.

Owens, N. W. (1975). Social play behaviour in free-living baboons (*Papio anubis*). *Animal Behaviour, 23,* 387–408.

Parker, S. T. (1977). Piaget's sensorimotor series in an infant macaque: A model for comparing unstereotyped behavior and intelligence in human and nonhuman primates. In S. Chevalier-Skolnikoff & F. Poirier (Eds.), *Primate bio-social behavior* (pp. 43–113). New York: Garland.

(1984) Playing for keeps: An evolutionary perspective on human games. In Peter K. Smith (Ed.), *Play in animals and humans* (pp. 271–293). London: Blackwell Publisher.

(1991) A developmental approach to the origins of self-recognition in great apes and human infants. *Human Evolution, 6,* 435–449.

(1993a). Comparative perspectives on the ontogeny of role-playing and imitation in the social interactions of an infant gorilla. Paper delivered at the Animal Behavior Society Meetings, Davis, CA.

(1993b). Evolved mechanisms for cognitive construction: Circular reactions and imitation as self-teaching mechanisms. *Human Development, 36,* 309–323.

Patterson, F. (1978). Conversations with a gorilla. *National Geographic, 154,* 438–465.

(1980). Creative and innovative uses of language by a gorilla: A case study. In K. Nelson (Ed.), *Children's language* (Vol. 2, pp. 497–561). New York: Gardner.

Perner, J. (1988). Developing semantics for theories of mind: From propositional attitudes to mental representation. In J. Astington, P. Harris, & D. Olson (Eds.), *Developing theories of mind* (pp. 141–171). Cambridge University Press.

Piaget, J. (1952). *The origins of intelligence in children.* New York: Norton.

(1954). *The construction of reality in the child.* New York: Basic Books.

(1962). *Play, dreams and imitation.* New York: Norton.

(1965). *The moral judgment of the child.* New York: Free Press.

(1985). *The equilibriation of cognitive structures.* Chicago: University of Chicago Press.

Povinelli, D., Nelson, K., & Boysen, S. T. (1992). Comprehension of role reversal in chimpanzees: Evidence of empathy? *Animal Behaviour, 43*(4), 633–640.

Premack, D. (1988). "Does the chimpanzee have a theory of mind" revisited. In R. Bryne & A. Whiten (Eds.), *Machiavellian intelligence* (pp. 160–178). Oxford University Press.

Premack, D., & Woodruff, G. (1978). Does the chimpanzee have a theory of mind? *Behavioral and Brain Sciences, 1,* 515–526.

Redshaw, M. (1978). Cognitive development in human and gorilla infants. *Journal of Human Evolution, 7,* 122–141.

Russon, A. (1990). Peer social interaction in infant chimpanzees. In S. T. Parker & K. R. Gibson (Eds.), *"Language" and intelligence in monkeys and apes* (pp. 379–419). Cambridge University Press.

Russon, A., & Galdikas, B. (1992). *Imitation in rehabilitant orangutans: Model and action selectivity.* Paper delivered at the 14th congress of the International Primatological Society, Strasbourg, France.

(1993). Imitation in free-ranging rehabilitant orangutans. *Journal of Comparative Psychology, 107*(2), 147–160.

Sachs, H. (1980). Button button who's got the button? *Sociological Inquiry, 50,* 318–27.

Savage-Rumbaugh, S., & MacDonald, K. (1988). Deception and social manipulation in symbol-using apes. In A. Whiten & R. Bryne (Eds.), *Machiavellian intelligence* (pp. 224–236). Oxford University Press.

Snow, C. (1990). Building memories: The ontogeny of autobiography. In D. Cicchetti & M. Beeghly (Eds.), *The self in transition: Infancy to childhood* (pp. 213–241). Chicago: University of Chicago Press.

Stern, D. (1977). *The first relationship.* Cambridge, MA: Harvard University Press.

(1985). *The interpersonal world of the infant.* New York: Basic Books.

Stevens, T. R. (1977). Cognitive structure in sports tactics: A preliminary investigation. In P. Stevens (Ed.), *Studies in the anthropology of play.* West Point, NY: Leisure Press.

Symons, D. (1978). *Play and aggression: A study of rhesus monkeys.* New York: Columbia University Press.

Tomasello, M. (1990). Cultural transmission in the tool use and communicatory signaling of chimpanzees? In S. T. Parker & K. R. Gibson (Eds.), *"Language" and intelligence in monkeys and apes: Comparative developmental perspectives* (pp. 274–310). Cambridge University Press.

Trevarthan, C., & Hubley, P. (1978). Secondary intersubjectivity: Confidence, confiders and cooperative understanding in infants. In A. Lock (Ed.), *Action, Gesture and Symbol* (pp. 183–229). New York: Academic Press.

Trevarthan, C., & Logotheti, C. (1987). First symbols and the nature of human knowledge. In J. Montangero, A. Tryphon, & S. Dionnet (Eds.), *Symbolism and knowledge* (pp. 59–86). Geneva: Cahiers de la Fondation Archive Jean Piaget, No. 8.

Visalberghi, E., & Fragaszy, D. (1990). Do monkeys ape? In S. T. Parker & K. R. Gibson (Eds.), *"Language" and intelligence in monkeys and apes* (pp. 247–272). Cambridge University Press.

Vygotsky, L. S. (1962). *Thought and language.* Cambridge, MA: MIT Press.

de Waal, F. (1982). *Chimpanzee politics.* New York: Harper & Row.

(1989). *Peacemaking among primates.* Cambridge, MA: Harvard University Press.

de Waal, F., & Hoekstra, J. A. (1980). Contexts and predictability of aggression in chimpanzees. *Animal Behaviour, 28,* 929–937.

Watson, J. S. (1984). Bases of causal inference in infancy: Time, space, and sensory relations. In L. Lipsitt & C. Rovee-Collier (Eds.), *Advances in infancy research* (Vol. 3, pp. 152–165). New York: Ablex.

(1972). Smiling, cooing and "the game." *Merrill-Palmer Quarterly*, *18*, 323–339.

Weisner, T., & Gallimore, R. (1977). My brother's keeper: Child and sibling caretaking. *Current Anthropology*, *18*, 169–190.

Whiten, A., & Byrne, R. (1988). The manipulation of attention in primate tactical deception. In R. Byrne & A. Whiten (Eds.), *Machiavellian intelligence* (pp. 211–223). Oxford University Press.

Part II

The development of self in human infants and children

John S. Watson

The comparative analysis of when and how animals become aware of themselves has at least two levels of focus. One has to do with a conceptual awareness. It implies some representational memory and is reflected in some self-referencing behavior that discriminates unique features of the individual (Gallup, 1970; Lewis & Brooks-Gunn, 1979). Lewis and Brooks-Gunn call this level the "categorical self." The other level of focus has to do with a perceptual sensitivity. It implies some detection mechanism and is reflected in some self-referencing behavior that discriminates self from nonself at least momentarily. Lewis and Brooks-Gunn call this level the "existential self." This paper will be limited to a consideration of how infants detect their existential selves. It is the simpler of selves, but it has been none the less elusive in the history of developmental psychology.

Theoretical concern for detection of self

Self-discovery in Freudian and Piagetian theories

The two major developmental theories of this century, Freud's and Piaget's, each assume that normal development in humans requires an initial investment in the task of differentiating the self from its external environment. The theories differ as to when and how the task is accomplished. As for timing, Freud proposed a relatively speedy accomplishment within the first few months, whereas Piaget proposed that the task would take two or three times as long (Wolff, 1960). As for how the distinction between self and nonself is accomplished, the two theorists differed fundamentally.

Freud (1911/1946) attended to why we would be motivated to do so. As basic need states arise, primary thought processes are assumed to hallucinate goal states; but, given the failure of these to satisfy the underlying physiological needs, secondary thought processes arise to cope with reality and the exigencies of the external world. However, just how internal and external are distinguished is not adequately provided. There is a proposal that tension-inducing stimuli arising outside of the body can be distinguished from those arising within on the basis of their contingent relation to bodily movement. Rappaport (1951) states the proposal succinctly:

131

[T]he still helpless organism [has] the capacity for making a first orientation in the world by means of its perceptions, distinguishing both "outer" and "inner" according to their relation to actions of the muscles. A perception which is made to disappear by motor activity is recognized as external, as reality; where such activity makes no difference, the perception originates within the subject's own body – it is not real. . . .

In summary, motility first serves as the channel of discharge for tensions due to needs of drive origin; later it becomes the tool of primordial reality-testing by distinguishing between inner and outer sources of stimuli, that is to say, between the "I" and "not I"; finally, it assumes the character of action, altering the external world for the purpose of gratification. (footnotes on pp. 323–333)

Although this contingency algorithm for distinguishing self from nonself has notable ancestral resemblance to one that I shall elaborate below, it is clearly deficient on two grounds. First, by defining self in terms of noncontingent stimuli, the external events that are unaffected by the infant's behavior would be categorized as internal. For example, imagine a young infant in a baby carriage on a busy urban sidewalk. The sounds of traffic and the varying flux of light and smell would impinge upon the infant in a virtually random relation to his or her motor activity. By Freud's rule, these stimuli, unaffected by motor activity, would be categorized as originating within the subject's body (the "I"). Moreover, additional confusion should arise for the categorization of sensory innervations that are intrinsic to motor action. The stimulation generated by movement of muscle, tendon, and bone are concurrent effects of motor activity. These stimuli, unlike traffic sounds, start and stop in direct relation to motor activity. By Freud's rule, however, these proprioceptive stimuli, being affected by motor activity, would be categorized as originating in the external world (the "not-I").

Piaget's theory, as I understand it, does not really offer a mechanism for detecting self versus nonself, even though a major concept in his developmental theory is that of egocentrism and its progressive replacement by an objective world view. For Piaget, the first step in breaking out of a complete solipsistic self-containment is with the onset of what he termed secondary circular reactions at about 4 months of age. As infants advance to the stage of coordinating secondary circular reactions at about 8 months, they progressively distinguish their causal action from its environmental effects. Unfortunately, Piaget's definition of the crucial constructive transition from primary circular to secondary circular functioning introduces an additional piece of circularity if one tries to use this distinction as the mechanism for differentiating self from environment. In Piaget's (1936/1952) words:

After reproducing the interesting results discovered by chance on his own body, the child tries sooner or later to conserve also those which he obtains when his action bears on the external environment. It is this very simple transition which determines the appearance of the "secondary" reactions; . . .

Of course, all the intermediaries are produced between the primary circular reactions and the secondary reactions. It is by convention that we choose, as criterion of the appearance of the latter, the action exerted upon the external environment. (pp. 154–155)

It would be broadly appealing if the circular reactions could support a mechanism of self-detection. Researchers have fruitfully applied the Piagetian concept of circular reactions in comparative analyses of primate intellectual development (Antinucci, 1990; Parker, 1977). The problem for its use in explaining self-detection, of course, is that it would be logically circular to explain the infant's acquiring a distinction between self versus the environment by reference to secondary circular reactions that are themselves defined by this same distinction.

Dreaming as self-reference algorithm in connectionist theory

Something closer to a viable mechanism for sorting out self from environment has recently been offered by theorists in the camp of what is called "new connectionism." Hinton and Sejnowski (1986) have offered a refinement of an earlier suggestion by Crick and Mitchison (1983) to the effect that REM sleep may provide a control computation for disambiguating stimulation arising from self and environment.

Basically their proposal is this: Neural networks are good at mapping the structure of stimulation they receive. The classic problem of solipsism arises, however, because the nets are subject to stimulation from their own activation as well as to stimulation from the external environment. To disentangle the internal from the external contributions to the resultant mapping, the system sleeps. In this state, shutting out stimulation from the environment, the nets develop a self-map that can be subtracted from the tangled composite. The result is a map of the external world.

Subtracting dream maps from wakeful maps will not provide, however, a clear distinction between world and body outside the net. The body can stimulate itself from outside its neural nets, as when behavior stimulates the distal senses (seeing one's body move or hearing its sounds) or when body parts touch each other. In short, although dream maps may be able to help us avoid confusing in-net from out-of-net experience, they appear doubtful as a solution for disentangling in-body versus out-of-body experiences. Put another way, dream maps may well help us accurately perceive the structure of the physical world, but they can provide no clear boundary between the part of that world we occupy and the part we do not.

Two time-based algorithms for self-detection

There would seem to be at least two potential options for distinguishing self from environment as based on the temporal relations between efferent and afferent neural activations. Each offers some measure of distinction between afferent sensory signals that are provoked by environmental events versus those that are provoked directly by bodily motoric activation. Thus, each might provide a connectionist theory with the needed algorithm for resolving out-of-body from merely out-of-net.

Delay of effect

One option would be simply reference to delay between efferent and afferent activity. We might be endowed with a temporal filter that accepts an event as environmental if and only if it is not preceded within some specified time by efferent activity. Twenty years ago I proposed this option in connection with an explanation of why young infants seemed to develop exuberant smiling to the contingent movement of one kind of mobile versus another (Watson, 1972). I suggested that perhaps the development of the smiling reaction in this contingency situation was dependent on the half-second delay between instrumental behavior and the electromechanical activation of the mobile. In the ethological spirit of my hypothesis that early contingency may specify the subsequent targets of social responses, I tacitly assumed that the evolution of such an imprinting mechanism might have included a delay filter to deflect the possibility of infants imprinting on themselves. A problem with this criterion of self is that stimulus events that happen to occur immediately following a behavior would be categorized as arising within the body.

Imperfect contingency

Another option is that imperfect temporal contingency between efferent and afferent activity implies out-of-body sources of stimulation, perfect contingency implies in-body sources, and noncontingent stimuli are ambiguous. In order to specify this mechanism, it is necessary to first specify what I mean by temporal contingency. What I mean is the temporal pattern between two events that potentially reflects the causal dependency between them. Before I am accused of begging the question, let me admit that I do not know the best philosophical definition of causality, but I am aware of four measures of temporal relations that have been offered to explain how animals and infants might detect a causal relation between their behavior and a subsequent event. I favor one over the rest; but on the basis of existing research, none can be ruled out as potentially relevant to the infant's detection of contingency.

Four options for detecting contingency

The four options can be labeled in terms of what may function as the focal relation between an infant's behavior and consequent stimulation. These potential focal relations are contiguity, correlation, conditional probability, and causal implication. I will give a brief overview and example of each of these in the effort to note their most distinctive implications.

The four theoretical options can be illustrated by considering how an infant might be sensitive to a contingency between kicking behavior and movement of a mobile hanging overhead when one of the baby's legs is tethered to the mobile by a ribbon. This situation, in fact, has been used often by Rovee-Collier and her colleagues in the study of infant learning and memory

(Rovee & Rovee, 1969; Rovee-Collier, 1987). As we consider the four different theories of detection, we will imagine that an infant might focus on any one of four distinct questions. The infant attending to contiguity would ask, "Does the mobile move with or soon after my kick?" The infant attending to correlation would ask, "Does the mobile movement vary over time in relation to the variation of my kicking?" The infant attending to conditional probability would ask, "Does the probability of a mobile movement given some time following a kick differ from the probability of mobile movement without consideration of kicking?" Finally, the infant attending to logical implication would ask, "Does my kicking versus not kicking have a logical implication for mobile movement versus no movement?"

Contiguity

Consider first the option that the infant attends to temporal contiguity (see Figure 8.1). Contiguity was a focal aspect of Guthrie's (1959) theory of learning, but Skinner (Ferster & Skinner, 1957; Skinner, 1938) has probably been the most influential proponent of the central role of contiguity in an organism's sensitivity to contingency. The notion is a simple one. If a contingent stimulus occurs shortly in time following a behavior, then subsequent behavior of the subject will be altered as a function of the reinforcement value of the stimulus.

If contiguity were the infant's sole means of perceiving contingency, then the underlying mechanism for that perception need be no more than some sensitivity to the passing of time, I have represented that mechanism in Figure 8.1 with a stopwatch in the infant's head. That can be viewed in contrast to the depiction in Figure 8.2 of an infant attending to the correlation of behavior and stimulus over time and employing some mechanism to evaluate this correlation. For metaphoric simplicity. I have represented that mechanism with a pocket calculator.

Temporal correlation

Although often merged with discussion of conditional probability, some modern theories of learning appear to imply that temporal correlation will be a major informing experience about the existence of causal relations (Tarpy, 1982). If more reward or reinforcement is received during or shortly following periods in which greater amounts of a particular behavior are performed, and lesser amounts of reward are received when lesser amounts of behavior are performed, then this correlation of behavior and reward might inform the subject of the existence of a contingency between these events.

The advantage of attending to correlation versus contiguity would be the capacity to avoid both false positive and false negative perceptions of temporal contingency. Correlation uses enough information to avoid the false positive superstitions that might be aroused by chance temporal conjunctions.

Moreover, if lagged correlation is employed, contingencies that involve delays can be detected even though contiguity is absent.

Conditional probability

Figure 8.3 illustrates an infant attending to conditional probability; I have again represented the underlying mechanism with a pocket calculator. Note, however, that this metaphoric device is given additional memory relative to the calculator of correlation. That is to allow a basis for the important distinction between the two modes of contingency evaluation. Conditional probability evaluation involves keeping track of instances in which the behavior occurs and the stimulus does not, versus instances when the stimulus occurs but the behavior does not. Correlation evaluation effectively ignores the difference between these two forms of disjunctive relation. For that reason, I assume conditional probability analysis involves greater differentiation of memory.

As I have contended previously (Watson, 1979, 1984), a complete picture of contingency structure requires calculation of two rather independent conditional probabilities. These are termed the *sufficiency index* and the *necessity index*. Sufficiency refers to the probability of the stimulus given some specified time following the behavior. Necessity refers to the probability of the behavior given some specified time preceding the stimulus. A simple example should illustrate their distinction from the correlation index.

Imagine two babies are each linked by separate ribbons to a single mobile overhead. Consider the situation from the vantage point of an infant who is evaluating conditional probability. Each time he kicks, the mobile moves; but it also moves if he does not kick, because the other infant kicks occasionally too. If both infants are kicking at about the same rate, then the following should be evident. First, unless both infants are kicking in unison, the correlation between our infant's kicks and the mobile movement will be decidedly less than perfect. Now consider the two conditional probabilities that our infant is calculating. The sufficiency calculation will find that a mobile movement occurs each time the infant kicks, a conditional probability of 1.00. The necessity calculation, however, will find in the present example that, on average, only one in two movements of the mobile have been immediately preceded by our infant's kick, a probability of .50.

This hypothetical example should illustrate at least that one of the two conditional probability indices of contingency can remain very high while the correlation index falls far short of perfection.[1] In this manner, the conditional probability calculations can protect the infant from making false negative evaluations of contingency.

Causal implication

Bower (1989) has recently offered an alternative option for how the human infant may perceive contingency. Rather than contiguity, correlation, or conditional probability, Bower proposes that the infant may attend directly to

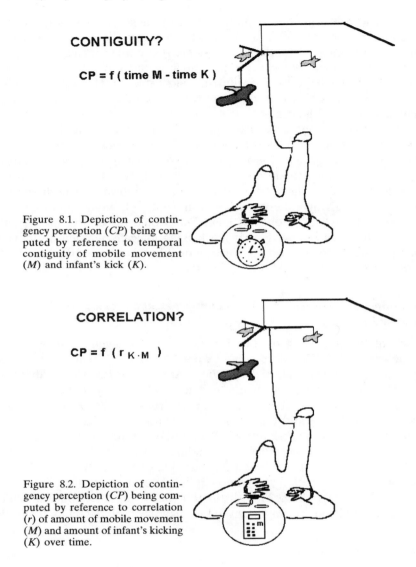

CONTIGUITY?

$$CP = f (\text{time M} - \text{time K})$$

Figure 8.1. Depiction of contingency perception (CP) being computed by reference to temporal contiguity of mobile movement (M) and infant's kick (K).

CORRELATION?

$$CP = f (r_{K \cdot M})$$

Figure 8.2. Depiction of contingency perception (CP) being computed by reference to correlation (r) of amount of mobile movement (M) and amount of infant's kicking (K) over time.

causal implication. By this he means that the infant observes the world with a natural inclination to formulate the potential cause–effect relations that would be *logically* consistent with the observed instances of behavior and stimulus events. In Figure 8.4, I have represented this logical sensitivity with, for want of a better metaphor, a "therefore" mechanism. It relies on the infant possessing a sense of implication.

Consider one of the examples used by Bower. An infant kicks and sees his mobile move. Bower proposes that this observation is placed in logical relation to the prior observation that in the absence of kicking there had been no mobile movement. Logically, these observations are consistent with either of two causal hypotheses. These are that either kicking implies mobile movement, or that not kicking implies no movement of the mobile. These are logically distinct hypotheses as illustrated in Table 8.1. Having the mobile not move following a kick is inconsistent with the former but not the latter, and having it move without a prior kick is inconsistent with the latter but not the former. Bower proposes that the infant accepts one of the hypotheses and then, in the spirit of a philosopher of science (namely Popper), the infant tries to disprove his hypothesis. If he has constructed the hypothesis that not kicking causes the mobile not to move, then, he must try to disprove it by seeking to observe the relation that would disconfirm it. The disconfirming instance for this hypothesis, as noted in Table 8.1, would be observing the mobile move when he did not kick. This would disconfirm the hypothesis because if it had been true that not kicking logically implied no movement of the mobile, then it would not be possible to observe movement of the mobile when kicking had not occurred.

A distinction between detecting contingency and seeking or avoiding it

Each of the preceding options for detecting contingency can work under some conditions. The more complex ones are valid over a wider range but they are cognitively more demanding. A point worth noting about each, however, is that each can work to detect contingency without necessarily involving the arousal (or suppression) of instrumental behavior. It is conceivable (as I will illustrate) that evidence of contingency detection may exist in the absence of a general rate change in the instrumental behavior. When detection does lead to arousal (or suppression) of instrumental behavior, then there is evidence of seeking (or avoiding) the contingency. From this perspective, then, seeking/avoiding implies detection, but detection does not imply seeking/avoiding. This distinction between detection and seeking/avoiding contingency is important to the analysis of self-detection that follows.

Studies of self-detection

Lewis and Brooks-Gunn (1979) speculate in a manner similar to Gibson (1966, 1979) that the earliest form of self–other distinction is probably based on some kind of response-contingent stimulation such as that produced when the infant sees the movements of his hand as he waves it in front of his face, or when he feels the proprioceptive stimulation of leg movement as he flexes his legs. Given the substantial body of evidence on the human infant's capacity to perceive response–stimulus contingencies by the age of 3 months or earlier, Lewis and Brooks-Gunn set this age period as the probable time of the first primitive discrimination of self and other.

CONDITIONAL PROBABILITY?

$$CP = f \ P(M/K_t)$$
and
$$CP = f \ P(K/_tM)$$

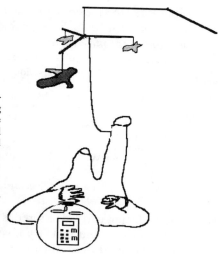

Figure 8.3. Depiction of contingency perception (*CP*) being computed by reference to the "sufficiency index" of conditional probability, $P(M/Kt)$, to be read as the probability of mobile movement (*M*) in a specified time span (*t*) following a kick (*K*) and by reference to the "necessity index" of conditional probability, $P(K/tM)$, to be read as the probability of a kick (*K*) in a specified time span (*t*) preceding a mobile movement (*M*).

CAUSAL IMPLICATION?

$$CP = f \ (\ K \ \text{implies} \ M \ )$$
and
$$CP = f \ (\ \text{not} \ K \ \text{implies} \ M \ )$$
and
$$CP = f \ (\ K \ \text{implies not} \ M \ )$$
and
$$CP = f \ (\ \text{not} \ K \ \text{implies not} \ M \ )$$

Figure 8.4. Depiction of contingency perception (*CP*) being deduced ("ERGO") by reference to the causal implication of observed relations between the four logical options for combining mobile movement (*M*) versus no movement (*not-M*) with kicking (*K*) versus no kicking (*not-K*) as represented in Table 8.1.

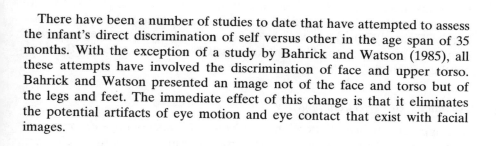

There have been a number of studies to date that have attempted to assess the infant's direct discrimination of self versus other in the age span of 35 months. With the exception of a study by Bahrick and Watson (1985), all these attempts have involved the discrimination of face and upper torso. Bahrick and Watson presented an image not of the face and torso but of the legs and feet. The immediate effect of this change is that it eliminates the potential artifacts of eye motion and eye contact that exist with facial images.

Table 8.1. *Observations for each of the four logical hypotheses about causal relation between kicking not kicking and mobile movement not movement*

Hypothesis	Observation			
	K + M	not K + M	K + not M	not K + not M
K implies M	C	C	I	C
Not K implies M	C	C	C	I
K implies not M	I	C	C	C
Not K implies not M	C	I	C	C

Abbreviations: C, consistent; I, inconsistent; K, kicking; M, mobile movement.

The infant sat facing two video screens. Each screen displayed the dynamic image of a pair of legs and feet. By inverting the video camera, the images produced approximated the retinal distribution of the subjects' normal view of their own legs and feet. That is, the left–right distribution was correct and the legs projected toward the upper portion of the visual field, as is so for infants' direct view of their legs and feet. In three of four experiments, the direct view was screened so that an infant's only visual information about leg movement was on one of the TV monitors (see Figure 8.5).

Three experiments were run to see if 5-month-olds would show discriminative attention to a choice between a live video image of their legs versus those of a peer or, in one experiment, a prior recording of themselves. The results confirmed the expectancy that 5-month-old infants would perceive the difference between the contingent self-image and the noncontingent peer image. Consistent with prior findings with facial images at this age, the subjects in each experiment displayed a significant preferential fixation on the image that was not their contingent self.

A fourth study was carried out with a sample of 3-month-olds. Examination of the distribution of individual percent fixation scores uncovered a significant bimodal distribution. That is, about half displayed strong preference for self and about half displayed strong preference for the noncontingent image. Few showed marginal preference. Thus it is possible that 3-month-olds are capable of discriminating self from other but are in a developmental phase involving a transition from attentional deployment toward self to attention away from self toward others. This interpretation fits with the findings of Field (1979) whose sample of 3-month-olds viewing faces showed a significant preference for the contingent self image as opposed to the noncontingent image.

Avoidance of perfect contingency

The apparent change of attentional preference for noncontingency over contingency after 3 months raises an important question. Does the preference

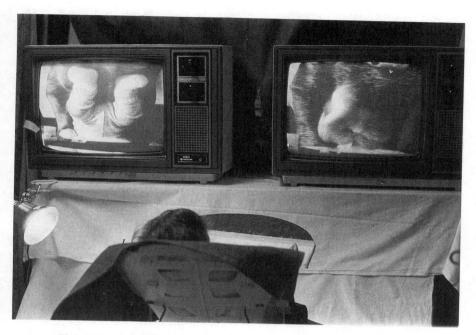

Figure 8.5. Testing situation used by Bahrick and Watson (1985) to assess infant's discrimination and preference for noncontingent versus perfectly contingent images of leg movement.

reflect a general lack of interest in all levels of contingency or just in perfect contingency? Lack of interest in all levels of behavior–stimulus contingency would seem counter productive for a species that relies as much on learning as humans do. It would seem more plausible, then, that the observed preference for noncontingent images by infants older than 3 months is a special avoidance reaction to perfect contingency. They detected the perfect contingency of the on-line image but selected to look at the noncontingent image. By this reckoning, we should expect that some level of imperfect contingency would not be avoided, but would indeed be arousing of behavior and attention. I have observed support for just such a relation between arousal and level of contingency in a series of learning studies with infants (Watson, 1984, 1985).

Some evidence of arousal from imperfect contingency

In these learning studies, infants were placed in a special chair that allowed them to control the running of a video tape by making small movements of their legs (see Figure 8.6). During an initial 1-min base period, kicking was recorded but had no effect on the video. This was immediately followed by a 6-min contingency period in which a kick would cause a woman's face to become animated with speech for about 3 sec. A "learning score" was derived by subtracting the behavior rate in the base period from the average rate observed in the final 2 min of the contingency period. Although unanticipated at the time, the infants experiencing immediate and "perfect" contingency in this study failed to show a significant increase in response rate. Reliable learning was observed only in infants who experienced moderately reduced contingency that involved imperfection in the probability their behavior would be immediately followed by the stimulus (the "sufficiency" index) *and* imperfection in the probability that a stimulus was immediately preceded by their behavior (the "necessity" index).

The final study of the series examined this curious effect by extending the range of contingency variation. It involved 120 16-week-olds. Five contingency conditions were selected based on previous results. One of the five conditions was the "perfect contingency" and another was the lowest-level contingency (approximately .10 on both indices). If any of the three intermediate conditions supported a significant increase in response rate, it should be either one or both of the conditions in which moderate imperfection existed in both the sufficiency index – the conditional probability of the stimulus following the behavior – and the necessity index – the conditional probability that a stimulus was preceded by behavior.

Learning scores were compared across conditions, with a breakdown by sex. Again the "perfect contingency" condition failed to arouse significant increment in the rate of kicking in either sex. In contrast, significant increase in rate of kicking occurred for both sexes in the intermediate contingency group, providing approximately .5 on both indices of contingency – one of the two higher intermediate levels as predicted. For both sexes this was the only condition to produce a reliable increase in instrumental behavior. Thus it would seem that 16-week-olds are less aroused by perfect control than by moderately imperfect control.

Detecting versus seeking

Did the infants in the perfect contingency not detect it? Or did they detect it and simply not seek it? To find out, we analyzed the momentary probability of kicking as a function of time from a rewarded kick, from an unrewarded kick, and from a noncontingent presentation of reward. These results are analyzed in detail in Watson (1984). For the purpose of this paper, it will suffice to note that the momentary reaction to contingent reward indicates that the perfect contingency was detected. Activation of the video, either contingently or noncontingently, tended to cause an immediate suppression of kicking for about 2 sec. This orienting reaction was very robust. Most notable, however, was that when the stimulus was perfectly contingent on kicking, the initial suppression was followed by a few seconds of significantly elevated probability of kicking again. There was no sign of a similar elevation in likelihood of repeating a kick in instances of unrewarded kicking from the base period, nor was there a significant elevation in kicking following

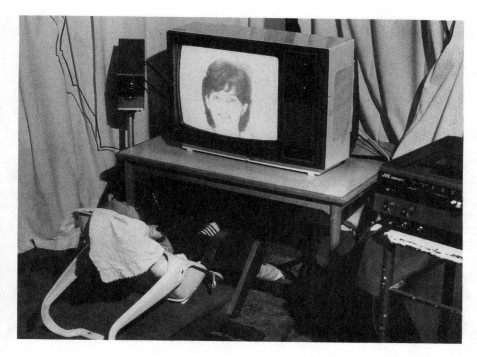

Figure 8.6. Testing situation used by Watson (1984, 1985) to assess infant's discrimination and behavioral arousal under various levels of contingency between kicking and animation of video.

noncontingent rewards from the contingency period of subjects exposed to imperfect contingency. The distinctive pattern of reaction to perfect contingency suggests that this contingency was detected even though the infants did not show a general increase in kicking rate.

A hypothesis of self-seeking followed by self-avoidance

Given this array of self-detection and learning data, I have been developing a hypothesis about the initial phases of human self-detection and related phases of self-seeking and self-avoidance. As it stands presently, I assume that the infant is endowed with an analytic mechanism that detects contingency. Based on the learning data cited above, I expect that this detection mechanism is some analog of the conditional probability option described above (but see Watson, 1979, for a complete discussion of computations in relation to chance expectancies). Until about age 3 months, the infant seeks the highest level of contingency it can detect; after that age, it would appear that a bias matures to the effect that perfect (or very nearly perfect) contingency is categorized as "self." This self-object is perceived as less attractive and arousing than objects whose reactive contingency is high but not perfect.

That bias affords a functional discrimination and categorization of self as distinct from other objects. It also orients the infant to engage the reactive causal properties of external objects in preference to generating recursive cycles of self-stimulation.

A potential function of initial self-seeking

Considering the preceding hypothesis, one must wonder why there would be an initial phase of self-seeking. If this is not a residual quirk of evolution, what might the phase serve? Some recent work of the connectionist Rumelhart (1990; see also Jordan & Rumelhart, 1991) offers an interesting possibility. He has tested various teaching strategies in simulations of a "neural net" learning to guide a mechanical arm in accurate reaching for objects. He describes the task as involving at least a neural net for mapping motor activation and arm movements (the motor net) and a neural net for mapping of perceived object placement and motor activation (the object-perception net). It appears that this double net learns the task most efficiently if it begins by organizing the motor net through a series of random self-activations prior to having output from the object-perception net serve as input to the motor-activation net. Rumelhart notes that Piaget seemed to have a similar strategy in mind in his notion of primitive sensory motor schemata becoming organized by their application on the self (as primary circular reactions) prior to their application on the external environment (as secondary circular reactions).

A speculation about deviant self-seeking in infantile autism and Rett syndrome

The hypothesis of normal development here formulated in regard to the human infant's detection of his or her empirical self provides a base for speculation about at least two forms of abnormal development. Given the slim empirical foundation of these speculations, I will not devote much space to them. Whether or not they eventually receive any degree of validation, their presentation here should help consolidate the ideas presented in this paper.

Infantile autism

One of the presenting symptoms of infantile autism is the appearance of unusual rhythmicities and odd stereotypies, such as twirling an object or flapping the arms. Another symptom is a low tolerance for change in schedule or routine. In the social realm, two notable peculiarities stand out. There is frequently a tendency to avoid looking at the face of another person, and there is a special deficiency in the child's sensitivity to the alterable dispositional states of others (e.g., understanding how perceptions may affect belief states).

Although surely an oversimplification, it is nonetheless of some interest that each of these symptoms is consistent with what might be expected to follow from an insufficient amount of resetting for the hypothesized target value of imperfect contingency seeking. The rhythmicities and stereotypies would stand as aberrant extensions of primary circular reactions that remain guided by their very high level of act–outcome efficacy.

The lowered capacity to adjust to dispositional change in others, even when controlled for general cognitive capacity (e.g., Perner, 1991), is accountable as the effect of pursuing high efficacy. Dispositional change over time is reflected in a lower contingency until the dispositional change is detected. For example, I may judge an acquaintance as being moderately predictable in his response to my requests. I may later notice that he is very accommodating whenever he has recently been humming, but if he has not been humming, he is rarely accommodating. He is now very predictable and I know when to ask for favors. However, if I had not attended to his moderate predictability (my moderate experience of contingency) his dispositional change in the course of the day might have been overlooked. The autistic's tendency to seek very high efficacy could in like manner obscure the chances of noticing and incorporating dispositional variation and thus severely limit the autistic's construction of a theory of mind.

The autistic's tendency to avoid eye contact is also consistent with the model of normal development proposed here. To make the connection, however, requires reference to an additional theoretical proposal about normal infants mentioned earlier. In that proposal, the arrival of exuberant smiling and cooing to the facial form at about 3 months of age was viewed as tied in with the infant's detection of efficacy in the interaction of playful games introduced by the caretaker (Watson, 1972). I would now propose that the autistic's perverse seeking of near-perfect efficacy diminishes the chance that normal interaction games will suffice to support exuberant response and attention to human faces. In its strongest form, I am proposing that the autistic's continued commitment to very high efficacy interferes with the process of imprinting as proposed for the normal 3-month-old, and also interferes with subsequent detection of mental state transformation in others.

Rett syndrome

In 1969, a neurophysiologist in Austria proposed a delineation of patients formerly classified as infantile autistic (Olsson & Rett, 1987). Within the last decade, the syndrome has received a great deal of attention worldwide under the label Rett syndrome (Hagberg, 1989; Acker, 1987). Major distinguishing features of the disorder are, as follows:

1. It appears to only occur in girls.
2. Development appears normal for a period of 6–16 months prior to a rapid loss of cognitive and manual function resulting in severe mental retardation.
3. The decline in cognitive capacity is accompanied by the onset of increasingly repetitive hand-on-hand and hand-to-mouth stereotypies.
4. These children tend to stare and smile at adult faces.

These characteristics, with the exception of the gender association, can be accounted for in terms of a simple disruption in the proposed model of normal regulation of contingency seeking. Assume that a child passes through the transition of normal target-level change. Perfect contingency is avoided and high but less-than-perfect contingency is sought after. Interaction games are engaged and faces become imprinted objects for later attention and release of social smiling. Cognitively, the child continues to expand secondary circular actions. Development appears normal. Then, for reasons presumably associated with defective alleles on the X chromosome, the target value of contingency seeking switches back to perfection. With this transformation a destructive course of development is initiated. In a manner analogous to a computer virus or "worm" the brain is literally reprogrammed to seek the repetition of behavior that has high act–outcome contingency. Perfect efferent–afferent contingencies of self-stimulation compete with established less-than-perfect secondary circular actions. Reaching gives way to hand wringing and mouthing.

These symptoms are not the simple spasms of a broken brain. In an intensive experimental intervention with one Rett child, my colleagues and I were able to briefly encourage a variety of secondary circular reactions and observed a concomitant decline in primary rhythmicities, but we did not succeed in halting the ultimate commitment to self-stimulation (Watson, Umansky, & Johnston, 1990).

The speculations offered here are surely far too simplistic to be claimed as adequate explanations of these disorders, or of the cognitive transitions of normal infancy, for that matter. They will have been worth consideration, in my judgment, if they have managed to help clarify the distinctions between self-detection, self-seeking, and self-avoiding as these may arise in the early development of humans and possibly of other species.

Other species

I am not sure these speculations are directly relevant to self-detection in other species. However, I can think of two reasons that they might be. First, whether or not an organism advances to the state of reflective self-awareness (as discussed by other authors in this volume), it will need to avoid having perception of its body elicit behavior evolved for acting upon the external world (particularly predatory behavior). Second, the forms of contingency analysis presented in this chapter appear applicable to any species capable of the implied computations. Although to date, logical inference has been attributed only to the human infant (Bower, 1989), many researchers have long been willing to attribute conditional probability analysis to many less complex species (Maier, Seligman, & Solomon, 1969; Rescorla, 1967; Seligman, 1975). Given the presumed value of detecting body boundaries, I would guess that many species other than ours take advantage of the algorithm of perfect contingency.

The assumption that many species may rely on contingency analysis and the algorithm of perfection leaves important questions for comparative developmentalists. Is there a connection between self-detection and self-awareness? Do other species, as they develop, change in the degree to which perfect contingency attracts their attention? If so, is it possible that the timing and magnitude of this change is critical for the effects it creates? More specifically, can the speculation above regarding contingency seeking in autism and Rett syndrome be modified to frame contrasts between those primates that do achieve self-awareness and those that do not? My guess is that the answer to each of these questions is yes.

Acknowledgment

I would like to thank the editors, especially Robert Mitchell, for the many suggested changes that I adopted in my effort to improve the clarity of this chapter.

Note

1. The present example illustrates the possibility of the correlation being diminished even though the sufficiency index is perfect. Alternatively, were we to assume only one infant connected to the mobile by a faulty mechanical device, then the situation could be one in which the correlation index and the sufficiency index are reduced, but in which the necessity index remains perfect because no matter how faulty the mobile connection it will always be true that when the mobile has moved it will have been preceded by a kick.

   If the infant uses both indices of contingency available in conditional probability analysis, then conflict may arise when one index specifies perfection and the other does not. On the other hand, I would expect that the algorithm of self-detection would require both indices to be near 1.0. An environmental event would be specified when either index was appreciable and greater than chance expectancy.

References

van Acker, R. M. (1987). *Stereotypic responding associated with the Rett syndrome: A comparison of girls with this disorder and matched subjects without the Rett syndrome.* Unpublished doctoral dissertation, Northern Illinois University.

Antinucci, F. (Ed.) (1990). *Cognitive structure and development in nonhuman primates.* Hillsdale, NJ: Erlbaum.

Bahrick, L. E., & Watson, J. S. (1985). Detection of intermodal proprioceptive-visual contingency as a potential basis of self-perception in infancy. *Developmental Psychology, 21,* 963–973.

Bower, T. G. R. (1989). *The rational infant.* Chicago: Freeman.

Crick, F., & Mitchison, G. (1983). The function of dream sleep. *Nature, 304,* 111–114.

Ferster, C. B., & Skinner, B. F. (1957). *Schedules of reinforcement.* New York: Appleton-Century-Crofts.

Field, T. (1979). Differential behavioral and cardiac responses of 3-month-old infants to a mirror and peer. *Infant Behavior and Development, 2,* 179–184.

Freud, S. (1911/1946). Formulations regarding the two principles in mental functioning. *Collected papers* (Vol. 4, pp. 13–21). London: Hogarth.

Gallup, G. G., Jr. (1970). Chimpanzees: Self-recognition. *Science, 167,* 86–87.

Gibson, J. J. (1966). *The senses considered as perceptual systems.* Boston: Houghton Mifflin.

(1979). *The ecological approach to visual perception.* Boston: Houghton Mifflin.

Guthrie, E. R. (1959). Association by contiguity. In S. Koch (Ed.), *Psychology: A study of a science,* (Vol. 2, pp. 158–195). New York: McGraw-Hill.

Hagberg, B. A. (1989). Rett syndrome: Clinical peculiarities, diagnostic approach, and possible cause. *Pediatric Neurology, 5*(2), 75–83.

Hinton, G. E., & Sejnowski, T. J. (1986). Learning and relearning in Boltzmann machines. In D. E. Rumelhart & J. L. McClelland (Eds.), *Parallel distributed processing* (Vol. 1, pp. 282–317). Cambridge, MA: MIT Press.

Jordan, M. I., & Rumelhart, D. E. (1991). *Forward models: Supervised learning with a distal teacher.* Occasional Paper #40, Center for Cognitive Science, Massachusetts Institute of Technology.

Lewis, M., & Brooks-Gunn, J. (1979). *Social cognition and the acquisition of self.* New York: Plenum.

Maier, S., Seligman, M. E., & Solomon, R. L. (1969). Fear conditioning and learned helplessness. In R. M. Church & B. Campbell (Eds.), *Punishment and aversive behavior.* New York: Appleton-Century-Crofts.

Olsson, B., & Rett, A. (1987). Autism and Rett syndrome: Behavioral investigations and differential diagnosis. *Developmental Medicine and Child Neurology, 29,* 429–441.

Parker, S. T. (1977). Piaget's sensorimotor period series in an infant macaque: A model for comparing unstereotyped behavior and intelligence in human and nonhuman primates. In S. Chevalier-Skolnikoff & F. E. Poirier (Eds.), *Primate biosocial development: Biological, social, and ecological determinants* (pp. 43–112). New York: Garland.

Perner, J. (1991). *Understanding the representational mind.* Cambridge MA: MIT Press.

Piaget, J. (1936/1952). *Origins of intelligence.* New York: Norton.

Rappaport, D. (1951). *Organization and pathology of thought.* New York: Columbia University Press.

Rescorla, R. A. (1967). Pavlovian conditioning and its proper control procedures. *Psychological Review, 74,* 71–80.

Rovee, C. K., & Rovee, D. T. (1969). Conjugate reinforcement in infant exploratory behavior. *Journal of Experimental Child Psychology, 8,* 33–39.

Rovee-Collier, C. (1987). Learning and memory in infancy. In J. D. Osofsky (Ed.), *Handbook of infant development* (pp. 98–148). New York: Wiley.

Rumelhart, D. E. (1990). The role of mental models in learning and performance: A connectionist account. Colloquium presented to Berkeley Cognitive Science Group.

Seligman, M. E. P. (1975). *Helplessness: On depression, development and death.* San Francisco: Freeman.

Skinner, B. F. (1938). *The behavior of organisms: An experimental analysis.* New York: Appleton-Century-Crofts.

Tarpy, R. M. (1982). *Principles of animal learning and motivation.* Glenview, IL: Scott, Foreman.

Watson, J. S. (1972). Smiling, cooing, and "the Game." *Merrill-Palmer Quarterly, 18,* 323–339.

(1979). Perception of contingency as a determinant of social responsiveness. In E. Thoman (Ed.), *The origins of social responsiveness* (pp. 33–64). Hillsdale, NJ: Erlbaum.

(1984). Memory in learning: Analysis of three momentary reactions of infants. In R. Kail & N. Spear (Eds.), *Comparative perspectives on the development of memory* (pp. 159–179). Hillsdale, NJ: Erlbaum.

(1985). Contingency perception in early social development. In T. M. Field & N. A. Fox (Eds.), *Social perception in infants* (pp. 157–176), Norwood, NJ: Ablex.

Watson, J. S., Umansky, R., & Johnston, C. (1990). Rett syndrome: An exploratory study. *Infant Behavior and Development, 13* (special ICIS issue; from abstracts of papers presented at the Seventh International Conference on Infant Studies, Montreal, April 1990, Abstract No. 663).

Wolff, P. H. (1960). The developmental psychologies of Jean Piaget and Psychoanalysis [Monograph 5]. *Psychological Issues, 2*(1), 3–181.

9 Social imitation and the emergence of a mental model of self

Daniel Hart and Suzanne Fegley

Self-awareness, and the object of its reflection, the self-concept, are central features of the conscious experience of adult humans in Euro-American cultures.[1] The phenomenal reality of these facets of experience have fascinated scholars interested in human nature for centuries. Philosophers, psychologists, sociologists, anthropologists, and literary critics have offered (often contradictory) perspectives on the essence of the sense of self. Perhaps the single most perplexing issue in the vast literature concerns the genesis of self-awareness and the self-concept: How can the zygote – which by consensus is without self-awareness or a self-concept – develop into an adult human within whom these qualities are apparent?

Many of developmental psychology's greatest theorists (e.g., Baldwin, 1902; Guillaume, 1926/1971; Luria, 1976; Mead, 1934; Piaget, 1932/1965) have attempted to explicate the nature of self-awareness and the self-concept, as well as the processes through which they are acquired. There is agreement among these early theorists that social interaction of one type or another (imitation, role taking, and language use have all been proposed) is central to the development of the sense of self. (For a partial review of these theories, see Hart, Kohlberg, & Wertsch, 1978.) The postulate of the social origins of self-awareness and the self-concept continue to be evident in more recent accounts as well (e.g., Mahler, Pine, & Bergman, 1975; Stern, 1985). Our goals in this chapter are, first, to examine the nature of the infant's self-awareness and self-concept, and then to consider the ways in which social interaction might be involved in the acquisition and developmental transformation of some facets of these constructs.

The nature of self-awareness and the self-concept

The sense of self is formed of many complexly interwoven experiences for which there is no single taxonomy. A review of divergent bodies of psychological work (Hart, 1992b), however, suggests that some distinctions emerge clearly among the varied experiences constituting the sense of self. The broadest of these is between *self-awareness* and the *self-concept;* as William James (1892/1961) pointed out a hundred years ago, a person not only has an

149

image of the self (a self-concept), but also knows that this image is the object of the process of self-reflection (self-awareness).

Although psychologists usually consider self-awareness a single category of phenomena, it has two facets that are usefully distinguished. First, self-awareness means that, among the infinite range of possible objects of attention, one is focusing on oneself. In this sense, self-awareness is but one of a class of awareness that might be imagined, such as "ball awareness," "food awareness," or "mother awareness." In the social psychological literature, this facet is called *objective self-awareness* (e.g., Duval & Wicklund, 1972), and there is now a large body of work with adults that demonstrates the importance of this construct for understanding many aspects of behavior.

There is a second facet to self-awareness, however, that is often overlooked by psychologists. Self-awareness is not only the focusing of attention on one set of stimuli, but also the awareness that this set of stimuli *is oneself* (*subjective self-awareness*). Nozick (1981, p. 73) points out that it is exactly this facet of self-awareness that is missing "when Oedipus sets out to find 'the person whose acts brought trouble to Thebes'. . . . He does not know that he himself is the culprit"; Oedipus' thoughts refer to himself – "the person whose acts brought trouble to Thebes" – and therefore there is evidence of objective self-awareness; but he does not realize that this self is his own. Subjective self-awareness is often manifested in (but not explained by) the network of emotions that are awakened when considering the facets of the self (for an interesting discussion of this issue, see Rosaldo, 1984). For instance, the patterning of affect is noticeably different when one's own goals appear thwarted than when one perceives the same occurring in another. Subjective self-awareness usually occurs in tandem with objective self-awareness, but the two are dissociable. For instance, subjective self-awareness is apparently lost in prefrontal lobotomies (see, for example, the cases reported in Freeman & Watts, 1942); and in some cases of psychopathology patients talk about their experiences and problems without the recognition (suggested in part by a lack of emotional arousal) that these experiences and problems belong to them (e.g., Rogers, 1951, p. 509).

There are some useful distinctions that can be made among the components of the self-concept as well. Research by cognitive psychologists (e.g., Medin, 1989) suggests that most categories and concepts are structured in identifiably different levels, ranging from specific to general; the self-concept shares this quality. At a very specific level, the adult's self-concept is composed of autobiographical memories like graduation from high school, a first date, or an angry argument with parents (see Brewer, 1986, for a discussion). There is a more general level at which a person thinks of the self in terms of representations and schemata such as "tall," "helpful," or "smart" (e.g., Markus, 1977). Finally, the self-concept is, in part, a mental model or theory, within which the various autobiographical memories and schemata are organized and synthesized into broad principles (e.g., Damon & Hart, 1988, 1992; Hart, 1992a). Because the mental model of self serves to integrate the knowledge deriving from many of the organism's interactions with the world, and in turn can be used by the organism to anticipate the consequences of future actions, it can be said to be a "mental model" in the sense used by many cognitive psychologists (e.g., Craik, 1943). Although these levels of the self-concept are psychologically real (in that they adhere to genuine cleavages in the self-concept), they do overlap to some extent. Furthermore, the levels are sure to interact. For instance, one might expect that an individual in possession of a mental model of self might be able to rapidly assimilate schemata about the self. Similarly, newly acquired personal memories might lead to revisions in self-schemata and in the mental model of self. Unfortunately, there has been relatively little research that investigates these interactions, so that our understanding of the directions of influence among the levels is limited.

Mirror self-recognition and the sense of self

Mirror self-recognition (MSR) is the only widely accepted assessment paradigm in the study of the infant's sense of self (Damon & Hart, 1992). To briefly summarize it (see Amsterdam, 1972; Johnson, 1983; and Lewis & Brooks-Gunn, 1979, for detailed descriptions of the procedure) infants are first seated in front of a mirror and their behavior observed; this serves as a baseline against which to judge the infant's behavior in the subsequent episode. Next, the mirror is turned away, and the infant's face is surreptitiously marked with rouge. The mirror is turned back, and the infant's behavior is observed once again. At approximately 18–24 months of age, a human infant in this second episode shows *mark-directed behavior*, touching the rouge on its face, which can be seen only by inspecting the image in the mirror. Interestingly, mark-directed behavior appears only in humans, chimpanzees, orangutans (Gallup, *SAAH*3), and gorillas (Patterson & Cohn, *SAAH*17). What does this behavior reveal about the sense of self?

Mark-directed behavior indicates that the infant is capable of objective self-awareness; that is, the infant can focus on a class of stimuli forming the self's mirror image and distinguish this class of stimuli from those forming images of others. (In control experiments, infants rarely touch their own faces when viewing images of another infant with rouge on its face; see Johnson, 1983.) Nevertheless, objective self-awareness is evident within the first several months of life: For instance, Watson (*SAAH*8) has demonstrated that infants can make use of contingency information to distinguish between televised images of their own moving legs and the moving legs of another infant. In addition, Neisser (1988) and Butterworth (1990) have described the powerful visual cues that cleave the world into self- and non–self-related stimuli, and Samuels (1986) has reviewed a range of clues that make objective self-awareness possible in the first months of life. So the finding that objective self-awareness can be detected at approximately 18 months of age in the mirror self-recognition task adds little to our knowledge of the development of the sense of self.

Of more significance is the evidence for subjective awareness; the infant knows that the image marred with rouge is of itself, and is sufficiently disturbed by the anomaly in the self's typical appearance to explore it tactually. Subjective self-awareness is also evident in the infant's emotional expressions when placed in front of the mirror. Lewis, Sullivan, Stanger, and Weiss (1989) report that infants who demonstrate mark-directed behavior tend to show evidence of embarassment in front of the mirror as well, with embarassment apparently absent in infants who do not touch the marks on their faces. Lewis and his colleagues argue that embarassment is a self-referential emotion caused by the perception that the self is exposed (in this case by the mirror), and consequently apparently requires subjective self-awareness.[2] To date, there is no evidence to indicate that subjective self-awareness can be detected at younger ages using another paradigm.

The most significant debate concerns what the task reveals about the nature of the self-concept: Some argue that mirror self-recognition demonstrates that the infant is in possession of a mental model or theory of self, while others argue that the task merely reveals the presence or absence of a schema or representation of the self's physical appearance. The perspective that MSR is a circumscribed skill, reflecting only a knowledge of the self's physical appearance, has been offered by Bertenthal and Fischer:

Historically, many investigators have used a rather diffuse definition of self and have assumed that one measure of a notion of self, such as self-recognition, reflects the entire category of behaviors relating to the self. We believe that the notion of self does not develop in a simple, unitary manner. The child develops many skills that relate to her or his own properties; self-recognition as we have measured it [with the mirror-and-rouge and other tasks] is only one of these many skills. (1978, p. 50)

If visual self-recognition is not a sign of a mental model of self but instead is but one of "many skills," then there is little reason to attribute special significance to the mirror self-recognition task.

There are many researchers who reject the idea that MSR assesses only a schema of the physical qualities of self. For instance, Gallup argues that the failure of monkeys to exhibit mark-directed behavior suggests that "the monkey's inability to recognize himself may be due to the absence of a sufficiently well-integrated self-concept" (1977, p. 334). If this position is true, one can infer that the development of visual self-recognition ought to be accompanied by a variety of other behaviors revealing of a mental model of self. Certainly it is possible to demonstrate important differences in social interaction, imitation, and deception – all behaviors that may require a mental model of self – between species that are capable or mirror self-recognition (e.g., chimpanzees) and those who apparently are not (e.g., macaques). The question here, however, is whether the emergence of mark-directed behavior at around 20 months of age coincides with the formation of a mental model of self.

We believe that a variety of findings from research on human infants can be offered in support of the hypothesis that mark-directed behavior is an index of a mental model of self (for an account that draws upon different studies for a similar goal, see Emde, 1983, who has described possible correlates of visual self-recognition in infants). Studies by Kagan and his colleagues (1981; Richman et al., 1983) have revealed that infants develop knowledge of their capabilities at about the same time that mark-directed behavior is first exhibited. These researchers modeled both simple and complex actions to infants of a variety of ages and found that a common response in infants 18–22 months of age was to imitate the simple actions and to exhibit distress after observing the complex actions. Kagan's interpretation of this finding is that infants at this age are aware of their own capabilities, recognize the implicit demand to imitate the actions of the experimenter and do so if they are able, but react negatively to invitations to imitate actions that they know exceed their capabilities. In other words, infants have knowledge of the self's abilities.

In a small sample of five infants, Kagan (1981) also found suggestions of an increase in self-descriptive language during free-play situations at this age, coinciding with the exhibition of distress to modeled actions. Lewis and Brooks-Gunn (1979) have reported that, at approximately the same age, infants respond to pictures of themselves by uttering their own names or personal pronouns. In their studies, these authors found that personal pronouns were used by infants only in reference to their own pictures, which suggests that personal pronouns are especially good indicators of self-awareness. Similarly, Zazzo (1982) has reported that there is a notable increase in the use of personal pronouns among infants of about 20–24 months of age, and Levine (1983) observed differences in peer interactions that were related to MSR and several other correlated indices of the sense of self (e.g., understanding pronouns, using "I" or "me" in conversation, and perceptual role taking).

Finally, the position that mirror self-recognition is only a measure of a schema of the physical qualities of self is undermined by research suggesting that the age at which mark-directed behavior emerges is unaffected by the extent of mirror exposure (e.g., Lewis & Brooks-Gunn, 1979); indeed, even infants without any experience with reflecting surfaces appear to recognize themselves in the mirror at about the same age (Priel & Schonen, 1986). Our interpretation is that infants with subjective self-awareness and a mental model of self are quickly or immediately able to assimilate visual images of themselves seen in mirrors and form schemata, and consequently demonstrate mark-directed behavior. If mark-directed behavior is an isolated skill, however, one might expect that its acquisition would require extensive practice.

Summary

To recapitulate, we have argued that the sense of self is composed of two types of self-awareness (objective and subjective) and three levels of self-concept (mental model, schemata, and personal memories). The most common assessment paradigm for the study of the sense of self in infancy, the

mirror self-recognition task, draws on both types of self-awareness and perhaps several levels of the self-concept. First, mark-directed behavior demonstrates the presence of objective self-awareness (the ability to distinguish
between self- and non–self-related classes of stimuli, a skill that is evident
even in early infancy). The patterning of emotional responses to the mirror
image that emerges at the same time as mark-directed behavior suggests that
MSR also draws upon subjective self-awareness (the knowledge that the object
of self-reflection is oneself). In terms of the self-concept, MSR certainly demands a schema or representation of the self's typical appearance, but we
have argued that a case can be made that a mental model of self is revealed
as well (although the studies marshaled in support of this position do not
demonstrate that this is so).

Social interaction and the sense of self

Despite a voluminous theoretical literature on social interaction and the
emergence and transformation of the sense of self, there has been very little
research on the topic in human infants. A few studies have attempted to
relate attachment status (securely attached or insecurely attached, as assessed
in the strange situation) to mirror self-recognition. Although we believe that
this research strategy is unlikely to yield much information, an overview of
a few studies provides a backdrop for our own work.

Attachment status, of course, refers to the quality of the affective relationship between infant and caretaker, which is thought to be determined in
large part by the responsiveness of the caretaker to the infant's needs (Damon,
1983). Responsive caretakers are more likely to have securely attached infants,
whereas caretakers having difficult relationships with their infants are more
likely to have insecurely attached infants. Although the attachment status
paradigm has served developmental psychology well, it is unclear how its set
of constructs are related to the sense of self. This confusion is amplified by
the conflicting results found in the literature. In one study of 19-month-old
infants, 9 of 10 securely attached infants recognized themselves in mirrors
while only 1 of 10 insecurely attached infants did so; this suggests that secure
attachment fosters self-recognition (Schneider-Rosen & Cicchetti, 1984; see
also Pipp, Easterbrooks, & Harmon, 1992). In another study, however, 6 of
10 infants (60%) insecurely attached at 12 months recognized themselves at
18 months, but only 11 of 27 infants (40%) securely attached at 12 months
did so (Lewis, Brooks-Gunn, & Jaskir, 1985; see also Tajima, 1984). The
conflicting pattern of results suggests that secure attachment may not be any
better than insecure attachment in supporting the development of mirror
self-recognition.

From our perspective, the problem with this line of research (besides its
failure to yield replicable findings) is that it offers no coherent conceptual
connection between the quality of the attachment relation and MSR. There
is no clear implication of attachment status for either type of self-awareness.

A secure attachment to the parent may result in the infant forming schemata of self such as "good" (as Sullivan, 1947, argued), but MSR activates most directly the infant's schemata of the physical qualities of the self. Moreover, there is no reason to believe that an emotionally secure relation by itself would facilitate (beyond an indirect contribution through its relation to the development of cognitive skills) the construction of a mental model of self in which the various schemata of self are interrelated. Consequently, if MSR does tap a mental model of self (as we have suggested), there is little reason to believe that its emergence is related to attachment status.

Social imitation and the emergence of a mental model of self

Our view is that imitation is of greater fundamental importance to the construction of a mental model of self than are the affective colorations deriving from individual differences in attachment. This perspective has its roots in Baldwin's account (1902) of the development of the self-concept across the transition from infancy to childhood, which remains one of the most influential. Guillaume (1929/1971), Mead (1909), and Piaget (1932/1965) all addressed Baldwin's insights, and more recent researchers such as Lewis and Brooks-Gunn (1979) and Meltzoff (1990) have returned to Baldwin's theory to one degree or another.

According to Baldwin, the development of self-awareness and the self-concept is the consequence of *imitation*. It will be useful here to distinguish between two products of imitation. The first concerns its cognitive qualities; imitation both reveals the limits of an infant's knowledge (one can reproduce another's actions only insofar as those actions are within the range of the self's competencies), and broadens it (new patterns of behavior are acquired). This process leads to the acquisition of new schemata of self, as the infant learns of its new capabilities. It is this aspect of imitation with which most recent researchers have been concerned.

However, as Užgiris (1983) argued, imitation has an important social facet as well. It is a means through which preverbal human infants can communicate with their caretakers. Imitation of the caretaker by the infant, and the caretaker's imitation of the infant's actions, result in a shared sense of mutuality and similarity. In turn, this mutuality and similarity provides the infant with information that permits the construction of a theory of the mind, as Royce has pointed out:

The psychological importance of imitation lies largely in the fact that in so far as a child imitates, he gets ideas about the inner meaning or intent of the deeds that he imitates, and so gets acquainted with what he really finds to be the minds of other people. (1910, p. 183)

The infant's theory of mind, then, is an exemplar based on the inferred relations among the thoughts, feelings, and actions of self and important others (see Smith & Zarate, 1992, for a discussion of exemplar and schema theories in social cognition). This exemplar of mind is refined, modified, and integrated into the evolving model of self.

Baldwin also asserted that imitation serves to accentuate subjective self-awareness by highlighting the distinction between one's own movements and those of another. The infant notices that despite the objective visual or auditory similarity between the actions of self and parent, it is only the self's actions that are accompanied by feelings of volition and proprioception.

To summarize, Baldwin's position is that imitation is a major force in the early development of self-awareness and the self-concept. Through imitation, the infant is able to:

1. incorporate into the self-concept features of others that become schemas of the self;
2. infer the connection between thoughts, emotions, and actions and consequently form a mental model of self; and
3. articulate differences between self and other.

Table 9.1 summarizes Baldwin's views on the influence of imitation on the sense of self, and notes as well some possible contributions of imitation to the facets of the experience of self Baldwin ignored.

A preliminary study

In the preliminary research that we have conducted, we have attempted to examine the relationship between the mental model of self and social imitation. One simple hypothesis derived from Baldwin's theory has been examined in our work: Infants who are more inclined toward social imitation ought to develop a mental model of self more quickly than those who are not. We have begun to examine this hypothesis in a small longitudinal sample of human infants and their mothers.

Method

Subjects. Initially, 23 infant–mother dyads were recruited for the study. Most were from middle-class or upper-middle-class backgrounds. The sample used in the analyses that follow is composed of those infants who recognized themselves, as indexed by the mirror-and-rouge task, by 24 months of age, and their mothers (19 dyads).

Procedure. The study was longitudinal in design, with sessions scheduled at approximately 6-week intervals, beginning when the infant was 14 months of age[3] and concluding 10 months later. The first four sessions (with mean ages in months and weeks of 14 : 0, 15 : 1, 16 : 3, and 18 : 1, respectively) are particularly relevant because it is during this time that the mental model of self is constructed.

Each session had several components: (1) assessments of the mental model

Table 9.1. *Influences of interpersonal imitation on self-awareness and the self-concept*

Objective self-awareness	Subjective self-awareness	Mental model or theory of self	Representations or schemata of self	Autobiographical episodes or personal memories
Imitation (except in neonates) results in an increase, because infants monitor their own actions to match those of another; sophisticated forms of imitation (of the types found only in humans and great apes) focus attention on the imitator's goals (Mitchell, 1987, 1993)	Imitation reveals a distinction between the actions of another and those of self; the latter are accompanied by particular intentions, affects, or sensations (Baldwin, 1902)	Imitation articulates the connection between actions of others and their intentions or affects; by imitating others, one is able to infer their states of mind, and to formulate a model of the relation of actions to cognition and affect; this model is both applied to, and influenced by, awareness of one's own actions, thoughts, and affects (Baldwin, 1902; Meltzoff, 1990; Royce, 1910)	Imitation leads to the acquisition of specific competencies, which are represented by schemata (Baldwin, 1902; Piaget, 1945/1962)	Imitation consolidates autobiographical episodes or personal memories, making them particularly accessible for reenactment and/or retrieval at a later time (Meltzoff, 1990)

of self, (2) cognitive imitation tasks, and (3) a free-play session. The sessions took place in each infant's home, and were videotaped.

Mental Model of Self. Two measures of the mental model of self were used;[4] the one of interest here is the mirror self-recognition task.[5] The age at which mark-directed behavior in the MSR task was first evidenced is of central importance in the analyses to follow.

Cognitive Imitation. Two cognitive imitation tasks were presented to the infants. These tasks were designed to assess the infant's capacity or ability to imitate complex actions. In terms of the distinction offered earlier, then, these are imitation tasks of the nonsocial variety. In the first task, the experimenter modeled short and long sequences of actions involving toy animals to the infant. For instance, a short sequence might be to hang a monkey on a tree, and a long one might involve having several different animals perform specific tasks like carrying food in order to prepare for an animal picnic. For these tasks, the dependent measure of interest was the number of steps, or component actions in the sequence, the infant was able to accomplish.

A second cognitive imitation task was borrowed from Case (Case & Khanna, 1981), and was intended to measure what he calls *central processing load*, a concept closely related to Pascual-Leone's (1970) construct of *m*-space. The experimenter modeled four increasingly difficult constructions out of blocks: The first involves placing one black on top of another, the second results in the formation of a right angle with three blocks, the third is the construction of a three-block arch, and the last involves the formation of a double arch construction using five blocks.

Social Imitation. Each mother–infant dyad played with a set of toys provided by the experimenter for 10 min. To ensure interest in the task, sets of toys not available in the child's home were selected (a zoo set, a Western town set, or a house). The videotapes were later coded for what we call *gestural imitation*: the reproduction of a novel action. The scoring criteria for gestural imitation are intended to emphasize the social qualities of imitation (e.g., Užgiris, 1983):

1. The imitating partner must observe the modeled behavior from beginning to end.
2. The modeling partner must pause at the completion of the behavior as an indication that the other is expected to perform the action, or communicate the expectation verbally (e.g., "you do it").
3. The modeled behavior must not have occurred in the previous minute.
4. The imitating partner must perform the action within one minute of the observation of the model.

Although gestural imitation by both the mother and the infant was coded, there was too little imitation by the mother across the sample to analyze these data. The index used in the analyses that follow is the percentage of actions modeled by the mother during the course of the play session that were imitated by the infant (interrater reliability for this index was calculated by having two raters independently code a randomly selected session for eight different dyads; the correlation between the two raters for the percentage of actions modeled by the mother that were imitated by the infant across a session was $r = .92$).

The play sessions were also coded for instances in which the mother did one of the following:

1. performed what her infant verbally requested,
2. accepted an object or toy offered to her by the infant, or
3. looked at the object to which her infant pointed.

All three behaviors were considered to be indices of maternal responsiveness. The percentage of actions during a session that met these criteria and to which the mother responded is the index used in the following analyses.

Results and discussion

The average age for mirror self-recognition ($M = 18 : 1$) is approximately the same as those reported by other investigators (e.g., Lewis & Brooks-Gunn, 1979), which suggests that the repeated testing did not result in mark-directed behavior emerging at an earlier age. More germane to the hypotheses of this study, the results indicate that a high level of gestural imitation by the infant in the free-play situation of age is correlated with mark-directed behavior in the mirror-and-rouge task at younger ages. The correlations between the age of self-recognition (as evidenced in the mirror-and-rouge task) and the percentage of maternal gestures imitated by the infant for the four sessions are .17, n.s.; $-.42$, $p < .05$ (one-tailed); $-.43$, $p < .05$ (one-tailed); and $-.60$, $p < .01$ (one-tailed). The last three correlations indicate that high scores for imitation are associated with self-recognition at younger ages. Because none of the infants had recognized themselves by the second session, the pattern of correlations presented here suggests that high levels of gestural imitation precede the acquisition of self-recognition, and do not merely accompany it. (If the correlations between age of self-recognition and gestural imitation at Sessions 3 and 4 are calculated *excluding* those infants who concurrently recognized themselves, the same pattern obtains: Session 3, $r = -.49$, $p < .06$ [one-tailed, 11 d.f.], and $r = -.82$ $p < .05$ [one-tailed, 5 d.f.].) These findings, then, are consistent with the hypothesis that high levels of infant social imitation between the ages of 15 and 19 months are related to the emergence of self-recognition behavior.

Figure 9.1 presents the average percentage of maternal gestures imitated by the infants at the sessions before, during, and following the session at which visual self-recognition was first evidenced. Note that age is not the metric of the x-axis; the reference point is the first session for which evidence of self-recognition was found. Consequently, if an infant recognized the self

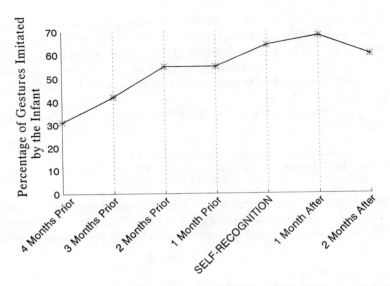

Figure 9.1. Percentage of the mother's gestures imitated at sessions preceding, during, and following the session at which visual self-recognition was first evidenced.

in the mirror at 18 months of age, the imitation data for the session preceding that session was collected at approximately 16.5 months; if the infant recognized the self at 16.5 months of age, the imitation data for the preceding session was collected at 15 months of age, and so on. This means that not all infants contributed data to each point on the graph (e.g., an infant who recognized the self at 16.5 months could not contribute data for the point representing three sessions preceding the session self-recognition was first evidenced, because the infants began participating in the study at 14 months of age).

The value of Figure 9.1 is that it provides a perspective on gestural imitation in relation to visual self-recognition. It appears that the infant's tendency to imitate its mother increases substantially in the months immediately preceding self-recognition, then remains relatively constant once self-recognition has been achieved.

Neither of the cognitive imitation tasks, nor the maternal responsiveness index, was a predictor of the age at which self-recognition was first evidenced. At first glance, the failure to find a relationship between cognitive tasks and visual self-recognition may appear to contradict previous findings suggesting that the cognitive factors (such as mental age) are good predictors of the acquisition of mark-directed behavior (e.g., in children with autism [Ferrari & Matthews, 1983] and Down's syndrome [Mans, Cicchetti, & Sroufe, 1978]). Although it seems certain that intact cognitive functioning is necessary for the development of a theory of self, attempts to identify concrete correspondences between the mental model of self and specific cognitive achievements such as object permanence have not been successful (Lewis & Brooks-Gunn, 1979; but see Mitchell, 1993).

In summary, the results of our preliminary investigation suggest that interpersonal imitation is related to the age at which mark-directed behavior is first manifested. Although it remains for future research to replicate these findings in a larger sample, and to contrast the effects of social imitation with other possible types of interactions such as pretend play and maternal mirroring, these initial results are consonant with the position that the construction of a mental model of self may have its origins in social imitation, much as Baldwin suggested.

Imitation as the process through which the mental model of self is formed is also supported by findings from comparative work on imitation in chimpanzees, orangutans, and monkeys. Chimpanzees and orangutans, which are capable of visual self-recognition, imitate others of their species frequently; monkeys, apparently without the ability to recognize the self in the mirror, rarely imitate others (Mitchell, 1993; Parker & Gibson, 1990; Visalberghi & Fragaszy, 1990).

One very significant problem remains with the Baldwinian account that we have offered: If interpersonal imitation is the central mechanism involved in the acquisition of the mental model of self, why does it take so long for infants to construct it? The work of Meltzoff (e.g., 1990) clearly reveals that infants have well-developed capabilities to imitate long before they first demonstrate visual self-recognition.

Our own tentative explanation is based on the widely accepted assumption that infants are, at first, incapable of almost every type of representational thought (e.g., Case & Khanna, 1981), perhaps due to their neurological immaturity (e.g., Fischer, 1987). The consequence is that the information about the self that is generated through imitation in the 1st year of life cannot be synthesized and integrated to form a mental model of the self. However, as representational skills emerge toward the end of the 1st year and through the middle of the 2nd, social imitation becomes more frequent because the self-relevant information that is generated can be integrated into a mental model of self.

Summary

Our goal in this chapter has been to examine the sense of self and its relation to one form of social interaction, imitation. The first part of the chapter articulated some important distinctions among the facets of the experience of self, and used these distinctions to explore the meaning of visual self-recognition.

Although a few investigators have tried to link MSR to attachment status, their studies have yielded conflicting or ambiguous results. For instance, some research seems to suggest that securely attached infants develop visual self-recognition at an earlier age than insecurely attached infants, while other investigations find just the opposite. The failure to find compelling connections, we suggested, reflects both theoretical and methodological problems. Baldwin's (1902) notion that interpersonal imitation is central to the development of the mental model of self appears to hold some promise. In the small longitudinal study of infants presented here, it appears to be the case that infants who frequently imitate their mothers develop visual self-recognition at younger ages than those who are less inclined to imitation. This suggests that infants utilize social imitation to articulate the differences and commonalities between themselves and others, information that is then integrated into a model of self.

Imitation, we have suggested, is most influential in the construction of a mental model of self, although we have outlined its possible connections to the other facets of the experience of self (Table 9.1). This should not be taken to imply that imitation is "the" form of social interaction through which the sense of self is acquired and transformed; our position is that imitation is of fundamental importance for only one facet of self (the mental model of self) at one developmental period (late infancy). Self-awareness and the self-concept are complex constructs that are subject to multiple influences across the life span, and nothing is gained by reducing them to caricatures.

We began this chapter with the observation that the ontogenesis of self-awareness and of the self-concept is a formidable mystery; the discussion and results presented here do not resolve it. Nonetheless, progress can be made toward understanding these topics, and we have tried to illustrate the process through which advances can be made. Knowledge about the development of self-awareness and the self-concept, and their relations to social interaction, can cumulate only if these aggregate constructs are separated into their constituents (i.e., the facets of self-awareness and the self-concept, and types of social interaction like imitation). With these distinctions, it becomes possible to move forward from fruitless arguments concerning when humans become "self-aware" or have a "self-concept," and discover the ways in which the facets of these phenomena are transformed through social interaction in the course of life.

Acknowledgment

The research described in this chapter was supported by an award from the Rutgers University Research Council. The chapter was written while the first author was a visiting scientist at the program on Conscious and Unconscious Mental Processes of the John D. and Catherine T. MacArthur Foundation.

Notes

1. There is considerable controversy concerning the extent to which self-awareness and the self-concept are characteristic of the experience of people in non-Western cultures. (For discussions of this issue see Damon & Hart, 1988; Hart & Edelstein, 1992; Rosaldo, 1984.)

2. Moreover, these researchers assert that the sense that the self is exposed requires a mental model of self, not merely a more limited schema of the physical qualities of self. Although no compelling evidence is presented in defense of this claim, the issue of the connection between self-awareness and the self-concept is central.
3. Because of technical difficulties, data are available for only 12 dyads at the earliest testing point. Scheduling difficulties and infant illnesses resulted in missing data for two dyads at the second session, and one dyad at each of the next two sessions.
4. The second measure, from Kagan (1981), was descibed earlier. Unfortunately, too few of our subjects evidenced the criterion behavior (distress when expected to imitate complex behaviors) to allow meaningful analyses of the relationship of this measure of the representational self-concept to mark-directed behavior or to social imitation.
5. As noted above, 19 of 23 infants exhibited mark-directed behavior by 24 months of age. This percentage is considerably higher than that observed in cross-sectional laboratory studies (e.g., Johnson, 1983). Our interpretation is that the infants in our study were much more comfortable with the procedure due to its longitudinal nature, combined with testing in the infant's own home, and consequently were more likely to manifest mark-directed behavior.

References

Amsterdam, B. (1972). Mirror self-image reactions before age two. *Developmental Psychobiology,* 5, 297–305.
Baldwin, J. M. (1902). *Social and ethical interpretations in mental life.* New York: Macmillan.
Bertenthal, B., & Fischer, K. (1978). Development of self-recognition in the infant. *Developmental Psychology, 14,* 44–50.
Brewer, W. (1986). What is autobiographical memory? In D. Rubin (Ed.), *Autobiographical memory* (pp. 25–49). Cambridge University Press.
Butterworth, G. (1990). Self-perception in infancy. In D. Cicchetti & M. Beeghly (Eds.), *The self in transition: Infancy to childhood* (pp. 119–137). Chicago: University of Chicago Press.
Case, R., & Khanna, F. (1981). The missing links: Stages in children's progression from sensorimotor to logical thought. In K. Fischer (Ed.), *Cognitive development* (pp. 21–32). San Francisco: Jossey-Bass.
Craik, K. (1934). *The nature of explanation.* Cambridge University Press.
Damon, W. (1983). *Social development.* New York: Norton.
Damon, W., & Hart, D. (1988). *Self-understanding in childhood and adolescence.* Cambridge University Press.
   (1992). Social understanding, self-understanding, and morality. In M. Bornstein & M. E. Lamb (Eds.), *Developmental psychology: An advanced textbook* (3rd ed., pp. 421–464). Hillsdale, NJ: Erlbaum.
Duval, S., & Wicklund, R. (1972). *A theory of objective self-awareness.* New York: Academic Press.
Emde, R. (1983). The prerepresentational self and its affective core. *Psychoanalytic Study of the Child, 38,* 165–192.
Ferrari, M., & Matthews, W. (1983). Self-recognition deficits in autism: Syndrome-specific or general developmental deficit? *Journal of Autism and Developmental Disorders, 13,* 317–324.
Fischer, K. (1987). Relations between brain and cognitive development. *Child Development, 58,* 623–632.
Freeman, W., & Watts, J. (1942). *Psychosurgery: Intelligence, emotion, and social behavior following prefrontal lobotomy for mental disorder.* Baltimore: Charles C. Thomas.
Gallup, G. (1977). Self-recognition in primates: A comparative approach to the bidirectional properties of consciousness. *American Psychologist, 32,* 329–338.
Guillaume, P. (1929/1971). *Imitation in children.* Chicago: University of Chicago Press.
Hart, D. (1992a). *Becoming men: The development of aspirations, values, and adaptational styles.* New York: Plenum.
   (1992b). *Self-awareness and the self-concept: An integrational model.* Unpublished paper; available from the author.

Hart, D., & Damon, W. (1986). Contrasts between the understanding of self and an understanding of others. In R. Leahy (Ed.), *The development of self* (pp. 151–178). New York: Academic Press.

Hart, D., & Edelstein, W. (1992). Self-understanding development in cultural context. In R. Lipka and T. Brinthaupt (Eds.), *The self: definitional and methodological issues* (pp. 291–322). Albany, NY: SUNY Press.

Hart, D., Kohlberg, L., & Wertsch, J. (1987). The developmental social-self theories of James Mark Baldwin, George Herbert Mead, and Lev Semenovich Vygotsky. In L. Kohlberg (Ed.), *Child psychology and childhood education: A cognitive-developmental view* (pp. 223–258). New York: Longman.

James, W. (1892/1961). *The principles of psychology*. New York: Harper & Row.

Johnson, D. (1983). Self-recognition in infants. *Infant Behavior and Development, 6*, 211–222.

Kagan, J. (1981). *The second year: The emergence of self-awareness*. Cambridge, MA: Harvard University Press.

Levine, L. (1983). Mine: Self-definition in 2-year-old boys. *Developmental Psychology, 19*, 544–549.

Lewis, M., & Brooks-Gunn, J. (1979). *Social cognition and the acquisition of self*. New York: Plenum.

Lewis, M., Brooks-Gunn, J., & Jaskir, J. (1985). Individual differences in visual self-recognition as a function of mother–infant attachment relationship. *Developmental Psychology, 21*, 1181–1187.

Lewis, M., Sullivan, M., Stanger, C., & Weiss, M. (1989). Self-development and self-conscious emotions. *Child Development, 60*, 146–156.

Luria, A. R. (1976). *Cognitive development: Its cultural and social foundations*. M. Lopez-Morillas & L. Solotaroff (Trans.). Cambridge, MA: Harvard University Press.

Mahler, M., Pine, F., & Bergman, A. (1975). *The psychological birth of the human infant*. New York: Basic.

Mans, L., Cicchetti, D., & Sroufe, A. (1978). Mirror reactions of Down's syndrome infants and toddlers: Cognitive underpinnings of self-recognition. *Child Development, 49*, 1247–1250.

Markus, H. (1977). Self-schemata and processing information about the self. *Journal of Personality and Social Psychology, 35*, 63–78.

Mead, G. H. (1909). Social psychology as counterpart to physiological psychology. *Psychological Bulletin, 6*, 401–408.

(1934). *Mind, self, and society*. Chicago: University of Chicago Press.

Medin, D. (1989). Concepts and conceptual structure. *American Psychologist, 44*, 1469–1481.

Meltzoff, A. (1990). Foundations for developing a concept of self: The role of imitation in relating self to other and the value of social mirroring, social modeling, and self practice in infancy. In D. Cicchetti & M. Beeghly (Eds.), *The self in transition: Infancy to childhood* (pp. 139–164). Chicago: University of Chicago Press.

Mitchell, R. W. (1987). A comparative-developmental approach to understanding imitation. In P. P. G. Bateson & P. H. Klopfer (Eds.), *Perspectives in ethology* (Vol. 7, pp. 183–215). New York: Plenum.

(1993). Mental models of mirror self-recognition: Two theories. *New Ideas in Psychology, 11*, 295–325.

Neisser, U. (1988). Five kinds of self-knowledge. *Philosophical Psychology, 1*, 35–59.

Nozick, R. (1981). *Philosophical explanations*. Cambridge, MA: Harvard University Press.

Parker, S., & Gibson, K. (Eds.) (1990). *"Language" and intelligence in monkeys and apes*. Cambridge University Press.

Pascual-Leone, J. (1970). A mathematical model for the transition role in Piaget's developmental stages. *Acta Psychologica, 32*, 301–345.

Piaget, J. (1932/1965). *The moral judgment of the child*. M. Gabain (Trans.). New York: Free Press.

(1945/1962). *Play, dreams, and imitation in childhood*. New York: Basic.

Pipp, S., Easterbrooks, A., & Harmon, R. (1992). The relation between attachment and knowledge of self and mother in one to three year old infants. *Child Development, 63*, 738–750.

Priel, B., & de Schonen, S. (1986). Self-recognition: A study of a population without mirrors. *Journal of Experimental Child Psychology, 41*, 237–250.

Richman, C., Novack, T., Price, C., Adams, K., Mitchell, R., Reznick, J., & Kagan, J. (1983). The consequences of failing to imitate. *Motivation and Emotion, 7*, 157–167.

Rogers, C. (1951). *Client-centered therapy.* Boston: Houghton Mifflin.

Rosaldo, M. (1984). Toward an anthropology of self and feeling. In R. Shweder & R. Levine (Eds.), *Essays on mind, self, and emotion* (pp. 137–157). Cambridge University Press.

Royce, J. (1910). *Studies of good and evil.* New York: Appleton.

Samuels, C. (1986). Bases for the infant's developing self-awareness. *Human Development, 29*, 36–48.

Schneider-Rosen, K., & Cicchetti, D. (1984). The relationship between affect and cognition in maltreated infants: Quality of attachment and the development of visual self-recognition. *Child Development, 55*, 648–658.

Smith, E., & Zarate, M. (1992). Exemplar-based model of social judgment. *Psychological Review, 99*, 3–21.

Stern, D. (1985). *The interpersonal world of the infant.* New York: Basic.

Sullivan, H. (1947). *Conceptions of modern psychiatry.* New York: Norton.

Tajima, N. (1984). Infants' temperamental disposition, attachment, and self-recognition in the first 20 months of life. *Annual report of the research and clinical center for child development, 82–83*, 71–80.

Užgiris, I. (1983). Imitation in infancy: Its interpersonal aspects. In M. Perlmutter (Ed.), *Minnesota symposium on child psychology* (Vol. 17, pp. 1–33). Hillsdale, NJ: Erlbaum.

Visalberghi, E., & Fragaszy, D. M. (1990). Do monkeys ape? In S. T. Parker & K. R. Gibson (Eds.), *Language and intelligence in monkeys and apes: Comparative developmental perspectives* (pp. 247–273). Cambridge University Press.

Zazzo, R. (1982). The person: Objective approaches. In W. W. Hartup (Ed.), *Review of child development research* (Vol. 6, pp. 247–290). Chicago: University of Chicago Press.

10 Minds, bodies, and persons: Young children's understanding of the self and others as reflected in imitation and theory of mind research

Alison Gopnik and Andrew N. Meltzoff

In the 1946 film noir *Lady in the Lake*,[1] the hero, Philip Marlowe, is represented by the camera's eye, and by a voice-over narration. We hear his thoughts and see his vision of others: the heroine's face coming into close-up for a kiss, the villain's scowl, his own feet upended as he falls to the ground from a Mickey Finn, his unshaven face briefly in a mirror after he regains consciousness.

The film graphically presents a common philosophical and psychological view of our understanding of ourselves and other people. On this view, we have direct access to our own internal mental states, our emotions, beliefs, desires, sensations, and so on, much like Philip Marlowe's internal monologue in the movie. However, we have only indirect and partial access to our appearance and behavior; our bodies, after all, are only peripherally part of our field of vision. Like Marlowe we can see only small parts of our selves. Conversely, we seem to have direct access to the physical appearance and behavior of others; but we have no access to their internal states, either their "feel" for their own bodies or their internal mental states. Although Marlowe can see the heroine's face and hear her words, he has no way of knowing whether her professed affection for him is genuine, or whether her internal monologue is like his (an important plot device in 1940s movies).

Yet the odd thing is that *Lady in the Lake* is so phenomenologically unreal. We don't feel that we consist of thoughts, feelings, feet, and a blurry edge of nose, while others consist of a bag of skin moving over ground. Instead we see both ourselves and others as persons, minds and bodies intertwined into a creature that both feels and behaves. This has seemed to many to pose a deep philosophical and developmental problem. How do we ever escape from solipsism? How do we defy Descartes' demon and leap out of the mirrored case of Leibniz's monads? In short, how do we accomplish the mapping from internal states to behavior that results in a unified sense of persons?

166

The problem seems to run in both directions. In one direction it concerns the way we perform this mapping for ourselves. How do we discover that our minds have bodies that have the same appearance as the bodies of others? The Gallup mirror self-recognition task (Gallup, 1970; Lewis & Brooks-Gunn, 1979) has seemed to many to be particularly significant because it provides a relevant developmental measure. It suggests that before the age of 18 months we do not, in fact, know that our internal proprioceptive body image is the same as the physical appearance we see in the mirror; we develop this knowledge in the 2nd year of life. Perspective-taking tasks, both in the classic Piagetian tradition and in recent "theory of mind" research, seem relevant to the opposite puzzle. How do we discover that other people's bodies have minds? The suggested solution, both in the classic Cartesian philosophical tradition and in the notion of "perspective-taking" in the developmental literature, is that we reason by a kind of analogy from ourselves, we see that our mental states cause our behavior and then infer that others with similar behavior have similar mental states. In both cases the assumption is that we have one sort of information given perceptually, minds for the self, bodies for the other, and another given by a kind of inference, bodily appearances for the self, minds for the other.

We want to suggest that this picture is radically mistaken. We do have information about persons both through perception and inference; but in both cases there is a kind of symmetry in the information we have about ourselves and others. Certain kinds of information about persons, particularly about the structure of their movements and gestures, are given innately by the perceptual system. However, this information includes both kinesthetic and proprioceptive sensations (relevant to the self) and visually perceived bodily expressions and movements (relevant to the other). These two types of information are mapped onto each other innately. Other kinds of information, particularly information about the representational character of mental states, are not given perceptually but must be constructed through a process of inference in the preschool years. They constitute a kind of theory of mind (ToM). This theory, however, is constructed equally on the basis of information about the self and about others, and is equally applicable to both. When this theory is wrong or incomplete, children are just as inaccurate in reporting their own mental states as they are in reporting the mental states of others. Moreover, we will suggest that our innate perceptual capacity to map our own internal bodily sensations onto the seen bodily appearance of others may form the foundation for our later more sophisticated theory of persons. Our deep-seated assumption that we are like other people in important ways carries over to our later theoretical understanding of ourselves and others.

Thus, both perceptually and in inference, mapping the self and other is a nonproblem. The purported psychological phenomenology of *Lady in the Lake* feels strange and unreal because it *is* strange and unreal, just as a sequence of our retinal images would look nothing like our perceived visual world. Our understanding of persons comes from both perception and inference, but both apply equally to ourselves and others.

Some of the most direct evidence for the first part of this claim – that there are innate mappings from self to others – comes from recent studies of imitation (see Meltzoff & Gopnik, 1993). Imitation is a particularly interesting and potent behavior because it automatically implies a certain recognition of the similarities between the self and the other. Moreover, the imitation of facial gestures, in particular, suggests a mapping from internally felt kinesthetic sensations to physical appearances and movements. To imitate another person poking out his tongue, for instance, I have to recognize the equivalence between the spectacle of his face, an oval with a protruding moving element in its lower portion, and the set of feelings that make up my own kinesthetically perceived but invisible tongue. Similarly, to recognize that another person is imitating you, you must map your gesture onto their movement. We will show that there is an innate capacity for facial imitation, and that recognition that others are imitating you is possible long before mirror self-recognition is.

The second part of our argument comes from recent studies of what has been called "children's theories of mind," particularly children's understanding of intentional states like belief and desire. An essential feature of persons is that they are characterized in terms of what is sometimes called "folk psychology." We believe that persons (though not objects) have beliefs and desires, emotions and intentions, that these internal mental states lead to their actions, and that these states have a representational character. How do we come to believe this? Recent evidence suggests that we develop a succession of different, progressively more complex and accurate theories of the mind during the preschool years (Gopnik & Wellman, 1992). More significantly, these theories seem to be equally applicable to the self and others. Even though we seem to perceive our own mental states directly, this direct perception is an illusion. In fact, our knowledge of ourselves, like our knowledge of others, is the result of a theory, and depends as much on our experience of others as on our experience of ourselves (see Gopnik, 1993).

So the director's concept in *Lady in the Lake*, and the philosophical view it embodies, is based on two mistakes. It assumes that our knowledge of the visual appearance of others is perceptual, but that our knowledge that we have a similar appearance is inferred. (It has been proposed that the latter is constructed at around 18 months of age.) In fact, the evidence suggests that some knowledge of our own visual appearance is innately mapped onto our internal kinesthetic sensations, right from the very earliest phases of infancy. On the other hand, the standard view of "perspective taking" assumes that the sort of knowledge of our own internal states that is required to solve ToM tasks is given directly, through a kind of internal perception, and that our knowledge of the mental states of others must be inferred. In fact, the evidence suggests that our knowledge of some aspects of our own mind is based on inference rather than perception.

Imitation, identification, and the developmental origins of theory of mind

The idea that infants have an innate rapport with other people is, of course, not new. It is part of certain psychoanalytic traditions and has been taken up in Trevarthen's (1980) idea of "primary intersubjectivity" and in extensions of this by Stern (1985), Bruner (1983), and Hobson (1989). It reflects our intuitive judgment that infants somehow behave as if they understand persons, and our sense that our own interactions with infants have a richness and intimacy that implies a deep interpersonal connection.

However, it has not been clear just what these claims amount to. The authors above have seen the roots of intersubjectivity in the early temporal coordination of infant and adult behavior, the "conversational dances" that infants and caretakers perform. Nevertheless, there seems no clear reason why these behaviors, by themselves, should lead infants to think of other people as similar to themselves in deep ways. Similarly, some believe that normal infants are innately endowed with a special attentiveness to the human face (Johnson & Morton, 1991; Meltzoff & Moore, 1993). In themselves, however, such facial pattern detectors also do not provide a link between the self and the other. The infant might see the adult face as a particularly interesting entity, but because infants cannot see their own facial features, why should they think of the adult as someone like themselves?

We will argue that infants' initial "like-me" experiences, their sense that they are the same as other people, are rooted in their perception of bodily movement patterns and postures. Infants become aware of their own body movements through the internal sense of proprioception and can detect cross-modal equivalences between those movements as felt and the movements they see others perform. Indeed, one reason normal infants preferentially attend to other people may be their sense that those entities are "like me." Without this, other humans might have salient visual or temporal features, but they would not have the special meaning they do in our world. It is this primordial connection between self and other that we wish to examine in more detail.

Before Meltzoff and Moore's (1977) work on neonatal imitation, there was no reason to suppose that young infants could apprehend cross-modal equivalences between body movements as felt in the self and body movements as seen in others. Indeed, orthodox developmental theories explicitly denied this capacity to young infants, describing them as "solipsistic," "radically egocentric," and even as being in a phase of "normal autism" (Mahler, Pine, & Bergman, 1975). These theories proposed that the imitation of facial actions emerged only at about 1 year of age (Piaget, 1962). Facial imitation was classically viewed as a late achievement because infants cannot see their own faces. How can infants possibly match a gesture they see with an action of their own that they cannot see? How can infants come to bridge the gap between visible and invisible experiences? The gap seemed so wide that most theorists embraced the idea that facial imitation first became possible at about 1 year. The implication is that some type of associative or inferential process must take place that allows one to map one's own felt face and the seen faces of others, just as one might need to learn by association and inference that a particular rattle makes a particular sound.

Meltzoff and Moore's (1977) discovery of early facial imitation alerted theorists that something was amiss in this conception of infants' initial understanding of persons. Meltzoff and Moore found that 12–21-day-old infants imitated tongue protrusion, mouth opening, and lip protrusion. Several features of these studies are particularly important. The imitative effects were surprisingly specific. Genuine imitation was demonstrated because infants responded differentially to two types of lip movements (mouth opening vs. lip protrusion) and two types of protrusion actions (lip protrusion vs. tongue protrusion). The response was not a global or a general reaction to the presence of a human face or even to the presence of protrusions or mouth movements. Rather, the infants differentially imitated specific types of lip movements and specific types of protrusions.

Meltzoff and Moore also provided evidence that early matching was not simply restricted to some sort of immediate reflexive mimicry, a kind of Gibsonian "resonance" in which perception of human acts somehow "directly" led to their motor production with no intervening mediation. In one study a pacifier was put in infants' mouths as they watched the display so that they could observe the adult demonstration but not duplicate the gestures. After the infant observed the display, the adult assumed a passive face pose and only then removed the pacifier. In fact, once the pacifier was removed, the infants imitated the earlier displays. The finding suggests that imitation, even this very early imitation, could be mediated by memory of the absent display (Meltzoff & Moore, 1977, Study 2, 1989).

Other data show that early imitation is not well characterized as a simple reflex. In particular, the imitative response was not simply triggered by the sight of the adult display. The first responses of the infants often involved the correct body part but only approximated the adult's act. Infants would move their tongues but not produce full tongue protrusions. Infants then appeared to home in on the detailed match, gradually correcting their responses over successive efforts to more exactly correspond to the details of the display (Meltzoff & Moore, in press). For these reasons and others it seems more accurate to think of early imitation as intentional matching to the target provided by the other, rather than as a rigidly organized, purely reflexive response (Meltzoff & Moore, 1992).

The subjects in the previous studies were 12–21 days old. At first glance this seems young enough to justify claims about an innate capacity; but perhaps neonates had been conditioned to imitate during the first weeks of life. Perhaps imitation was dependent upon early mother–infant interaction. An experimentalist's job is never done . . . well, almost never. To resolve the point,

Meltzoff and Moore (1983) tested 40 newborns in a hospital setting. The average age of the sample was 32 hours; the youngest infant was only 42 min old. The results showed that the newborns imitated both of the gestures shown to them, mouth opening and tongue protrusion. Nativist claims are, of course, commonplace in the recent literature, but there are few cognitive capacities that have actually been demonstrated to be present at birth. You can't get much younger than 42 min old. These data directly demonstrate that a primitive capacity to imitate is part of the normal child's biological endowment.

The discovery of early infant imitation was initially considered surprising. The findings have now been replicated and extended in well over 20 different studies from a dozen independent labs. Positive results have been reported in the United States, Canada, France, Switzerland, Greece, Sweden, Israel, and even in babies tested in rural Nepal. In brief, the basic phenomenon reported by Meltzoff and Moore has now been documented by independent investigators, in different settings, using a variety of different procedures. At a behavioral level, the findings are secure. Attention has now shifted from debates about the existence of early imitation to explorations of the mechanisms underlying this behavior and its role in development.

Meltzoff and Moore suggested that imitation is based on infants' capacity to register equivalences between the body transformations they see in other people and the body transformations they feel themselves make. On this account early imitation involves a kind of cross-modal matching. Infants can, at some primitive level, recognize an equivalence between the acts they see others do and the acts they do themselves. There appears to be a primitive and foundational "body scheme" that allows the infant to unify the seen acts of others and their own felt acts into one framework. The infant's own facial gestures are invisible to them, but they are monitored by proprioception. Conversely, the adult's acts are not felt by proprioception but they can be seen. Infants' cross-modal capacities allow them to make this linkage.

Meltzoff and Moore have argued that early imitation is one manifestation of a larger capacity, an amodal representational system, that can also be tapped by other tasks including cross-modal object tests and work on lip–voice correspondences for speech (Bower, 1982; Kuhl & Meltzoff, 1982; Meltzoff, 1990a, 1993; Meltzoff & Borton, 1979). In the Meltzoff and Borton experiment, for example, 1-month-old infants were given one of two pacifiers to suck, a smooth ball or a knobby one. The infants could not see the pacifiers, but merely feel them. The infants were then presented visually with replicas of the two shapes (so that at olfactory cues were eliminated); they could see them but not feel them. Infants consistently looked longer at the shape that matched the one they had sucked, which suggests that infants are equipped to detect correspondences between tactile and visual information. This cross-modal coordination is the underlying mechanism for imitation.

In order for infant imitation to play a central role in interpersonal development, infants would need not only to map bodies to bodies, but also to map larger units – like adults' behavior with objects – from the seen world to their own actions. They would also need to keep adult models in mind for long periods, so that they could imitate not only while the adult is present but also after significant delays. Meltzoff has recently explored deferred imitation of some actions on objects, and the results here, too, suggest that imitation from memory is more powerful and is developed earlier than classical theory supposed.

Piaget claimed that this type of deferred imitation was not possible until around 18 months, during sensorimotor Stage 6 (Piaget, 1962). However, the new data demonstrate that infants as young as 9 months can imitate actions on objects after a 24-hour delay (Meltzoff, 1988a). Other studies used an even longer time interval, 1 week. Importantly, the studies did not involve just familiar actions, but also the imitation of novel acts after this delay. For example, in Meltzoff (1988b) infants saw a small wooden box with a trans- lucent orange plastic panel for a top surface, and then saw the experimenter bend forward and touch the panel with the top of his forehead. After a week they returned to the laboratory and were given the objects one at a time to play with. The results provided clear evidence for deferred imitation. Of the 12 children in the imitation group, 11 duplicated three or more target behaviors on Day 2, whereas only 3 of 24 control subjects did so ($p < .0001$). The data showed that these infants duplicated the novel act from memory. In the control conditions not one infant produced the novel act, in contrast to 67% of the infants who had seen it modeled a week earlier ($p < .0001$).

We have suggested that early imitation bridges the behavior of the other with the infant's own behavior and internal states. If this bridge exists, it should bear two-way traffic: Infants should not only imitate adults, but should also recognize when they themselves are being imitated by adults. It is, after all, true in this case that the infant's behavior and the adult's are equivalent. A series of experiments were conducted in which an adult purposely imitated the child, with the goal of determining if the child could recognize that his or her own behavior was being mimicked (Meltzoff, 1990b). In a first experiment two adults sat across a table from each subject (all of whom were 14 months old). All three participants were provided with replicas of the same toys. Everything the child did with his toy was directly mimicked by one of the adults who had been assigned as the imitator. The second adult sat passively, holding the toy loosely on the table top. Children preferred to look at the imitating adult and also smiled at him more. Moreover, children would in- vestigate this relationship between the self and other by experimenting with it. For example, children might modulate their acts by performing sudden and unexpected movements to check if the imitating adult was still conform- ing to their actions. This is a way of "catching the adult out," a way of experimenting with the relationship between self and world.

It is, of course, possible that the children simply preferred an active experi- menter to a passive one, or even preferred an experimenter whose actions were contingent on the child's own actions. To distinguish these alternatives a second study was conducted in which the purely temporal aspects of the contingency were controlled by having both experimenters act at the same time. This was achieved by having three predetermined pairs of target actions. Both experimenters sat passively until the infant performed one of the target actions on this list. If and only if the infant exhibited one of these target actions, both experimenters began to act in unison. The imitating adult performed the infant's act, and the control adult performed the other behavior that was paired with it from the predetermined target list. The infants looked, smiled, and directed more testing behavior at the adult who imitated them. Thus, even with temporal contingency information controlled, infants can recognize the structural equivalence between self and other. In a very real sense, infants can recognize the reflection of themselves in another.

Note also that these children engaged in these behaviors well before the age at which they would have passed the classical mirror self-recognition task (Lewis & Brooks-Gunn, 1979). Nevertheless, in a sense what the children have done here is exactly to recognize the correspondence between themselves and a "mirror," though the mirror in this case is another person mirroring their own movements. To use another movie reference, these children are like Chico Marx in *Monkey Business*,[2] seeing Groucho mimicking his every move through an empty doorframe. In fact, like Chico they produce "testing behaviors" to see if the match will continue to be maintained. Is Chico's recognition of his correspondence to Groucho different from his recognition of his correspondence to a real mirror? We will suggest that there are some differences later. However, the essential link between the internally perceived self and a physically perceived appearance is in place long before children master the mirror task.

Imitation and theory of mind

The kinds of research that we have discussed so far all involve equivalences between the child's own body and the bodies of others. Now we will turn to the question of the relations between minds and behaviors. This second problem involves the recognition that other people have the same internal mental states that I do. Is it helpful to think of infants' understanding of bodily movements as the foundation for "like me" judgments? Does assuming that I and others have similar bodies lead me to think that we also have similar minds?

Although we tend to think of bodies as physical objects, a body scheme that includes internal proprioceptive sensations actually serves as an interesting halfway point between purely physical objects and mental states. One classical characteristic of mental states that distinguishes them from physical objects is their perceived spatial location. Mental states are located inside the skin, whereas physical objects, including the bodies of others, are located outside it. Similarly, the paradigmatic example of behavior is the body movements of others. The work on early imitation shows that even infants recognize some equivalences between externally perceived behavior, that is, perceived body movements, and literally internal proprioceptive states. Moreover, such proprioceptive sensations, in addition to being spatially located "inside," would seem to have much of the character of mental states. In particular, they are not publicly observable and are private experiences. Indeed, on many philosophical accounts, pains and other internal sensations, which are phenomenologically similar to proprioceptive sensations, are the mental states par excellence.

One fundamental assumption of mentalism, that external, visible behaviors are mapped onto phenomenologically internal mental states, is apparently given innately. Clearly infants have much to learn about the nature of the mind, but apparently they need not learn that it, or something like it, exists, or that it is shared by themselves and others. Ironically, given the great Platonic–Cartesian philosophical tradition of divorcing bodies and minds, it may, quite literally, be our knowledge of the body that leads us to knowledge of the mind. From a developmental viewpoint, knowing that we inhabit bodies similar to others, and assuming that they share our internal bodily states, might be an important precursor to inferring that they share more abstract mental states as well. A person is, after all, both a body and a mind, and for young infants these two aspects of personhood may be deeply intertwined.

Theories of mind

We've suggested that children innately map certain of their own internal mental states, particularly their kinesthetic sensations, onto the bodily movements of others. What evidence is there about children's understanding of the relations between the self and the other for more exalted mental states such as beliefs, desires, and perceptions? In the past few years there has been a veritable explosion of interest in children's ideas about these more sophisticated mental states (see Astington, Harris, & Olson, 1988; Perner, 1991; Wellman, 1990; see also Astington & Gopnik, 1991b; Gopnik, 1990, for reviews). Most of this work, however, has centered on children's understanding of the minds of others, rather than their own minds. We have a fairly clear picture of how children come to understand the minds of others. The crucial question, then, is how this knowledge is related to children's knowledge of their own minds. We will review the data on children's understanding of other minds, and then present our own research on children's understanding of their own minds. As we will see, the pattern of results for the self and the other is extremely similar, which we use to support our thesis that knowledge of self and other develop in tandem.

Understanding other minds

We have suggested that even in infancy children have some primordial sense of being like another, and have a first mapping of bodies and minds. The findings of infant imitation indicate that young infants are not as solipsistic as we might have thought; but we also suggest that children's later understanding is rather different from the standard picture. In classical Piagetian theory, infants emerge from an early solipsistic universe only to be trapped in an egocentric one. They become aware of others, but are doomed to ignore their perspectives. Just as infancy research has forced us to revise the view of infants as solipsists, so the recent research in theory of mind has forced us to revise the view of preschoolers as egotists.

Recent evidence suggests that by the time they are 3 years old, children know many things about their own minds and the minds of others. They seem to know that mental entities are different from physical entities, and understand that images and pretenses are different from reality (Flavell, Flavell, & Green, 1987; Wellman & Estes, 1986). They know that imagining an apple is different from seeing the apple, and that someone may pretend to be a dog without actually being one. Moreover, they have an initial capacity to engage in a certain kind of nonegocentric visual-perspective taking; they know they may see something that someone else does not see. For example, if they sit opposite another person at a table with an opaque screen in the middle, they know that the other person will not be able to see an object on the child's side of the screen (Flavell, Everett, Croft, & Flavell, 1981).

These are impressive abilities in such very young children. However, the inabilities of these children are equally impressive. Three-year-old children consistently fail to understand certain other problems, or rather understand them in a way that is very different from adult understanding. These problems all require that the child understand the complex representational process that relates real objects and mental representations of those objects. The evidence suggests that, unlike the perceptually given representations of body movements, our understanding of these aspects of the mind is constructed through a process of inference and theory formation.

One widely investigated ToM task has been called the false-belief task (Hogrefe, Wimmer, & Perner, 1986; Perner, Leekam, & Wimmer, 1987; Wimmer & Perner, 1983). In one version of the false-belief task children are asked to predict how a deceptive object will appear to others. For example, children see a candy box that turns out to be full of pencils (Perner et al., 1987). They are asked what someone else will think when they first see the box. Three-year-old children consistently say that the other person will think there are pencils in the box. They apparently fail to understand that the other person's beliefs may be false. This finding has proved to be strikingly robust. Children make this error in many different situations, involving many different kinds of objects and events. They continue to make this error even when they see the other person respond to the box with surprise, or when they are explicitly told about the other person's false belief (Moses & Flavell, 1990; Wellman, 1990). Moreover, they make incorrect predictions about the other person's actions, which reflect their incorrect understanding of the other person's beliefs (Perner et al., 1987).

Children also have difficulty understanding the sources of their beliefs. O'Neill, Astington, and Flavell (1992) performed an experiment testing whether children understood that only certain kinds of information could be obtained from particular types of sources. For example, children were shown two objects that could be differentiated only by touch, and two others that could be differentiated only by sight; they were then shown a person who either saw or touched the objects. Three-year-olds had difficulty predicting which objects the other person would be able to differentiate.

These tasks concern children's understanding of beliefs. An interesting question, only recently being investigated, concerns children's understanding of desires. One particularly suggestive finding concerns children's understanding of differences in judgments of value, which are more closely related to desires than to beliefs (Flavell, Flavell, Green, & Moses, 1990). Children appear to be more able to make the correct judgment in these cases than in the standard false-belief task. They appeared more willing to say, for example, that they might think a cookie was "yummy" whereas another person could think it was "yucky," than that they might think the box was full of candy whereas the other thought it was full of pencils. Nevertheless, a fair minority of 3-year-old children made errors on these tasks. Understanding these aspects of desires appears to be somewhat less difficult than understanding false beliefs, though still more difficult than understanding images, pretenses, and simple perceptions.

Four-year-olds have little difficulty solving any of these problems. Moreover, in addition to these tasks, there are a number of related tasks that show significant changes between ages 3 and 4. In this period children begin to make distinctions between appearance and reality (Flavell, Green, & Flavell, 1986); they begin to understand that an object that looks one way from one perspective may look quite different from another (Flavell et al., 1981), and they begin to understand subjective probability, that is, that people may have different degrees of certainty or conviction about their beliefs (Moore, Pure, & Furrow, 1990). Moreover, there is evidence for an association among these tasks: Children who do well on one of these tasks are also likely to do well on the others (Flavell et al., 1986; Gopnik & Astington, 1988; Moore et al., 1990).

How do we interpret these results? There is some consensus among investigators that there is a general shift in the child's conception of the mind, and in particular in the child's epistemological concepts (those of the relation between mind and world) at around $3\frac{1}{2}$ years. The precise characterization of this shift has been a matter of some debate. However, the essential idea is that the 3-year-old believes that objects or events are directly apprehended by the mind, or to use Chandler's phrase, that objects are bullets that leave an indelible trace on any mind that is in their path (Chandler, 1988). Different theorists have captured this idea in different ways: Astington and Gopnik (1991a) call this a "Gibsonian model"; Flavell (1988) talks about cognitive connections; Wellman (1990) talks about a "copy theory of representation";

and Perner (1988) talks about connections to situations. By age 4 or 5, children have developed a different notion of mind – something more like a "representational model of mind" (Flavell, 1988; Forguson & Gopnik, 1988; Perner, 1991; Wellman, 1990). In this view almost all psychological functioning is mediated by representations. These mental states all involve representations of reality rather than realities themselves.

Understanding your own mind

So far, we have suggested that there is at least plausible evidence for a deep change in children's understanding of the psychological states of other people somewhere between ages 3 and 4. The crucial question is whether this change in the conception of the mind is applicable only to others or to the self as well. Returning to the film *Lady in the Lake*, it might seem that the problems we have described should apply only to children's understanding of the minds of others, but not to themselves. Children's knowledge of their own minds should be direct, whereas their knowledge of the minds of others, like their knowledge of their own bodily appearance, must be inferred. In fact, some might want to argue that the changes in the understanding of the mind we have described so far reflect the difficulties inherent in applying knowledge of one's own mental states to other people, and indeed this argument has been made in the literature (Harris, 1991; Johnson, 1988). Piaget's notion of childhood egocentrism implies a similar picture (Piaget, 1929).

Certainly, it seems children assume that their current beliefs about the world will be shared by others. There is however, a difficulty here, for our concept of what is real is constituted by our current beliefs. Children's errors might come from two different sources. Children might think that everyone else believes as they do, or they might think that everyone believes what is actually the case – where what is the case, for children as well as adults, is specified by current beliefs. In the latter case, the problem with false beliefs would not be that the child assumes her particular beliefs are shared, but simply that she assumes others believe what is so – what she knows, or thinks she knows, to be the actual state of affairs.

We can differentiate between these possibilities by looking at children's understanding of their immediately past beliefs or psychological states, states they no longer hold. If children's problems genuinely stem from a kind of egocentrism, they should not have similar difficulties understanding their own immediately past beliefs. These beliefs are, after all, as much their own as are their current beliefs.

Our research program has compared children's performance on "other minds" tasks like those we have just described, to their performance on tasks that require them to report their own immediately past mental states. The basic finding is that children make errors about their own immediately past states that are similar to the errors they make about the states of others. This is true even though children ought to have direct first-person psychological experience of these past states.

In our original experiment (Astington & Gopnik, 1988; Gopnik & Astington, 1988) we presented children with a variety of deceptive objects, such as the candy box full of pencils, and allowed them to discover the true nature of the objects. We then asked the children about their own immediately past false beliefs about the box: "When you first saw the box, before we opened it, what did you think was inside it?" One-half to two-thirds of the 3-year-olds said that they had originally thought there were pencils in the box. They apparently failed to remember their immediately previous false beliefs. Moreover, children's ability to answer the false-belief question about their own belief was significantly correlated to their ability to answer the question about the false belief of another, and to answer the appearance–reality question, even with age controlled. This result was recently replicated by Wimmer and Hartl (1991) and by Moore et al. (1990), who also found significant correlations to children's understanding of subjective probability.

The children in Gopnik and Astington (1988) also received a control task. They saw a closed container (a toy house) with one object inside it. The house was then opened, the object was removed, and a different object was placed inside. Children were asked, "When you first saw the house, before we opened it, what was inside it?" This question had the same form as the belief question. However, it asked about the past physical state of the house rather than asking about a past mental state. Children were included in the experiment only if they answered this question correctly, and so demonstrated that they could understand that the question referred to the past, and remember the past state of affairs.

In a second set of experiments (Gopnik & Graf, 1988; O'Neill & Gopnik, 1991) we investigated children's ability to identify the sources of their own beliefs, elaborating on a question first used by Wimmer, Hogrefe, and Perner (1988). This question parallels the understanding of the sources of other's beliefs investigated by O'Neill et al. (1992). Children found out about objects that were placed in a drawer in one of three ways: They either saw the objects, were told about them, or figured them out from a simple clue. Then we asked, "What's in the drawer?" and all the children answered correctly. Immediately after this question we asked about the source of the child's knowledge: "How do you know there's an x in the drawer? Did you see it, did I tell you about it, or did you figure it out from a clue?" Again, 3-year-olds made frequent errors on this task. Although they knew what the objects were, they could not say how they knew. They might say, for example, that we had told them about an object, when they had actually seen it. In a follow-up experiment (O'Neill & Gopnik, 1991) we used different and simpler sources (tell, see, and feel) and presented children with only two alternative possibilities at a time. We also included a control task that ensured the children understood the meaning of "tell," "see," and "feel." Despite these simplifications of the task, the performance of the 3-year-olds was similar to their performance in the original experiment.

In a more recent experiment we investigated whether children could understand changes in psychological states other than belief (Gopnik & Slaughter, 1991). When the child pretends that an object is another object, or imagines an object, there need be no understanding of the relation between those representations and reality. For young children, pretenses and images are unrelated to reality: They cannot be false – or true, for that matter. And as we have seen, these mental states are apparently well understood by 3-year-olds (Flavell et al., 1987; Wellman & Estes, 1986). Similarly, we have seen that even without an understanding of the representational process, 3-year-olds can tell that someone else may not see something that they see themselves, and vice versa. Can 3-year-old children also understand changes in their own pretenses, images, and perceptions?

To test this we placed children in a series of situations in which they were in one mental state, and their mental state was then changed, situations comparable to the belief-change case. For example, we asked children to pretend that an empty glass was full of hot chocolate. The glass was then "emptied" and the child was asked to pretend that it was full of lemonade. We then asked them, "When I first asked you. . . . what did you pretend was in the glass then?" We also asked them to imagine a blue dog and then a pink cat and asked them, "When I first asked you. . . . what did you think of then?" In both of these cases, as in the belief change case, the child is first in one mental state and then in another, even though nothing in the real world has changed. In these cases, however, unlike in the belief case, the mental states need not be interpreted as involving any representational relation to the real world. In a perceptual task, paralleling the perspective-taking task of Flavell et al. (1981), we placed the children on one side of a screen from which an object was visible, and then moved them to the other side, from which another object was visible, and asked them to recall their past perception. The 3-year-olds were able to solve these tasks; only 1 out of 30, for example, failed to remember her earlier pretense. However, a majority of these same 3-year-olds were unable to solve the belief-change task.

We also tested children's understanding of changes in desires. In three different tasks we presented children with situations in which their desires were satiated, and so changed. In all three tasks a sizable minority of 3-year-olds (30–40%) reported that they had been in their final state all along. Thus, for example, hungry children were asked "Are you hungry?" at snack time. We then fed them crackers until they were no longer hungry and asked, "Were you hungry before we had the snack?" A third of them reported that they were not. Nevertheless, just as in the Flavell task, which measured children's understanding of the desires of others ("Does Ellie think the cookie is yummy or yucky?"; Flavell et al., 1990), so in our similar task ("Were you hungry before?") the children were better at reporting past desires than past beliefs. Indeed, the absolute levels of performance were strikingly similar in our task and Flavell's.

In short, the evidence suggests that there is an extensive parallelism between children's understanding of their own mental states and their understanding

Table 10.1. *Children's knowledge of their own mental states and those of others*

States	Other	Self
Easy		
Pretense	Before age 3 (Flavell et al., 1987)	Before age 3 (Gopnik & Slaughter, 1991)
Imagination	Before age 3 (Wellman & Estes, 1986)	Before age 3 (Gopnik & Slaughter, 1991)
Perception (Level 1)	Before age 3 (Flavell et al., 1981)	Before age 3 (Gopnik & Slaughter, 1991)
Intermediate		
Desire	Age 3–4 (Flavell et al., 1990)	Age 3–4 (Gopnik & Slaughter, 1991)
Difficult		
Sources of belief	After age 4 (O'Neill et al., 1992)	After age 4 (Gopnik & Graf, 1988)
False belief	After age 4 (Wimmer & Perner, 1983)	After age 4 (Gopnik & Astington, 1988)

of the mental states of others (see Table 10.1). In each of our studies, children's reports of their own immediately past psychological states are consistent with their accounts of the psychological states of others. When they can report and understand the psychological states of others, in the cases of pretense, perception, and imagination, they report having had those psychological states themselves. When they cannot report and understand the psychological states of others, in the case of false beliefs and source, they do not report that they had those states themselves. Moreover, and in some ways most strikingly, the intermediate case of desire is intermediate for self and other.

Just as the imitation data suggest an early mapping of internal sensations onto the bodily movements of others, so these results suggest an early mapping of sophisticated mental states onto the behaviors of others. In these examples, however, the mapping seems to be based on inference rather than on perception. In particular, that children are consistently wrong in their reports of their own mental states suggests that these states are not, in fact, known directly through some sort of perceptual process. As adults in these tasks, our impression is that we find the correct answer simply by remembering our immediately past state, rather than by reconstructing or inferring it; but the developmental evidence suggests that this is not true of 3-year-olds.

The conclusion that emerges from these findings is that, contrary to what one might have imagined, children appear to show a striking synchrony in their understanding of themselves and others. Rather than having to make different inferences about themselves and others – that they have bodies, that others have minds – they instead seem to make these inferences in parallel. At each stage of development they appear to assume deep commonalities between their own minds and bodies and those of others.

In fact, we would suggest that the primitive understandings that underlie early imitation may themselves be the basis for the theory of mind ascriptions of older children. The early assumption that the self and others have the same internal states and that behavior maps onto internal experience, may be quickly extended. As children construct more abstract explanations for the behavior of others, they may assume that these explanations also apply to themselves, and may even "experience" the abstract entities those explanations propose.

Contingency, self–other differentiation, and making sense of mirrors

It would seem, then, that the problem might not be one of mapping the self onto the other so much as differentiating the self from the other. Again, this is an old idea in the psychoanalytic tradition. However, just as imitation suggests a mechanism by which an initial identification between the infant and mother may take place, so it also suggests a mechanism by which this differentiation might occur. Watson (1979; Bahrick & Watson, 1985) has demonstrated the power of contingency analysis in the infant. Watson argues that different patterns of contingency are at the root of the differentiation of the self and others. We would echo this idea, but would also underscore that the relations may be stronger than temporal contingency.

For the self, the relation between proprioceptive and visual information will be one of perfect structural as well as temporal matching. Whatever information the proprioceptive system gives us about our own body movements will be perfectly matched by the information the visual system gives us. For the other, this match will not be perfect. Even in mutual imitation games, for example, variations in the form of the mother's imitative responses are common. In addition, of course, there will be a perceptible gap between the infant's own felt movements and the mother's seen movements. These differences could help the infant identify the mother as someone who is both like and not like himself. Mutual imitation games are exercises in both self–other equivalence and differentiation.

Phenomena like *hand regard* in 3-month-olds – the hours that young infants spend gazing at their hand as they wave it to and fro in the visual field – may also reflect children's discovery of these different contingency patterns, and their own attempts to investigate the distinctive properties of the self. Unlike Piaget (1962), we do not think that hand regard provides the original basis for cross-modal matching: That ability is clearly in place much earlier. It may, however, be a way of working through the different contingency patterns of the proprioceptive–visual match for the self and the other.

The mirror task may also reflect this understanding of contingencies. Clearly, the task does not simply index an ability to map internal information about the self and appearances; this is in place much earlier. What is interesting about this task is that it requires precisely the sort of information about physical appearance that is not given by the early cross-modal mapping. In fact, the experiment is carefully designed so that the child has no way of using tactile or proprioceptive information to identify the visual mark. At the same time, it seems plausible that children use the visual–proprioceptive correspondences between their own bodies and the movements of their image in the mirror (correspondences that are given by the innate cross-modal system) to recognize the correspondences between the visual mark and their own physical appearances (correspondences that are not innate). In particular, they may use the perfect structural contingencies between their own proprioceptive image and the visual image in the mirror as a clue to the fact that even unfelt aspects of appearance, spots, eye color, or hairstyle are part of the self (Meltzoff, 1993; Mitchell, 1993). Mirror experience may allow us to identify the purely visual aspects of our appearance with our body schemes. The spot on my face, my eye color, my hair, are not given to me by experience, and yet mirror experience allows me to identify them with myself.

Some of our adult phenomenology reflects this. The queasy ambivalence of personal vanity comes precisely from this sense that these aspects of our physical appearance, the ones that strike us in the mirror – the ones, moreover, that we can hope to alter or modulate – have this strange quality of being a part of us and yet not part of us at the same time. Other aspects of our appearance – what in the 18th century was called our "countenance," our spontaneous expressions, our movements, the way we smile, frown, or look intense – are precisely those things that we don't see gazing in the mirror, and that speak most directly to others and reflect our deepest sense of ourselves. The vertiginous experience of "perm shock," the existential horror one experiences looking in the mirror after even a successful permanent wave, or the similar case of "beard shock" when a long-established beard is shaved off, are particularly striking examples of this phenomenon, the sense that the image in the mirror is both a part and not a part of ourselves.

Conclusion

The significance of early interpersonal understanding for later development of a theory of mind may be particularly apparent in cases where this innate capacity is missing. There is reason to believe that the disorder of autism is such a case. Children with autism seem to lack the fundamental appreciation of the commonality between themselves and others. This lack is apparent in their language and in their profound deficits in developing social skills, such as referential pointing and social referencing, that normally developing children acquire even in infancy (Baron-Cohen, 1991; Hobson, 1989; Sigman, 1989). These children also display a specific inability to accomplish later ToM tasks (Baron-Cohen, Leslie & Frith, 1985). Moreover, there is some evidence of imitative deficits in these children (Meltzoff & Gopnik, 1993; Mitchell, 1993;

Rogers & Pennington, 1991). It's as if the basic assumption that the self and other are the same is missing in these children, and this makes them incapable of developing a more elaborated commonsense psychology. These children may really be in the position of Philip Marlowe in *Lady in the Lake*, trapped in a solipsistic universe, seeing others as objects rather than persons, and having to laboriously construct even a basic understanding of the relations between themselves and others.

Most children, however, never live in that universe. We are given a sense of the person that automatically includes both our own internal states and the behavior of others. Indeed there may even be some evidence that this sense has emerged at a particular point in evolutionary history. Nonhuman primates show a similar pattern of behavior on both generative imitation and theory of mind tasks. Monkeys show little capacity in either of these areas, whereas great apes have been demonstrated to have some primitive versions of both imitative and ToM abilities (Meltzoff, 1988c, 1988d; Mitchell, 1987; Povinelli, Nelson, & Boysen, 1990; Premack & Woodruff 1978; Tomasello, Kruger, & Ratner, 1993; Tomasello, Savage-Rumbaugh, & Kruger, in press; Visalberghi & Fragaszy, 1990; Whiten & Ham, 1991).[3]

Initially the range of entities that are subsumed in the innate concept of the person is rather limited – body schemes, simple emotions, and perhaps aspects of attention. The development of a more sophisticated theory of mind, however, builds on this foundation and keeps this general structure. The elaborated sense of persons we have as adults, with all the added knowledge that comes from our cultural and social experience, retains this quality of transcending boundaries between bodies and minds, between ourselves and others. Cross-modal matching is nature's solution to the mind–body problem, and nature's way of exorcising the solipsistic demons of philosophy.

Notes

1. *Lady in the Lake*, dir. Robert Montgomery (Metro-Goldwyn-Mayer, 1946).
2. *Monkey Business*, dir. Norman MacLeod (Paramount, 1931).
3. It is also interesting that success on Gallup's (1970) mirror self-recognition test has been found to follow the same phylogenetic lines (Gallup, 1983; Mitchell, 1993, in press; Povinelli, 1987) suggesting that all three may be related.

References

Astington, J. W., & Gopnik, A. (1988). Knowing you've changed your mind: Children's understanding of representational change. In J. W. Astington, P. L. Harris & D. R. Olson (Eds.), *Developing theories of mind*. Cambridge University Press.

(1991a). Developing understanding of desire and intention. In A. Whiten (Ed.), *Natural theories of mind: Evolution, development and simulation of everyday mindreading* (pp. 39–50). Cambridge, MA: Blackwell Publisher.

(1991b). Theoretical explanations of children's understanding of the mind. *British Journal of Developmental Psychology*, 9, 7–31.

Astington, J. W., Harris, P. L., & Olson, D. R. (Eds.) (1988). *Developing theories of mind*. Cambridge University Press.

Bahrick, L. E., & Watson, J. S. (1985). Detection of intermodal proprioceptive–visual contingency as a potential basis of self-perception in infancy. *Developmental Psychology, 21*, 963–973.

Baron-Cohen, S. (1991). Precursors to a theory of mind: Understanding attention in others. In A. Whiten (Ed.), *Natural theories of mind: Evolution, development and simulation of everyday mindreading* (pp. 233–251). Cambridge, MA: Blackwell Publisher.

Baron-Cohen, S., Leslie, A. M., & Frith, U. (1985). Does the autistic child have a "theory of mind"? *Cognition, 21*, 37–46.

Bower, T. G. R. (1982). *Development in infancy* (2nd ed.). San Francisco: Freeman.

Bruner, J. (1983). *Child's talk: Learning to use language.* New York: Norton.

Chandler, M. (1988). Doubt and developing theories of mind. In J. W. Astington, P. L. Harris, & D. R. Olson (Eds.), *Developing theories of mind.* Cambridge University Press.

Flavell, J. H. (1988). The development of children's knowledge about the mind: From cognitive connections to mental representations. In J. W. Astington, P. L. Harris, & D. R. Olson (Eds.), *Developing theories of mind* (pp. 244–267). Cambridge University Press.

Flavell, J. H., Everett, B. A., Croft, K., & Flavell, E. R. (1981). Young children's knowledge about visual perception: Further evidence for the Level 1–Level 2 distinction. *Developmental Psychology, 17*, 99–103.

Flavell, J. H., Flavell, E. R., & Green, F. L. (1987). Young children's knowledge about the apparent-real and pretend-real distinctions. *Developmental Psychology, 23*, 816–822.

Flavell, J. H., Flavell, E. R., Green, F. L., & Moses, L. J. (1990). Young children's understanding of fact beliefs versus value beliefs. *Child Development, 61*, 915–928.

Flavell, J. H., Green, F. L., & Flavell, E. R. (1986). Development of knowledge about the appearance–reality distinction. *Monographs for the Society for Research in Child Development* (Serial No. 212, 51, No. 1).

Forguson, L., & Gopnik, A. (1988). The ontogeny of common sense. In J. W. Astington, P. L. Harris, & D. R. Olson (Eds.), *Developing theories of mind* (pp. 226–243). Cambridge University Press.

Gallup, G. G., Jr. (1970). Chimpanzees: Self-recognition. *Science, 167*, 86–87.

(1983). Toward a comparative psychology of mind. In R. Mellgren (Ed.), *Animal cognition and behavior* (pp. 473–510). New York: North Holland.

Gopnik, A. (1990). Developing the idea of intentionality: Children's theories of mind. *Canadian Journal of Philosophy, 20*, 89–114.

(1993). How we know our minds: The illusion of first-person knowledge of intentionality. *Behavioral and Brain Sciences, 16*, 1–14.

Gopnik, A., & Astington, J. W. (1988). Children's understanding of representational change and its relation to the understanding of false belief and appearance–reality distinction. *Child Development, 59*, 26–37.

Gopnik, A., & Graf, P. (1988). Knowing how you know: Young children's ability to identify and remember the sources of their beliefs. *Child Development, 59*, 1366–1371.

Gopnik, A., & Slaughter, V. (1991). Young children's understanding of changes in their mental states. *Child Development, 62*, 98–110.

Gopnik, A., & Wellman, H. M. (1992). Why the child's theory of mind really is a theory. *Mind and Language, 7*, 145–172.

Harris, P. (1991). The work of the imagination. In A. Whiten (Ed.), *Natural theories of mind: Evolution, development, and simulation of everyday mindreading* (pp. 283–304). Cambridge, MA: Blackwell Publisher.

Hobson, R. P. (1989). Beyond cognition: A theory of autism. In Geraldine Dawson (Ed.), *Autism: Nature, diagnosis, and treatment* (pp. 22–48). New York: Guilford.

Hogrefe, G. J., Wimmer, H., & Perner, J. (1986). Ignorance versus false belief: A developmental lag in attribution of epistemic states. *Child Development, 57*, 567–582.

Johnson, C. N. (1988). Theory of mind and the structure of conscious experience. In J. W. Astington, P. L. Harris, & D. R. Olson (Eds.), *Developing theories of mind* (pp. 47–63). Cambridge University Press.

Johnson, M., & Morton, J. (1991). *Biology and cognitive development: The case of face recognition.* Cambridge, MA: Blackwell Publisher.

Kuhl, P. K., & Meltzoff, A. N. (1982). The bimodal perception of speech in infancy. *Science, 218,* 1138–1141.

Lewis, M., & Brooks-Gunn, J. (1979). *Social cognition and the acquisition of self.* New York: Plenum.

Mahler, M. S., Pine, F., & Bergman, A. (1975). *The psychological birth of the human infant.* New York: Basic.

Meltzoff, A. N. (1988a). Infant imitation and memory: Nine-month-olds in immediate and deferred tests. *Child Development, 59,* 217–225.

(1988b). Infant imitation after a 1-week delay: Long-term memory for novel acts and multiple stimuli. *Developmental Psychology, 24,* 470–6.

(1988c). Imitation, objects, tools, and the rudiments of language in human ontogeny. *Human Evolution, 3,* 45–64.

(1988d). The human infant as *Homo imitans.* In T. Zentall and J. Galef (Eds.), *Social learning: Psychological and biological perspectives* (pp. 319–341). Hillsdale, NJ: Erlbaum.

(1990a). Towards a developmental cognitive science: The implications of cross-modal matching and imitation for the development of representation and memory in infancy. In A. Diamond (Ed.), *The development and neural bases of higher cognitive functions.* Special issue of *Annals of the New York Academy of Sciences, 608,* 1–31.

(1990b). Foundations for developing a concept of self: The role of imitation in relating self to other and the value of social mirroring, social modeling and self practice in infancy. In D. Cicchetti & M. Beeghly (Eds.), *The self in transition: Infancy to childhood* (pp. 139–164). Chicago: University of Chicago Press.

(1993). Molyneux's babies: Cross-modal perception, imitation, and the mind of the preverbal infant. In N. Eilan, R. McCarthy, & B. Brewer (Eds.), *Spatial representation: Problems in philosophy and psychology* (pp. 219–235). Oxford: Blackwell Publisher.

Meltzoff, A. N., & Borton, R. W. (1979). Intermodal matching by human neonates. *Nature, 282,* 403–404.

Meltzoff, A. N., & Gopnik, A. (1993). The role of imitation in understanding persons and developing a theory of mind. In S. Baron-Cohen, H. Tager-Flusberg, & D. Cohen (Eds.), *Understanding other minds: Perspectives from autism* (pp. 335–366). New York: Oxford University Press.

Meltzoff, A. N., & Moore, M. K. (1977). Imitation of facial and manual gestures by human neonates. *Science, 198,* 75–78.

(1983). Newborn infants imitate adult facial gestures. *Child Development, 54,* 702–709.

(1989). Imitation in newborn infants: Exploring the range of gestures imitated and the underlying mechanisms. *Developmental Psychology, 25,* 954–962.

(1992). Early imitation within a functional framework: The importance of person identity, movement, and development. *Infant Behavior and Development, 15,* 479–505

(1993). Why faces are special to infants: On connecting the attraction of faces and infants' ability for imitation and cross-modal processing. In B. de Boysson-Bardies, S. de Schonen, P. Jusczyk, P. MacNeilage, and J. Morton (Eds.), *Developmental neurocognition: Speech and face processing in the first year of life* (pp. 211–225). Boston, MA: Kluwer.

(in press). Imitation, memory, and the representation of persons. *Infant Behavior and Development.*

Mitchell, R. W. (1987). A comparative-developmental approach to understanding imitation. In P. Klopfer & P. Bateson (Eds.), *Perspectives in ethology* (Vol. 7, pp. 183–215). NY: Plenum.

(1993). Mental models of mirror self-recognition: Two theories. *New Ideas in Psychology, 11,* 295–325.

(in press). The evolution of primate cognition: Simulation, self-knowledge, and knowledge of other minds. In D. Quiatt and J. Itani (Eds.), *Hominid culture in primate perspective.* Denver: University Press of Colorado.

Moore, C., Pure, K., & Furrow, D. (1990). Children's understanding of the modal expression of speaker certainty and uncertainty and its relation to the development of a representational theory of mind. *Child Development, 61,* 722–730.

Moses, L. J., & Flavell, J. H. (1990). Inferring false beliefs from actions and reactions. *Child Development, 61,* 929–945.

O'Neill, D. K., Astington, J. W., & Flavell, J. H. (1992). Young children's understanding of the role that sensory experiences play in knowledge acquisition. *Child Development, 63,* 474–490.

O'Neill, D. K., & Gopnik, A. (1991). Young children's ability to identify the sources of their beliefs. *Developmental Psychology, 27,* 390–7.

Perner, J. (1988). Developing semantics for theories of mind: From propositional attitudes to mental representations. In J. W. Astington, P. L. Harris, & D. R. Olson (Eds.), *Developing theories of mind* (pp. 141–172). Cambridge University Press.

——— (1991). *Understanding the representational mind* (pp. 171–172). Cambridge, MA: MIT Press.

Perner, J., Leekam, S., & Wimmer, H. (1987). Three-year-olds' difficulty with false belief: The case for a conceptual deficit. *British Journal of Developmental Psychology, 5,* 125–137.

Piaget, J. (1929). *The child's conception of the world.* New York: Harcourt Brace.

——— (1962). *Play, dreams, and imitation in childhood.* New York: Norton.

Povinelli, D. J. (1987). Monkeys, apes, mirrors, and minds: The evolution of self-awareness in primates. *Human Evolution, 2,* 493–507.

Povinelli, D. J., Nelson, K. E., & Boysen, S. T. (1990). Inferences about guessing and knowing by chimpanzees (*Pan troglodytes*). *Journal of Comparative Psychology, 104,* 203–210.

Premack, D., & Woodruff, G. (1978). Does the chimpanzee have a theory of mind? *Behavioral and Brain Sciences, 4,* 515–526.

Rogers, S. J., & Pennington, B. F. (1991). A theoretical approach to the deficits in infantile autism. *Development and Psychopathology, 3,* 137–162.

Sigman, M. (1989). The application of developmental knowledge to a clinical problem: The study of childhood autism. In D. Cicchetti (Ed.), *The emergence of a discipline: Rochester symposium on developmental psychopathology* (pp. 165–187). Hillsdale, NJ: Erlbaum.

Stern, D. N. (1985). *The interpersonal world of the infant.* New York: Basic.

Tomasello, M., Kruger, A., & Ratner, H. (1993). Cultural learning. *Behavioral and Brain Sciences, 16,* 495–552.

Tomasello, M., Savage-Rumbaugh, S., & Kruger, A. (in press). Imitative learning of actions on objects by children, chimpanzees, and enculturated chimpanzees. *Child Development.*

Trevarthen, C. (1980). The foundations of intersubjectivity: Development of interpersonal and cooperative understanding. In D. Olson (Ed.), *The social foundation of language and thought: Essays in honor of Jerome Bruner* (pp. 316–342). New York: Norton.

Visalberghi, E., & Fragaszy, D. (1990). Do monkeys ape? In S. Parker & K. Gibson (Eds.), *Comparative developmental psychology of language and intelligence in primates* (pp. 247–273). Cambridge University Press.

Watson, J. (1979). Perception of contingency as a determinant of social responsiveness. In E. Tohman (Ed.), *The origins of social responsiveness* (pp. 33–64). Hillsdale, NJ: Erlbaum.

Wellman, H. M. (1990). *The child's theory of mind.* Cambridge, MA: MIT Press.

Wellman, H. M., & Estes, D. (1986). Early understanding of mental entities: A reexamination of childhood realism. *Child Development, 57,* 910–923.

Whiten, A., & Ham, R. (1991). On the nature and evolution of imitation in the animal kingdom: Reappraisal of a century of research. In P. Slater, J. Rosenblatt, C. Beer, & M. Milinski (Eds.), *Advances in the study of behavior* (pp. 239–283). New York: Academic Press.

Wimmer, H., & Hartl, M. (1991). Against the Cartesian view on mind: Young children's difficulty with own false beliefs. *British Journal of Developmental Psychology, 9,* 125–138.

Wimmer, H., Hogrefe, J.-G., & Perner, J. (1988). Children's understanding of informational access as a source of knowledge. *Child Development, 59,* 386–396.

Wimmer, H., & Perner, J. (1983). Beliefs about beliefs: Representation and constraining function of wrong beliefs in young children's understanding of deception. *Cognition, 13,* 103–128.

Part III

Self-awareness in great apes

11 Social and cognitive factors in chimpanzee and gorilla mirror behavior and self-recognition

Karyl B. Swartz and Siân Evans

The investigation of animal cognition is currently an active area of research in comparative psychology. Such cognitive processes as representation of lists (Swartz, Chen, & Terrace, 1991), complex strategies of spatial memory (Cook, Brown, & Riley, 1985), use of numbers (Boysen & Berntson 1989), and acquisition of concepts (D'Amato & Van Sant, 1988; Herrnstein & Loveland, 1964; Oden, Thompson, & Premack, 1990) have been addressed in primate and nonprimate species. One of the most exciting topics relevant to both comparative and developmental psychology has been that of mirror self-recognition (MSR), with its implications for comparative cognition (see Gallup, 1982, 1991; Gallup & Suarez, 1986; Mitchell, 1993; Parker, 1991; Swartz, 1990).

This phenomenon, first demonstrated in chimpanzees by Gallup (1970), involves two empirical observations that have been interpreted as evidence of self-recognition. When first exposed to a mirror, chimpanzees showed socially appropriate behaviors directed toward the mirror, as though the reflection were another conspecific. As mirror exposure continued, the chimpanzees' behavior changed. Social behaviors waned and self-directed or self-referred behaviors appeared. Self-directed behaviors involve using the mirror to guide movements to areas of the body not visible without the aid of the mirror; self-referred behaviors involve manipulation of the environment while watching the action in the mirror. The appearance of self-directed behaviors was interpreted as indicative of self-recognition on the part of the chimpanzee (Gallup, 1970). The use of the mirror to inspect areas of the body that were previously invisible, and the presence of *self*-directed rather than *mirror image*-directed behaviors, the argument goes, suggest that the chimpanzee must have the idea that the mirror image is a reflection of its own body.

In order to support this interpretation of self-directed behaviors in the presence of the mirror, Gallup (1970) developed a more objective test of self-recognition using the mirror. He anesthetized the animals and placed a red mark on their faces above the eyebrow, where it could be seen only with the aid of a mirror. When the chimpanzees recovered from the anesthesia, they were presented with the mirror. Touches to the mark on the face occurred

189

quickly, with the aid of the mirror to guide the hand to the animal's face rather than to the mirror image. This behavior has been interpreted to suggest that the subjects recognized themselves in the mirror, recognized the change in their appearance manifested by the presence of the mark, and demonstrated by the touching of their own bodies their knowledge of the locus of the mark. For more than 20 years this was considered to be a robust phenomenon in captive chimpanzees, occurring in all socially housed adult chimpanzees tested (Calhoun & Thompson, 1988; Gallup, 1970; Lethmate & Dücker, 1973; Suarez & Gallup, 1981). However, not all chimpanzees demonstrate this phenomenon (Swartz & Evans, 1991; see also Povinelli, *SAAH*18).

In our study (Swartz & Evans, 1991), 11 chimpanzees were given mirror exposure and tested with the mark test. Only one of the 11 chimpanzees touched the mark during the test, although several showed self-directed behavior using the mirror to guide their movements. In all previous studies to date, each animal that has shown mirror recognition has also shown self-directed behavior. Conversely, animals who do not pass the mark test typically have not shown self-directed behavior prior to the test (Gallup, 1977a; Gallup, McClure, Hill, & Bundy, 1971; Gallup, Wallnau, & Suarez, 1980; Ledbetter & Basen, 1982; Suarez & Gallup, 1981). The results of the Swartz and Evans (1991) study are counter to the preceding statement, and clearly question the presumed relationship between self-directed behavior and mark-directed behavior.

These results are extremely provocative in light of species differences in MSR observed in nonhuman primates. Although chimpanzees and orangutans have been reported to demonstrate this phenomenon rather consistently using the mark-test criterion, gorillas have not (see Anderson, 1984a; Ledbetter & Basen, 1982; Suarez & Gallup, 1981). Despite heroic efforts to demonstrate this phenomenon in gorillas, only two to date have provided evidence of self-recognition: Koko, through her spontaneous behavior with mirrors (Patterson & Cohn, *SAAH*17, who present provocative but inconclusive evidence for self-recognition in another gorilla, Michael), and King, a gorilla living at Monkey Jungle, in Miami, who recently passed the mark test (Siân Evans, unpublished data). The present chapter addresses the issue of mirror self-recognition from the perspective of failures to find evidence for this phenomenon in some species, and in some individuals within species that are reported to have the capacity.

Failures to pass the mark test

The question here is what does it mean to find failures to pass the mark test? We will first deal with the issue of individuals within a species (chimpanzees) who fail to provide evidence for MSR, and then move to a discussion of species differences in this phenomenon. When there is a failure to replicate a well-accepted empirical phenomenon, the first response is to question the methodology of the study that failed to achieve replication. In our 1991 paper, we addressed several subject and procedural variables that we felt were not sufficient explanations for the failure of ten chimpanzees to show evidence of mirror self-recognition. At the Sonoma Conference on Self-recognition and Self-awareness, Gallup presented an expanded discussion of methodological considerations in conducting mirror self-recognition studies, some of which are also raised in his chapter in this volume (*SAAH*3).

In the event that methodological differences are not sufficient to explain the discrepant results, the phenomenon at issue becomes more complex and subject to additional theoretical consideration. It is our position that the latter event has occurred, and that MSR should undergo some theoretical enrichment (see Mitchell, 1993, who has begun this task with an extensive theoretical discussion of MSR). In order to support our position that our failure to replicate is theoretically interesting and important, we will address some of these issues. The three major areas of concern are subject variables, mirror-exposure variables and mark test considerations and controls. These will each be discussed in turn, addressing these variables specifically with respect to the Swartz and Evans (1991) study.

Subject variables

A number of subject variables must be addressed to eliminate individual physical or psychological limitations as explanations of a failure to pass the mark text (Gallup, *SAAH*3). These include visual capacity, health status, maturational constraints, social rearing history, and initial interest in the mirror. Although there is no direct evidence supporting the absence of visual anomalies in the chimpanzee population included in the Swartz and Evans (1991) study, neither is there any evidence suggesting their presence. All subjects were successfully living in social groups in housing facilities that promoted physical activities requiring good vision (i.e., jumping, climbing, social interaction, and visual inspection and manipulation of objects). Further, all subjects were marked on the wrist as well as the face during mark tests. In *all* cases (even during repeated tests) the subjects investigated the mark on the wrist. Although the possible presence of visual anomalies must be acknowledged in individuals of any population who do not demonstrate a popularly accepted visually mediated phenomenon, it does not appear to be an explanation for the failure of ten apparently normal chimpanzees to touch the mark on their faces, nor has it been suggested as a possible explanation for the failure of gorillas to pass the mark test (Gallup, 1991; Suarez & Gallup, 1981).

Similarly, the general health status of the subjects in our study and their experimental history have been challenged by Gallup (*SAAH*3). Because of the anomalous nature of our results relative to those published previously, we were very sensitive to this issue with respect to our subjects. From all reports, these animals were in fine health. They received regular medical examinations and were all reported by competent veterinary staff to have no health problems.

As in humans, there is likely a developmental progression for mirror self-recognition in chimpanzees (Lin, Bard, & Anderson, 1992; Parker, 1991; Povinelli, *SAAH*18; Robert, 1986); however it is unlikely that maturation was a problem in the Swartz and Evans (1991) study. Two previous studies report mirror-guided mark-directed behavior in chimpanzees aged 3–6 (Calhoun & Thompson, 1988; Gallup et al., 1971), and two, in chimpanzees under 3 years (Hill, Bundy, Gallup, & McClure, 1970; Lin et al., 1992). Makata, the youngest subject in our study, was estimated to be 4 years of age based on weight and tooth eruption. His weight is consistent with Gallup's subjects who were also wild-born. Indeed, age estimates of Gallup's original subjects using age–weight data from comparable chimpanzees in the CIRMF population provide median age estimates of 5.7 and 5.6 years for the two females and 5.1 and 3.6 years for the two males, with an overall range of 3.2–6.4 years. (See Note at the end of this chapter for more detail.) It should be noted that Gallup's youngest subject produced the largest number of mark-directed responses. It is unlikely that Makata's failure to pass the mark test was a function of maturational constraints; however, even if this is an acceptable explanation, it does not explain the failure of the older animals in our study to evidence MSR.

Social rearing history has been shown to be a major factor in failures to demonstrate MSR (Gallup et al., 1971; Hill et al., 1970). Isolation-reared chimpanzees failed to pass the mark test (after failing to respond with self-directed behaviors during the mirror exposure period); however provision of social experience with other chimpanzees leads to this capacity. Our wild-born chimpanzees were all socially living and had been so for years prior to the study; the implication is that that situation was true for Gallup's (1970) original chimpanzee subjects, as well.

Mirror-exposure variables

Such variables as size and placement of the mirror, and temporal spacing of mirror exposure, may lead to differences in how quickly mirror self-recognition is demonstrated (Gallup, *SAAH*3), but if this phenomenon does indeed reflect a capacity inherent in all members of a species, such minor experimental variables should not interfere with an individual's ability eventually to demonstrate the phenomenon. As with the subject variables of age, sex, and previous social experience, such experimental factors as mirror size, position, or temporal spacing of the mirror exposure were insufficient to explain the difference between the Swartz and Evans (1991) findings and previous ones.

Mark-test considerations and controls

Gallup (*SAAH*3) has suggested that repeated mark tests could have decreased the subjects' interest in the face mark in our study. Indeed, in the single animal (Berthe) who evidenced MSR, the latency to touch the mark and the vigor with which the mark was investigated waned across the three mark tests given. However, in all cases Berthe *did* touch the mark, even with minimal reexposure to her unmarked mirror image between mark tests. Further, the interest in the wrist mark placed on all subjects for all mark tests maintained across repeated tests. Although a better design for the Swartz and Evans study would have included variations in the placement and color of the mark across tests, the foregoing factors suggest that the repeated mark tests were not the basis for the failure of ten chimpanzees to show mirror self-recognition.

An additional concern expressed by Gallup (*SAAH*3) is the potential effect that exposure to cagemates that had been marked might have on a particular animal's tendency to touch the mark on his/her own face. Gallup's suggestion that prior to being tested some of our subjects had had "extensive exposure" to marked social partners is simply not the case; but even had that been so, it is unclear why a subject *who investigates the mark on another chimpanzee's face* would not investigate a similar mark on its own face *if the animal had recognized himself/herself in the mirror*. Further, human infants have been observed to touch their own faces when they observed a mark on another person's face (Johnson, 1983; Lewis & Brooks-Gunn, 1979). If chimpanzees were inclined to touch their own faces in the presence of other chimpanzees who were marked, that should increase the proportion of our subjects who passed the mark test. However, such a result would call into question the rationale underlying the mark test (see Mitchell, 1993, for a discussion of this). We did not observe this behavior in the few instances when a marked animal was reintroduced to the social group.

In our earlier paper, we concluded that procedural differences were unlikely to be responsible for the differences between our findings and earlier findings. However, although we addressed the issue of social rearing, we did not specifically address the social context in which mirror exposure and testing were conducted, a factor that we now think may be important in the demonstration of mirror self-recognition in individuals. In his original study Gallup (1970) tested four young chimpanzees in isolation (although they were presumably group living prior to testing) in a small cage placed in a corner of a larger empty room. Mirrors were presented to them in this situation in order to "enforce self-confrontation." In contrast, the chimpanzee subjects of our study remained in their social group and received their mirror exposure on a daily basis when they were locked in their individual feeding cages as part of their daily routine. During mirror exposure they could hear, smell, and to a limited extent see other chimpanzees and their caretakers, who were continuing with their normal husbandry practices. The difference in the social environment for mirror exposure and text may be especially relevant in view of the observations reported by Gallup (*SAAH*3) that he and Povinelli have found that chimpanzees exposed to mirrors in large cages containing other animals have much more to do and thus spend very little time looking at themselves in the mirror. Distraction by their cagemates and other extraneous events capture their attention and pull it away from the mirror. It is possible that because mirrors were present during temporary interruptions from their normal active social life, some of the chimpanzees in our study were not as interested in them as chimpanzees tested in social isolation. Thus, the social context of mirror presentation may be very important, with some individuals more sensitive to it as evidenced by their responses (or lack thereof). Further, as we will suggest later in this chapter, there may be species differences in the manner in which the social context of mirror exposure affects interest in and responsiveness to the mirror.

Indeed, as Gallup suggests, initial interest in the mirror is an issue with respect to MSR. Chimpanzees (or any organisms) that do not sample the mirror sufficiently to determine its qualities cannot be expected to evidence self-recognition in a later mark test. We (Swartz & Evans, 1991) eliminated one of our chimpanzee subjects for this reason: She was much more interested in looking at the human observers than at her own reflection in the mirror. However, this raises an important point with respect to reports of chimpanzee mirror self-recognition. Although we suggest earlier in this chapter that the social context during mirror exposure may explain some variation in relative interest in the mirror, there may also be individual differences in the degree to which chimpanzees may be interested in the mirror. Is it the case that previous reports of self-recognition in chimpanzees have included only those animals that have shown some interest in the mirror? There is no suggestion in the published reports of these studies that such subject selection has occurred. That some manner of subtle subject selection may have occurred in the past to provide such consistent robust findings of MSR in chimpanzees is supported by Povinelli's (SAAH18) report that of approximately 90 chimpanzees he has tested, only 5% under 7 years of age and 30% over age 7 show evidence of MSR.

Whether or not previous subject selection with chimpanzee subjects has occurred, the situation is that we now know that not all chimpanzees display an active interest in the mirror. It is important to determine

1. why that might be so,
2. what experimental variables (such as the social context of mirror exposure) might increase an individual animal's interest in the mirror, and
3. whether then that animal will demonstrate mirror self-recognition.

Gorillas, who fail to pass the mark test and fail to show a great deal of self-directed behavior in the presence of the mirror, also show little visual interest in the mirror (see Figure 11.1). Low levels of interest in the mirror may underlie a failure to pass the mark test, but the basis for this lack of interest remains another unanswered question with respect to mirror behavior. The failure of some chimpanzees to show sufficient interest in the mirror to support self- or mark-directed behavior does not explain the subject in our study (Makata) that showed interest in the mirror and demonstrated self-directed behaviors in its presence, but nevertheless failed to pass the mark test.

In the context of previous theoretical treatments of MSR, Makata's behavior provides a real puzzle with respect to this phenomenon. However, in the context of more recent theoretical approaches (Mitchell, 1993; Parker, 1991; Thompson & Contie, *SAAH*26), his behavior may be seen as showing some ability to make use of mirror information that may or may not eventually lead to a demonstration of mirror self-recognition with the mark test. The relationship between forms of self-referred and self-directed behaviors other than mark-directed behavior (e.g., Menzel, Savage-Rumbaugh, & Lawson, 1985) and behaviors using mirror information (e.g., Anderson, 1986) is only now being addressed. If one accepts self-directed behavior using the mirror image as a guide to indicate self-recognition, then chimpanzees Makata, Henri, and Nestor (Swartz & Evans, 1991), and several gorillas (Parker, *SAAH*19; see also discussion of additional cases in Patterson & Cohn, *SAAH*17) can be said to have demonstrated MSR. If one requires demonstrations of additional uses of the mirror with respect to the body (see Thompson & Contie, *SAAH*26), then the issue of individual differences and species differences is still unresolved.

Species differences

In addition to the observed individual differences obtained in mirror self-recognition, species differences have been well documented (Gallup, 1991; Ledbetter & Basen, 1982; Suarez & Gallup, 1981), with gorillas typically showing little interest in mirrors, demonstrating few self-directed behaviors, and not passing the mark test. In an attempt to study this failure of gorillas in detail, we collected data regarding the mirror behavior and mark test behavior of two gorillas at the same time that data were collected with chimpanzee subjects. The following is a comparison of the mirror behavior of our two gorilla subjects with that of the chimpanzees we studied.

Subjects

Two lowland gorillas (*Gorilla gorilla gorilla*) participated in the study. They were housed at the Centre International de Recherches Médicales de Franceville (CIRMF) in Franceville, Gabon, Africa. The first subject, Zoé, was a 5-year-old female who had been orphaned and brought to the primate facility at CIRMF when she was approximately 2 years old. At the time of the study, she was living in a social group with two younger male gorillas with similar histories. The second subject, Etoumbi, was a 14-year-old wild-born male who had been captured during infancy and had lived in a captive gorilla colony in England for ten years before arriving at CIRMF. At the time of the study he was living in a family group with an adult female and their 2-year-old female infant. The experimental histories of these animals contained no pathological, toxicological, or invasive procedures beyond periodic anesthetization and standard health care. They were, by all available information, healthy, disease free, visually unimpaired, and behaviorally intact.

Apparatus

Two mirrors were used. The smaller mirror, presented to Etoumbi, was a 0.6-m-square unbreakable Plexiglas mirror mounted on an iron pedestal with the bottom of the mirror 0.9 m from the floor. The mirror was placed approximately 0.45 m from the cage. Etoumbi was given mirror exposure and was tested in a raised feeding cage attached to his home cage. The mirror was positioned a little below the animal's eye level. The feeding cage was sufficiently large to allow movement, and to allow Etoumbi various views in the mirror of his own body in various areas of the cage, as well as a clear view of his face when he was at the front of the cage. He was well acclimated to the daily caging procedure, and often was kept much longer in the cage than required to ingest the food provided. The mirror and the method of mirror presentation for Etoumbi was identical to that used for ten of the chimpanzees in Swartz and Evans (1991).

The second mirror was the same mirror used for Makata in the Swartz and Evans (1991) study. Zoé was tested using this mirror in a small (1.2 m in all dimensions) test cage with a 0.6-m-wide and 1.2-m-tall mirror contained in one wall of the cage. The mirror was full length, directly accessible, and was not obscured by cage bars.

Procedure

The procedure was similar to that used by Swartz and Evans (1991). For purposes of mirror exposure both subjects were removed from their social groups and presented with the mirror. Zoé was given mirror exposure for 12 one-hour periods, during which she was observed constantly. Following 12 hours of mirror exposure, she was given a 1-hour mark test.

Etoumbi was provided with 80 hours of mirror exposure, in daily sessions that lasted $\frac{1}{2}$–1 hour. During the course of each midday mirror exposure session, Etoumbi was fed. He was observed half an hour a day during mirror exposure. Following 80 hours of mirror exposure he was anesthetized and marked on the brow and upper forehead as well as the wrist with pink greasepaint.

Specifics of procedure differed for the two subjects. Etoumbi was provided with mirror exposure and tested in his feeding cage. As a standard maintenance procedure at CIRMF, all adult animals were individually fed twice a day in their feeding cages. During this feeding time the mirror was presented to Etoumbi. He was provided with two 1-hour sessions of mirror exposure a day, in the morning and at midday. Observations were conducted at the beginning of the mirror exposure sessions, and lasted half an hour or 15 min each. In all cases, he was observed at least half an hour per day throughout the mirror exposure period. During the daily observation sessions, the observer sat in front of the cage, to the side of the mirror, in a position so that she was directly observable to Etoumbi and could always see the animal, although she could not see into nor be seen in the mirror. Behavior was recorded every 30 sec (instantaneous sampling) on a checklist. In addition, amount of time spent looking into the mirror was recorded cumulatively during each session. Behavioral responses while looking into the mirror were divided into three categories: social behavior, self-directed behavior, and other. Social behaviors were displaying at the mirror, hooting while looking at the mirror, grimacing, bouncing swinging, and, in the case of Zoé, physical contact with the mirror (mouthing, touching, swiping, ventral–ventral contact, hitting, hanging from, or approaching). Self-directed behaviors included any behaviors directed at the body or any manipulations of the body that occurred while looking into the mirror or while using the mirror to guide behavior or observe its outcome. These behaviors were touching, grooming, or scratching areas of the body visible only with the aid of the mirror (i.e., the face and back areas), making faces, making noises with the mouth, blowing bubbles, and manipulating food or water with the mouth, all done while watching in the mirror. Behaviors categorized as "other" were twirling, scratching, manipulating toes, and staring into the mirror. These behaviors were classified as "other" because they were considered ambiguous. Clearly self-directed behavior involves either (1) manipulation of the body using the mirror to guide the body-directed movements (self-directed behaviors) or (2) using the mirror to investigate the effects of certain behaviors, e.g., blowing bubbles, making faces, or making noises with the mouth while looking into the mirror (self-referred behaviors). These descriptions, used also by Swartz and Evans (1991), were developed from observations with chimpanzees and from the behaviors listed in previous studies of mirror behavior of great apes (Gallup, 1970; Suarez & Gallup, 1981).

Zoé's 12 daily one-hour mirror exposure sessions were provided in a cage, described above, separated from her living area but connected by a tunnel. At the beginning of each session she was lured into the cage through the tunnel and was then confined in the test cage. During Zoé's mirror exposure and mark test, her behavior was recorded with an ad-lib sampling procedure by both observers. To make the data comparable to all other animals tested, these data sheets were scored as though an instantaneous sampling procedure had been conducted, with only those behaviors occurring at each 30-sec interval being recorded. Both methods of sampling produced the same picture of Zoé's behavior.

Mark tests were scheduled to coincide with regular health exams. The animals were anesthetized with a mixture of ketamine HCI (5–10 mg/kg) and acepromazine-maleate (0.05–0.10 mg/ml) administered intramuscularly. Immediately following the health examination, while the animal was still deeply anesthetized, a mark was placed on the left brow ridge, the right ear, and the right wrist. The marks were pink, made by a combination of red and white greasepaint (Stein Cosmetics, New York) and dusted with corn starch (Suarez & Gallup, 1981). Based on the first author's experience with marks applied to her eyebrows, forehead, and ears, the marks were tactually undetectable to a human.

Following the marking procedure, each animal was returned to the mirror-presentation cage. They were informally observed periodically following their return to the feeding cage, and a note was made if they were observed to attend to the mark applied to the wrist. When recovery from the anesthesia was such that the animal was sitting up and seemed alert, a 30-min control period was begun. This period began 2.5–3 hours following administration of anesthesia. During the control period the mirror was in its standard place, with the nonreflecting side facing the gorilla. All mark-directed responses were recorded. Immediately following the control period the mirror was turned so that the reflecting side faced the animal, and the mark test was conducted.

Results

Figure 11.1 presents total time spent looking at the mirror during the course of mirror exposure for the two gorilla subjects (Panel a), and for chimpanzee subjects from previous studies (Panels b and c; Gallup, 1970; Swartz & Evans, 1991). As shown in Figure 11.1a, the amount of time spent looking into the mirror decreased over the 80 hours of mirror exposure for Etoumbi, and did not recover during his mark test. Zoé maintained high levels of looking at the mirror, relative to Etoumbi, but not relative to chimpanzees. Panels (b) and (c) present looking time for Gallup's (1970) original chimpanzee subjects and for chimpanzees included in the Swartz and Evans (1991) study. For purposes of comparison these data are divided into two groups: Panel (b) presents mean looking time for Gallup's (1970) original four chimpanzee subjects compared with the two Swartz and Evans (1991) chimpanzees, Berthe and Makata, who were most responsive to the mirror. Berthe was the single subject in our study who passed the mark test. Makata was extremely responsive to the mirror and showed a high level of self-directed behaviors (Figure 11.2b), but he did not pass the mark test. Figure 11.1c presents looking time data for the remaining six animals who completed our study (Amélie, N'tébé, Henri, Koula, Masuku, and Nestor), collapsed across subjects. These chimpanzee subjects showed low levels of self-directed behaviors during the mirror exposure period, and there was a clear difference in the degree of interest shown in the mirror by these subjects compared to the other chimpanzee subjects. Of note for the present discussion, the looking times obtained for Zoé and Etoumbi fall in the range of those demonstrated by the unresponsive chimpanzees shown in Figure 11.1c, and are consistently lower than the responsive chimpanzees presented in Panel (b).

Figure 11.2 presents self-directed and social behaviors during mirror exposure for gorilla subjects (Panel a) and chimpanzee subjects (Panels b and c). Figure 11.2a demonstrates that Etoumbi showed no social behavior directed

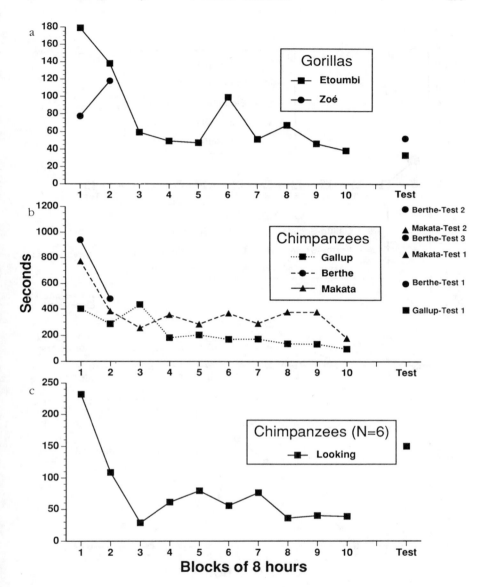

Figure 11.1. Mean looking time at mirror during mirror exposure and mark tests. Data are presented in 8-hour blocks to allow comparison with previously published data. (a) Gorilla subjects (Etoumbi and Zoé); (b) responsive chimpanzee subjects (Berthe and Makata) from Swartz and Evans (1991) and chimpanzees from Gallup (1970); (c) nonresponsive chimpanzee subjects (Amélie, N'tébé, Henri, Koula, Masuku, and Nestor) from Swartz and Evans (1991).

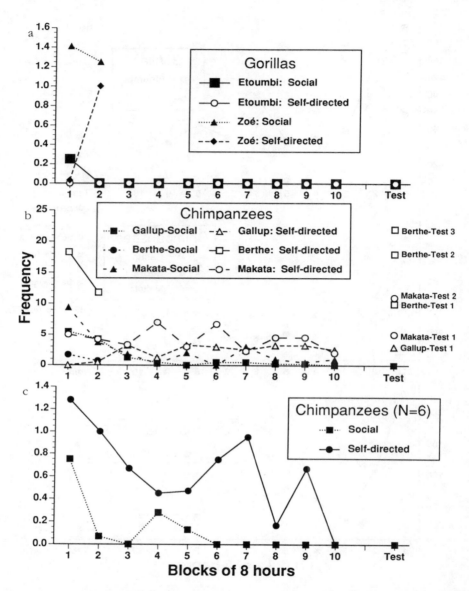

Figure 11.2. Mean social and self-directed behaviors during mirror exposure and mark tests. Data are presented in 8-hour blocks to allow comparison with previously published data. (a) Gorilla subjects (Etoumbi and Zoé); (b) responsive chimpanzee subjects (Berthe and Makata) from Swartz and Evans (1991) and chimpanzees from Gallup (1970); (c) nonresponsive chimpanzee subjects (Amélie, N'tébé, Henri, Koula, Masuku, and Nestor) from Swartz and Evans (1991).

toward the mirror, and only occasional self-directed behavior. Although Zoé showed social behavior comparable to that of nonresponsive chimpanzees (Fig. 11.2c), the frequency was extremely low compared to that of responsive chimpanzees (cf. Fig. 11.2b and note differences in scale on the ordinates). Neither gorilla touched the mark during the mark test.

Sources of individual and species differences

The important points to consider from the data are these:

1. Gorillas are less responsive to the mirror than are chimpanzees who pass the mark test.
2. Looking times, social behaviors, and self-directed behaviors of chimpanzees who do not pass the mark test are consistent with those behaviors in gorillas, showing that both may show lack of interest in the mirror.

Two factors may be operating here. First, the social context of testing may influence the degree of interest in the mirror during mirror exposure, with chimpanzees requiring a restricted social environment to produce interest and gorillas requiring a more supportive and rich social environment to induce attention to the mirror. (See Boccia, *SAAH*23, for a similar discussion in reference to social conditions that may facilitate mirror self-recognition in macaques.)

That social context is an important factor for gorillas was demonstrated clearly by King during mirror exposure (Siân Evans, unpublished data). King was approximately 22 years old at the time of test, having been a stage and circus performer before he was brought to Monkey Jungle in 1979. He was provided with 10 min of mirror exposure in daily 15-min sessions for 2 months. During all mirror exposure sessions, his trainer was present. Mirror exposure was broken up into 5-min segments: The back of the mirror was presented during the first 5 min, the mirror alone was presented during the second 5 min, and the trainer held the mirror during the last 5 min. When the trainer held the mirror, she did not interact with King or call attention to the mirror. She was careful not to speak to him about mirror-related events that would ordinarily prompt a response from her, but she was passive with respect to the mirror. King showed most self-directed behavior when the trainer held the mirror; behaviors included opening his mouth, chewing, and wiggling his fingers and toes, all while watching his image.

To prepare King for a mark test, he was wiped on the brow with a wet Q-tip each day for a month prior to the test. King accepted this as part of his daily routine and did not touch his brow or face following "water marking." On the day of King's mark test the water was replaced with a nontoxic pink paint that had been determined by the experimenters to be odorless and to provide no tactile cues when worn on the brow. During the 10-min nonmirror control period, King did not touch his brow. When the mirror was presented, he touched his brow twice within the 10 min. Following the first touch, he smelled his fingers and brought them to his mouth. King also showed additional behaviors that in their subtlety may reflect the species differences between chimpanzees and gorillas. He raised his brows several times while watching in the mirror, and he played "hide-and-seek" with his mark by placing his head so that the mark was hidden behind his cage bars and then moving his head just enough to reveal the mark in the mirror, all the while watching closely in the mirror. He also rubbed the mark on the cage bars. Some of these behaviors are consistent with those reported for Michael (see Patterson & Cohn, *SAAH*17, and discussion later in this chapter), who did not demonstrate many overt responses to the mark. Although King showed the classic behavior reported of many chimpanzees who pass the mark test (touching the mark and then smelling his fingers), he also showed some very subtle mark-directed behaviors. Note further that, in contrast to Koko, King's clear evidence of MSR appears in the absence of a history of early language training. As both Gallup (*SAAH*3) and Povinelli (*SAAH*18) suggest, early cognitive challenges may have enriched Koko's abilities sufficiently to allow demonstration of mirror self-recognition. King very likely had a human-supported and training-intensive early background, but it is unlikely that he was provided the same cognitively rich and structured environment. It may be the case that although early experience certainly expands the cognitive abilities of primates, that may not be the necessary condition for the demonstration of MSR by gorillas.

Consider the social context of the two gorillas (Koko [Paterson & Cohn, *SAAH*17] and King [Siân Evans, unpublished data]) who passed modified mark tests. Although neither of these gorillas was part of a social group, they both received extensive contact with human caretakers. Both King and Koko received mirror exposure in a familiar setting, and their preferred caretaker presented the mirrors to them and remained with them to record their responses. Encouragement rather than enforcement may be extremely important for the development of MSR in gorillas.

Testing gorillas in isolation with only an unfamiliar observer present, as is typically done, may not facilitate their interest in mirrors. The differences in the personalities of chimpanzees and gorillas are well known. Yerkes and Yerkes (1929) described gorillas as undemonstrative creatures compared with chimpanzees and recognized that they may possess capacities and potentialities that they decline to exhibit. Consistent with this suggestion is the report of the inconclusive results with the gorilla Michael reported by Patterson and Cohn (*SAAH*17). They suggest that Michael's equivocal behavior during the mark test may have been a function of the novelty of the social situation during the test, although they attribute his behavior to the distraction provided by the presence of two experimenters and they suggest that a better test of Michael's abilities would be to isolate him. We suggest that the best test for Michael (and any other human-oriented gorilla) would be to present the mirror in a familiar social setting in the presence of a caretaker who could provide undistracting social support and encourage interest in the mirror without directly influencing the test.

Species differences in responsiveness to mirrors may be affected by the social context of mirror exposure, but there may be a second reason for the gorilla's relative lack of interest in mirrors. Gallup (1991) has proposed a sampling error hypothesis to explain the failure of gorillas to show mirror MSR. He suggests that there may be species differences in the distribution of self-recognition and concedes that maybe a small percentage (5%) of gorillas can recognize themselves in mirrors compared with a much larger percentage of chimpanzees (80%; but see Povinelli, *SAAH*18). Indeed, although chimpanzees and gorillas are sympatric in some places (e.g., Gabon), their social structures are different. ·Gorillas live in stable family groups, whereas chimpanzees have a more fluid social structure (Tuttle, 1986.) Further, wild chimpanzees spend considerable time in tool-mediated extractive foraging (Goodall, 1968; Parker & Gibson, 1977) whereas wild gorillas engage only in hand-mediated extractive foraging (Byrne, 1991; Schaller, 1963). Because gorillas fail to use tools in the wild, they may be less interested in certain aspects of the environment, and may not be inclined to investigate the properties of the mirror. Our results with both Etoumbi and Zoé confirm that gorillas show little interest in mirrors. Further, Zoé's mirror-directed behavior did not show the pattern typical of chimpanzees. Although Zoé's social behaviors (which were all aggressive) did wane with mirror exposure, they were not replaced by self-directed behaviors (compare patterns of social and self-directed behaviors by Zoé and Makata, Figure 11.2a–b). These different tendencies of the two species to use information from the environment may have been exacerbated by the social context of mirror exposure and mark tests in previous studies, providing a seeming all-or-none dichotomy in the evidence available for mirror-guided, mark-directed behavior in great apes. As the mirror responses of more apes are investigated in social contexts judged most suitable for facilitating self-recognition for each species, perhaps meaningful estimates of the distribution of self-recognition, and the mechanism underlying these differential distributions, can be determined.

Self-recognition as demonstrated by the mark test may be preceded by a series of mirror responses that indicate an increasingly sophisticated use of the information provided by the mirror such that, eventually, the organism is able to use that information to address its own body and demonstrate what is currently interpreted to be self-recognition (Lin et al., 1992; Mitchell, 1993; Parker, 1991; Thompson & Contie, *SAAH*26; see also Lewis & Brooks-Gunn, 1979). This suggestion is related to Gallup's (1977b) idea that the presence of the mirror does not *cause* the acquisition of self-recognition in chimpanzees but *allows the demonstration* of a capacity inherent in the animal. It would appear that, starting with Gallup's (1970) initial demonstration of this phenomenon, the results of all studies of mirror behavior were interpreted in relation to self-recognition. In keeping with alternative theoretical treatments (Mitchell, *SAAH*6; Parker & Milbrath, *SAAH*7; Thompson &

Contie, *SAAH*26), it is clearly important to address the phenomenon of self-recognition from a comparative viewpoint and to separate it from mirror behavior. Animals from a number of species may show different abilities to use mirror information that may or may not include self-recognition. Self-recognition may be manifested qualitatively and quantitatively differently across species, and it should be investigated using methods in addition to mirrors.

Important questions remaining in this area are those of the developmental and phylogenetic differences of the use of mirror information, the relationship between cognitive abilities and the use of mirror information (independent of the demonstration of self-recognition), the role of social context in tests of mirror self-recognition, and the relationship between self-directed behaviors and mark-directed behaviors. Additional consideration of self-recognition, including the development of other measures of self-recognition, is warranted. Further, Gallup's (1991, *SAAH*3) suggestion that additional measures of social intelligence should be developed provides a clear step toward understanding this phenomenon. Addressing these questions and developing additional measures may provide a more extensive empirical framework for the development of theories of self-recognition and its cognitive correlates (cf. Anderson, 1984a,b; Gallup, 1979, 1982, 1991; Gallup & Suarez, 1986; Mitchell, 1993; Parker, 1991; Swartz, 1990).

Acknowledgments

We thank Gordon G. Gallup, Jr., for providing us with the raw data from Gallup (1970) included in Figures 11.1 and 11.2.

We thank the editors of this volume, Sue Parker, Robert Mitchell, and Maria Boccia, for their comments on an earlier version, which significantly strengthened the final one.

Preparation of this chapter was funded in part by grant No. S06 GM08225 from the National Institutes of Health awarded to Karyl B. Swartz. We thank Sharon Himmanen for her help in preparing the figures.

References

Anderson, J. R. (1984a). Monkeys with mirrors: Some questions for primate psychology. *International Journal of Primatology, 5*, 81–98.
(1984b). The development of self-recognition: A review. *Developmental Psychobiology, 17,* 35–49.
(1986). Mirror-mediated finding of hidden food by monkeys (*Macaca tonkeana* and *M. fascicularis*). *Journal of Comparative Psychology, 100,* 237–242.
Boysen, S. T., & Berntson, G. G. (1989). Numerical competence in a chimpanzee (*Pan troglodytes*). *Journal of Comparative Psychology, 103,* 23–32.
Byrne, R. (1991). Hand preferences in the skilled gathering tasks of mountain gorillas. *Cortex, 27,* 521–546.
Calhoun, S., & Thompson, R. L. (1988). Long-term retention of self-recognition by chimpanzees. *American Journal of Primatology, 15,* 361–365.
Cook, R. G., Brown, M. F., & Riley, D. A. (1985). Flexible memory processing by rats: Use of

prospective and retrospective information in the radial maze. *Journal of Experimental Psychology: Animal Behavior Processes, 11,* 453–469.

D'Amato, M. R., & Van Sant, P. (1988). The person concept in monkeys (*Cebus apella*). *Journal of Experimental Psychology: Animal Behavior Processes, 14,* 43–55.

Gallup, G. G., Jr. (1970). Chimpanzees: Self-recognition. *Science, 167,* 86–87.

(1977a). Absence of self-recognition in a monkey (*Macaca fascicularis*) following prolonged exposure to a mirror. *Developmental Psychobiology, 10,* 281–284.

(1977b). Self-recognition in primates: A comparative approach to the bidirectional properties of consciousness. *American Psychologist, 32,* 329–338.

(1979). Self-recognition in chimpanzees and man: A developmental and comparative perspective. In M. Lewis & L. A. Rosenblum (Eds.), *The child and its family* (pp. 107–126). New York: Plenum.

(1982). Self-awareness and the emergence of mind in primates. *American Journal of Primatology, 2,* 237–248.

(1991). Toward a comparative psychology of self-awareness: Species limitations and cognitive consequences. In G. R. Goethals & J. Strauss (Eds.), *The self: An interdisciplinary approach* (pp. 121–135). New York: Springer-Verlag.

Gallup, G. G., Jr., McClure, M. K., Hill, S. D., & Bundy, R. A. (1971). Capacity for self-recognition in differentially reared chimpanzees. *The Psychological Record, 21,* 69–74.

Gallup, G. G., Jr., & Suarez, S. D. (1986). Self-awareness and the emergence of mind in humans and other primates. In J. Suls & A. Greenwald (Eds.), *Psychological perspectives on the self* (Vol. 3, pp. 3–26). Hillsdale, NJ: Erlbaum.

Gallup, G. G., Jr., Wallnau, L. B., & Suarez, S. D. (1980). Failure to find self-recognition in mother–infant and infant–infant rhesus monkey pairs. *Folia Primatologica, 33,* 210–219.

Goodall, J. (1968). The behavior of free-ranging chimpanzees in the Gombe Stream Reserve. *Animal Behavior Monographs, 1,* 161–311.

Herrnstein, R. J., & Loveland, D. H. (1964). Complex visual concept in the pigeon. *Science, 146,* 549–551.

Hill, S. D., Bundy, R. A., Gallup, G. G., Jr., & McClure, M. K. (1970). Responsiveness of young nursery-reared chimpanzees to mirrors. *Proceedings of the Louisiana Academy of Science, 33,* 77–82.

Johnson, D. B. (1983). Self-recognition in infants. *Infant Behavior and Development, 6,* 211–222.

Ledbetter, D. H., & Basen, J. A. (1982). Failure to demonstrate self-recognition in gorillas. *American Journal of Primatology, 2,* 307–310.

Lethmate, J., & Dücker, G. (1973). Untersuchungen zum Selbsterkennen im Spiegel bei Orang-utans und einigen anderen Affenarten. *Zeitschrift für Tierpsychologie, 33,* 248–269.

Lewis, M., & Brooks-Gunn, J. (1979). *Social cognition and the acquisition of the self.* New York: Plenum.

Lin, A. C., Bard, K. A., & Anderson, J. R. (1992). Development of self-recognition in chimpanzees (*Pan troglodytes*). *Journal of Comparative Psychology, 106,* 121–127.

Menzel, E. W., Jr., Savage-Rumbaugh, E. S., & Lawson, J. (1985). Chimpanzee (*Pan troglodytes*) spatial problem solving with the use of mirrors and televised equivalents of mirrors. *Journal of Comparative Psychology, 99,* 211–217.

Mitchell, R. W. (1993). Mental models of mirror self-recognition: Two theories. *New Ideas in Psychology, 11,* 295–325.

Oden, D. L., Thompson, R. K. R., & Premack, D. (1990). Infant chimpanzees (*Pan troglodytes*) spontaneously perceive both concrete and abstract same/different relations. *Child Development, 61,* 621–631.

Parker, S. T. (1991). A developmental approach to the origins of self-recognition in great apes. *Human Evolution, 6,* 435–449.

Parker, S. T., & Gibson, K. R. (1977). Object manipulation, tool use and sensorimotor intelligence as feeding adaptations in cebus monkeys and great apes. *Journal of Human Evolution, 6,* 623–641.

Robert, S. (1986). Ontogeny of mirror behavior in two species of great apes. *American Journal of Primatology, 10,* 109–117.

Schaller, G. (1963). *The mountain gorilla: Behavior and ecology.* Chicago: University of Chicago Press.

Suarez, S. D., & Gallup, G. G., Jr. (1981). Self-recognition in chimpanzees and orangutans, but not gorillas. *Journal of Human Evolution, 10,* 175–188.

Swartz, K. B. (1990). The concept of mind in comparative psychology. *Annals of the New York Academy of Science, 602,* 105–111.

Swartz, K. B., Chen, S., & Terrace, H. S. (1991). Serial learning by rhesus monkeys: I. Acquisition and retention of multiple four-item lists. *Journal of Experimental Psychology: Animal Behavior Processes, 17,* 396–410.

Swartz, K. B., & Evans, S. (1991). Not all chimpanzees (*Pan troglodytes*) show self-recognition. *Primates, 32,* 483–496.

Tuttle, R. H. (1986). *Apes of the world.* Park Ridge, NJ: Noyes.

Yerkes, R. M., & Yerkes, A. W. (1929). *The great apes.* New Haven, CT: Yale University Press.

Note added in proof

Age estimates for Gallup's (1970) original subjects were derived from the population of wild-born chimpanzees living at CIRMF during 1979–86. Age estimates for the CIRMF population were based on tooth eruption and were constantly revised as the animals developed and more data became available. Using the most recently determined (and hence most accurate) age estimate for each animal, age–weight charts were constructed for all individuals. Because weights were taken at irregular intervals, not all animals were represented at all weight levels; however, there was a wide age and weight range represented overall. To ensure comparability with Gallup's subjects, data were used only from chimpanzees in the CIRMF population who had been in the laboratory for at least two years and who had come into the facility before age 3. This provided a pool of seven males and seven females. The age at which individual CIRMF chimps of the same gender attained the weight reported for each of Gallup's subjects was determined, and mean, median, and ranges were calculated. The obtained age estimates for each of Gallup's subject were as follows: Marge (no. 1825; 22.5 kg), mean = 5.4 years, median = 5.7, range = 4.6–6.2; Roxie (no. 355; 23.75 kg), mean = 5.6, median = 5.6, range = 5.0–6.4; Mack (no. 1831; 20.0 kg), mean = 5.0 years, median = 5.1, range = 4.4–5.4; Raymond (no. 1854; 14.75 kg), mean = 3.6, median = 3.6, range = 3.3–4.0 years. Six animals were used to estimate the ages of Roxie and Mack, seven for Marge, and two for Raymond.

The CIRMF population provides an ideal comparison group for Gallup's original subjects, with the exception of subspecies. The CIRMF animals used to obtain the above estimates were all *Pan troglodytes troglodytes*, which are smaller than *Pan troglodytes verus*. It is almost certain that Gallup's subjects were *Pan troglodytes verus*, as they were captured in Sierra Leone or Liberia, an area populated only by *P. t. verus* (International Animal Exchange, personal communication, September 21, 1993). If they were *P. t. verus*, then the above age estimates should be adjusted downward.

12   The comparative and developmental study of self-recognition and imitation: The importance of social factors

Deborah Custance and Kim A. Bard

Both mirror self-recognition (MSR) and "true" or "representational" imitation are among a host of different abilities that emerge in human infants between 18 and 24 months of age, when mental representation develops (Piaget, 1952). Mental representation is the highest cognitive achievement of the sensorimotor period. We believe that this is the basis for the expression of both self-recognition and imitation; by considering comparative and developmental evidence we shall discuss some of the cognitive conditions for each. Several researchers have suggested that the ability to imitate is a necessary condition for the development of self-recognition in human infants (e.g., Baldwin, 1903; Kaye, 1982; Lewis & Brooks-Gunn, 1979; see also Gopnik & Meltzoff, *SAAH*10; Hart & Fegley, *SAAH*9). However, contrary to this position, one hypothesis of this chapter is that imitation is not necessary for the development of self-recognition.

We believe that the developmental and comparative approaches must be used together in order to specify cognitive conditions for self-recognition and imitation (Mitchell, 1987; Parker, 1990, 1991). It is particularly difficult to determine the cognitive prerequisites by studying children alone, because many complex behaviors appear concurrently within a very narrow time frame. Developmental study of closely related primate species may be required to isolate the factors that are important in the development of cognitively complex behaviors such as self-recognition and imitation (Antinucci, 1989; Parker, 1991). Therefore, we shall utilize developmental data collected from our closest genetic relatives, chimpanzees, to compare with developmental data from human primates.

We are using Piaget's interpretation of the term "mental representation" (Piaget, 1952, 1962). Mental representation involves more than simply remembering past events or using a memory to direct one's present actions. Piaget described a dynamic process in which mental actions, images, or memories are *actively* manipulated in order to solve problems. In fact, the Piagetian term "mental representation" is close in meaning to Köhler's term "insight" or intelligence (Köhler, 1925). Köhler tested chimpanzees' intelligence by

207

presenting them with the novel situation of food hanging out of reach with several boxes scattered in the immediate vicinity. Intelligence was exhibited when the solution of stacking the boxes in order to reach the food was thought out, prior to acting. It was as if the subject created a mental model of the situation and then manipulated that mental model to solve the problem. All the trial-and-error problem solving took place internally. Therefore, for a behavior to indicate the presence of mental representation, there must be an indication that activity upon a mental model has occurred.

We begin with the hypothesis that mental representation is the cognitive foundation for the expression of both self-recognition and true imitation (see also Chapman, 1987). We will discuss cognitive conditions for the development of self-recognition, present data from a developmental study with chimpanzees, and then do the same for imitation. The final section of this chapter is a brief discussion of why we do not believe imitation is a necessary prerequisite for self-recognition.

The development of self-recognition

Self-recognition is the only aspect of the self that has been extensively studied in human and nonhuman primates (Anderson, 1984; Brooks-Gunn & Lewis, 1982; see review by Parker, Mitchell & Boccia, *SAAH*1; Suarez & Gallup, 1981). The standard test for self-recognition is the mirror mark test (Amsterdam, 1972; Gallup, 1970). Odorless, colored dye or rouge is surreptitiously placed on the face of a subject who is then presented with a mirror. The subject is judged to exhibit self-recognition if upon seeing itself in the mirror it touches the mark on its own face rather than directing its behavior toward its image in the mirror. Some argue that it is the process of deduction required in order to recognize that the mirror image is a reflection of one's own body that involves mental representation (see Mitchell, 1993, for more detailed discussion of deductive and inductive processes related to self-recognition).

The major goal of the study by Lin, Bard, and Anderson (1992) was to screen young chimpanzees of different ages in order to determine the age at which self-recognition occurs. A secondary aim was to compare the ages at which self-recognition and contingent behaviors occurred, to clarify the relationship between these behaviors in chimpanzees.

Methods

Subjects. Twelve chimpanzees (*Pan troglodytes*), $1\frac{1}{2}$–5 years old, were our subjects (nine males and three females). All subjects were born at the Yerkes Regional Primate Research Center of Emory University, and were raised in the great ape nursery. In order to study age-related developmental changes, the subjects were divided into four age groups: Shirlie and Brandy composed the 5-year-old group; Kasey, Carl, Justin, and Tank made up the 4-year-old group; Winston and Keith were the $2\frac{1}{2}$-year-old group; and Donald, Jarred, Lamar, and Fritz composed 2-year-old group.

Previous mirror and social experience. Experience with mirrors (Anderson, 1984) and with social partners (Gallup, McClure, Hill, & Bundy, 1971; Hill, Bundy, Gallup, & McClure, 1970; Lewis & Brooks, 1978) are crucial to the expression of self-recognition. Standard nursery conditions at the Yerkes Center include conspecific peer groups (as early as 1 month of age) and group access to play rooms containing permanent reflective metal plates. All the subjects were provided with experience both with mirrors and with conspecifics sufficient to allow for the expression of self-recognition if the chimpanzees had developed the cognitive capacity.

Apparatus. Tests were conducted either indoors or outdoors. Outdoors, a one-way mirror was placed close to the chain-link fence. The video camera was positioned behind the mirror, and a blind was constructed so that the one-way mirror effect was achieved. The experimenter recorded data from live observations from beside the mirror. For the indoor tests, a mirror was placed about 1 ft from the fence, and the video camera was positioned at an angle beside it. The experimenter recorded data from live observations from a position behind the camera.

Procedures. All sessions were conducted with subjects remaining in their peer groups, because separating an individual from the group can result in distressed behavior counterproductive to any assessment of cognitive ability. All subjects received three mirror sessions: a "pretest," which established a baseline of general mirror responses, and two test sessions that involved marking the subject. Subjects were marked on the brow ridge with a white odorless makeup cream. Two mark sessions were given to maximize the likelihood that mark-directed behavior could be displayed.

Mutually exclusive and exhaustive codes were constructed to categorize behavior both from live observations and from videotaped observations. *Inter*observer and *intra*observer reliability was assessed: Percentage agreement was 85% for both; Cohen's K was .76 for interobserver agreement and .78 for intraobserver agreement, scores that are considered good to excellent (Bakeman & Gottman, 1986).

Results

Mirror-related behavior can be interpreted as a measure of general interest in the mirror and its visual effects. Mirror-related behaviors include all those directed to the mirror (*mirror-directed* behaviors), including those directed to the mirror image (e.g., mirror-directed social displays), as well as *contingent* behavior, and *mirror-guided* behaviors (including those that are self-directed and *mark-directed*). Not surprising, the older chimpanzees spent more total time and a greater percentage of time engaged in mirror activities, compared to younger chimpanzees. Specifically, the 5-year-olds spent an average of 327 sec, and the 4-year-olds 263 sec on average engaged in behavior with the mirror. In comparison, the $2\frac{1}{2}$-year-olds spent an average of 149 sec, and the 2-year-olds an average of 118 sec engaged in mirror-related behavior. The total time chimpanzees spent engaged in mirror-related behaviors showed age-related trends.

Of course, we're really interested in the percentage of time spent in the following subcategories of mirror-related behaviors:

1. mirror-directed behavior, the least cognitively complex;
2. contingent movement; and
3. mirror-guided behavior, the most cognitively complex.

Mirror-directed behaviors were defined generally as those directed toward the mirror itself and toward the mirror image as a social partner. Thus, mirror-directed behaviors included looking either at the mirror or at the mirror-image, reaching toward and touching the mirror, and searching behind the mirror. Mirror-directed social behaviors included prosocial (e.g., playful) attempts to interact with the mirror image and displays of fear, dominance, submission, or aggression directed toward the mirror image (see Lin et al., 1992, for data). In this review, we will present only the results concerning contingent behavior and mark-directed behavior, one subset of mirror-guided behavior.

Contingent behaviors were defined, in general, as explorations of the one-to-one correspondence between the subject's action and the action of the mirror image (Lewis & Brooks-Gunn, 1979; Parker, 1991). Specifically, contingent behaviors consisted of moving while watching the movements of the mirror image. We recorded contingent behaviors involving the face separately from contingent behaviors involving the body.

We found that there were age-related trends in the average percentage of mirror time spent engaged in contingent behavior. The youngest subjects engaged in the smallest percentage of contingent behavior (about 15%), the $2\frac{1}{2}$-year-olds engaged in contingent behavior 25% of the time, and the 4-year-olds approximately 32% of the time. The oldest subjects engaged in the largest percentage of contingent behavior (about 45%).

Contingent behaviors were among the most frequently exhibited mirror behaviors in all age groups (Table 12.1). The 2-year-old group and $2\frac{1}{2}$-year-old groups exhibited the fewest contingent behaviors (an average of less than 10 in a session). The 4- and 5-year-old groups typically engaged in more than 15 contingent behaviors in each session. It is interesting to note that there was only one age group that exhibited a marked increase in contingent responses across test days: the 2-year-olds, the youngest age group.

The amount of body-directed contingent behavior exceeded the amount of face-directed contingent behavior in each subject and in each age group (Table 12.1). The ratio of body- to face-directed contingent behaviors, over all 3 test

Table 12.1. *The average number of contingent and mark-directed responses by young chimpanzees of four different ages*

	Baseline	Mark test 1	Mark test 2
Contingent: Body			
2-year-olds ($n = 4$)	3.7	12.0	4.0
$2\frac{1}{2}$-year-olds ($n = 2$)	14.5	5.0	4.5
4-year-olds ($n = 4$)	20.5	12.0	10.5
5-year-olds ($n = 2$)	13.5	12.0	12.5
Contingent: Facial			
2-year-olds ($n = 4$)	0	1.0	0.5
$2\frac{1}{2}$-year-olds ($n = 2$)	0.5	0	2.0
4-year-olds ($n = 4$)	1.8	2.3	1.0
5-year-olds ($n = 2$)	1.0	3.0	5.5
Contingent: Total body & facial			
2-year-olds ($n = 4$)	3.7	13.0	2.0
$2\frac{1}{2}$-year-olds ($n = 2$)	15.0	5.0	6.5
4-year-olds ($n = 4$)	22.2	14.0	11.5
5-year-olds ($n = 2$)	14.5	15.0	18.0
Mark-directed responses (mirror-guided)			
2-year-olds ($n = 4$)	—	0.5	0.25
$2\frac{1}{2}$-year-olds ($n = 2$)	—	0.5	4.0
4-year-olds ($n = 4$)	—	1.5	10.5
5-year-olds ($n = 2$)	—	4.0	4.5

sessions, were 13 : 1 for the 2-year-olds, 10 : 1 for the $2\frac{1}{2}$-year-olds, 8 : 1 for the 4-year-olds, and 4 : 1 for the 5-year-olds. In general, there were few instances of facial contingent behaviors.

Mirror-guided behaviors were defined as those directed toward an object, usually the self, while monitoring the reflected mirror image. Thus mirror-guided behaviors included *object reach* – locating an object visible only by means of the reflected image – and *self-directed behaviors* – behavior directed either to the chimpanzee's own body or face, exclusive of the mark. *Mirror-guided, mark-directed behavior* was the highest level of behavior and was coded with a separate category. Mirror-guided, mark-directed behavior involved touching the mark while monitoring the reflected image of the self, and is considered the definitive indication of self-recognition (see Gallup, 1970, and *SAAH*3).

The methodology typically used with nonhuman primates is the application of a dye to the forehead when the subject is anesthetized (e.g., Gallup, 1970). Self-recognition is assessed, upon recovery from the anesthesia, by comparing the amount of mark-directed behavior with and without the use of a mirror. When the subject is a human infant, he or she is marked with a spot of rouge placed on the side of the nose. Self-recognition is judged to be present if:

1. the infant touches the nose,
2. looks at the mark in the mirror, or
3. answers the mother's question of "Who's that?" by naming the image (e.g., Amsterdam, 1972).

In human infants, self-recognition is expressed as present or absent; in fact, the amount of mark-directed is rarely even reported. In this study, we have compromised between the methodologies: The chimpanzees were marked while awake, but the mark was placed along the brow and forehead rather than alongside the nose. In terms of the analysis the compromise consists of reporting the presence or absence of self-recognition for each individual, as is done with human subjects, and reporting the amount of mark-directed behavior with the mirror versus the amount without it, as is typically done with nonhuman primates (see Lin et al., 1992, for further details).

The number of mirror-guided, mark-directed behaviors are presented in Table 12.1. While looking at the mirror image, the chimpanzee, Scott, touches the mark on his own forehead (Figure 12.1). Scott wiped dirt off his face with a cloth; in this case, when he sees the white mark on his forehead, he takes a piece of cloth, and while looking at himself in the mirror, wipes off the mark. The cases in which the mark is touched without the use of the mirror are recorded with a separate category of behavior (Figure 12.2). Again, this is Scott wiping away the dirt with a cloth. This behavior is not counted as mirror-guided, mark-directed behavior. (Scott was a subject in the imitation study described in the next section.)

The 4- and 5-year-olds show mark-directed behavior with the use of the mirror at the highest levels, an average of over 4 instances in each session. There were relatively high frequencies of very clear instances of mark-directed behavior in the $2\frac{1}{2}$-year-old group, which led us to conclude that they do exhibit self-recognition. This was not the case with the youngest subjects. For the 2-year-olds, we concluded that they do not exhibit self-recognition, because of the lack of clear instances of mark-directed behavior with the use of the mirror. The age at initial self-recognition, as we have described here, is probably a conservative estimate.

In summary, self-recognition in chimpanzees appears to be consolidated between the ages of 2 and $2\frac{1}{2}$ years. Clear evidence for self-recognition was found in both 5-year-olds, in three out of four 4-year-olds, and in both $2\frac{1}{2}$-year-old chimpanzees. There was no compelling evidence for self-recognition in any of the four 2-year-olds. Young chimpanzees of the current study exhibit frequencies of mark-directed behavior at levels comparable to those found in adults of previous studies (Gallup, 1970; Suarez & Gallup, 1981). Young chimpanzees "pass" the mark test around the age of $2\frac{1}{2}$ years, slightly later than 18–24 months, the age at which most human infants pass the rouge test (Amsterdam, 1972; Brooks-Gunn & Lewis, 1982; Chapman, 1987; Hart & Fegley, SAAH9; Johnson, 1983). The overall results indicate that the age of onset of self-recognition in chimpanzees is only slightly later than that reported for human infants.

Figure 12.1. Scott, a 4½-year-old chimpanzee demonstrates self-recognition by using the mirror to guide his movements with a cloth to wipe the mark off his face.

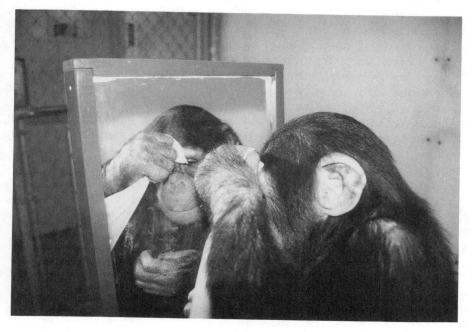

Figure 12.2. After looking at his marked mirror image, Scott turns away and wipes the mark off without looking at the result in the mirror.

With regard to our second question, whether there is a developmental relationship between contingent movement and mark-directed behavior, we answer, tentatively, "yes." All subjects exhibited contingent behavior. Young chimpanzees, even 2-year-olds, produced contingent movements. Our strongest indication of a developmental link between contingent movement and mark-directed behavior is that the 2-year-olds exhibited increased contingent movement rather than mark-directed behavior on the first mark test. Thus, it does appear that contingent behavior developmentally precedes self-recognition (Johnson, 1983; Lewis & Brooks-Gunn, 1979).

Mirror self-recognition demands knowledge that you are the agent of the actions of the mirror image, in addition to knowledge of the contingent aspect. You act and the image acts; in order to know that you are the origin of the image, you must recognize that your movements cause the image to move (e.g., Bigelow, 1981). In other words, you need an appreciation of causality. An understanding of causality may be a necessary condition for self-recognition (see Parker, 1991). Piagetian Stage 5 causality allows for the discovery of the mark through trial-and-error processes. The immediate search for the cause of the mark – the search on the face and not on the mirrored image – demonstrates Piagetian Stage 6 causality. Parker (1991) points out that cebus monkeys display some aspects of Stage 5 causality (i.e., in manual object-manipulation modality but not in the facial or gestural modality). Because cebus do not exhibit self-recognition (see Collinge, 1989) Parker therefore argues that achieving Stage 5 causality in other than the facial modality is not sufficient to cause the manifestation of mirror self-recognition. Alternatively, a Stage 6 level of causality might be required for MSR, that is, an ability to infer a casual agent from observing an effect, or the ability to predict an effect from observing the cause (Piaget, 1954). In other words, mental representation in the causality domain is required for MSR. If only the effect is observed (the mirror image has a mark) then it must be the result of a causal agent acting on the individuals' own face. If the infant knows the agent has acted (the mother having touched the infant's nose) then the mirror is utilized to observe the resultant effect, but the infant already knows that an effect exists as a result of the actions of the casual agent. Monkeys have not been found to display Piagetian Stage 6 in the causality domain, or in any modality for that matter (see Antinucci, 1989; Natale, Antinucci, Spinozzi, & Poti', 1986; Parker, 1991; Parker & Gibson, 1977; Visalberghi & Fragaszy, 1990). We can concur with Parker (1991) that an understanding of causality is required for the development of self-recognition, but the studies conducted to date cannot provide definitive conclusions regarding whether Stage 5 or Stage 6 levels are required.

The development of imitation

Piaget (1962) discussed sensorimotor imitation in terms of a sequence of six successive stages from birth through 2 years of age. It is not until Stage 6, with the appearance of delayed imitation, that Piaget believed there was evidence of mental representation. In delayed imitation, an individual watches a model perform a novel action and then at a later time, in the absence of a model, she reproduces the behavior. Piaget argued that the infant referred to a stored mental image of the modeled actions because she had no opportunity to learn it through simpler learning processes such as trial and error. Deferred imitation is representational because one actively uses the stored mental image in order to reproduce the novel behavior oneself. It is the mental activity involved in translating the stored mental image of another's actions into a set of motor acts, so that one can reproduce the same behavior, that distinguishes Stage 6 imitation as a representational ability (Bruner, 1970; Whiten & Byrne, 1988; Whiten & Ham, 1992). During Stage 6 the infant also performs immediate and accurate imitations of complex, novel behaviors, which Piaget (1962) argued also involved mental representation because "accommodation had taken place internally and without external experimentation" (p. 66).

Contingency is an important factor in both self-recognition and imitation, but the contingencies involved in the two processes differ in significant ways. For example, the contingency involved in mirror self-recognition is reversed from what we ordinarily think of as imitation; that is, the subject performs an action and the mirror "imitates" it. Moreover, the contingency between the observer's and the mirror image's actions is perfect, while there is rarely, if ever, perfect contingency found between two living organisms (see Watson, *SAAH*8). In MSR the subject can use the perfect contingency with the mirror to help guide the hand to the mark on the face, and locate the mark by seeing the hand touch the mark in the mirror image. Conversely, in imitation the subject is presented with the goal action and it must somehow work out how to replicate the model. Hence, the nature of the contingent feedback involved in self-recognition and imitation is very different indeed.

Social experience: Scaffolding

In our opinion, the most important question is why a chimpanzee, or a human infant for that matter, should imitate actions that have no direct function or goal. Adult human caregivers play a crucial role in the development of imitation in human infants (e.g., Kugiumutzakis, 1992; Meltzoff & Moore, 1977; Trevarthen, 1979). Adults shape and direct their babies' behavior so that the infants' imitative responses become more systematic (Kaye, 1982). Very early in infancy, turn-taking imitative exchanges emerge (e.g., Brazelton, Koslowski, & Main, 1974). Human parents constantly provide new models that are slightly more advanced than the actions their infant already performs (the notion of "zone of proximal development"; see Vygotsky, 1978). The system of development in which parents provide the structure and direction of the infant's learning agenda is called *scaffolding* (Wood, Bruner, & Ross, 1976).

We suggest that imitation requires scaffolding. This could explain why the strongest experimental evidence for true imitation in nonhuman primates has come from apes who were raised by human adults from a young age (Bard, in press-a; Galef, 1988; Russon & Galdikas, 1993a; Tomasello, Kruger, & Ratner, 1993). However, some convincing examples can be found from field research. There are two observations of demonstrational teaching of nut cracking techniques by wild chimpanzees (Boesch, 1991) and many observations of tutorials in the learning of communicative signals (Bard, in press-a; van de Rijt-Plooij & Plooij, 1989). Presumably there is little point in demonstrating an action to an infant if he or she cannot imitate. Mother chimpanzees perform many kinds of supportive behaviors that facilitate the learning of skills, such as nut cracking. Thus, young chimpanzees receive a degree of scaffolding, the structure that supports learning commensurate with existing ability. However, the turn taking exchanges and scaffolding received by young mother-reared chimpanzees do not equal those that human adults provide human infants (see Bard, 1993, in press-b; Parker, 1993; Plooij, 1979). Hence, the imitative response in wild chimpanzees appears to be less robust than imitation among humans and among human-reared chimpanzees.

Development of imitation in chimpanzees

We conducted a study of imitation in two chimpanzees that had extensive human contact but no experience with imitation. This study can be viewed as an exploratory investigation into the social experiences that are important in order to imitate.

Initial tests of imitation were begun with a "do as I do" experiment with two chimpanzees who were approximately $4\frac{1}{2}$ years of age. The imitation project was begun with these two chimpanzees in particular, because they were very cooperative, they were old enough to be cognitively capable of imitation, and they were older than chimpanzees who passed the self-recognition mark test (e.g., Lin et al., 1992). Scott and Katrina's rearing is typical of one set of conditions at the Yerkes Research Center. Their mothers did not have sufficient maternal behaviors to provide adequate care, so they were placed in the nursery within hours of birth (see Bard, in press-a, in press-c for more details of maternal competence in chimpanzees). Although given regular contact with humans, they were raised with same-aged peers from 6 weeks and 4 weeks of age, respectively. Chimpanzees are raised in the great ape nursery with the explicit goal of attaining chimpanzee species-typical behaviors. As young chimpanzees leave the nursery, they are gradually integrated into groups containing adult chimpanzees. While Scott and Katrina were in the great ape nursery they participated in much of the recent developmental research conducted there (Bard, in press-d; Bard, Hopkins, & Fort, 1990; Bard, Platzman, & Gardner, 1991; Bard, Platzman, Lester, & Suomi, 1992; Hopkins, Bard, Gardner, & Bennett, 1993).

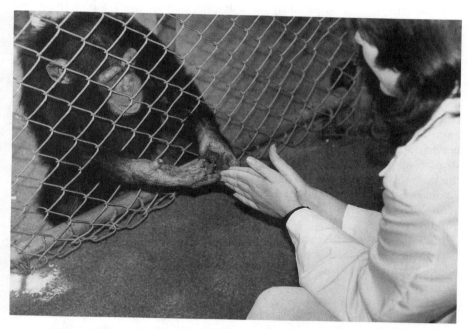

Figure 12.3. Scott, when 4½-years old, is naive to imitation and prefers to coact with a demonstrator's actions.

In these initial screens, the experimenter said "do this" and demonstrated one of eight arbitrary actions (Custance, Whiten, & Bard, 1993). Each of the actions was demonstrated repeatedly at different times of day; over the course of two weeks of initial pilot testing the chimpanzees' responses were recorded. When the "imitation of simple actions" task began, Scott and Katrina quickly learned to respond with some general actions, but they rarely reproduced the demonstrated action. One chimpanzee responded correctly to three items that involve touching the experimenter. The other chimpanzee responded correctly to two of these touching demonstrations.

We decided to institute a teaching procedure similar to that outlined by Hayes and Hayes (1952; Custance, Whiten, & Bard, in press). Learning took many steps: Scott and Katrina first learned to replicate the actions through a pattern of coacting with the demonstrator; then they learned turn taking; finally, attention to the model's action was emphasized.

As Scott learned "praying hands," it seemed that he conceived of the task as touching fingertips with the experimenter (Figure 12.3). Scott, in particular, showed a great desire to direct actions toward the demonstrator, trying to touch her or reach out to perform the action within her space. Through a gradual shaping process, performance was changed to making "praying hands." The same fading procedure was used for the actions, "slap the floor," "raise one arm," and "raise two arms."

Some actions were inspired by the chimpanzees' behavior, which the demonstrator then imitated. In this way, we included familiar actions already in the chimpanzees' repertoire within the actions to be imitated. "Wipe floor" was inspired by Scott wiping a lettuce leaf back and forth on the floor. The first time the demonstrator showed Scott the action, Scott did the same thing, but he did not repeat it again. So the demonstrator planned to improvise by demonstrating a combination of an old action ("slap the floor") with a new twist (from side to side). However, Scott began to do the action with his palm facing up. The demonstrator paused in the action and clearly turned her hand over. Scott hesitantly turned his over the same way. The demonstrator was so surprised that Scott imitated the action by demonstration that the demonstrator turned her hand back the other way; again he imitated it. Again the demonstrator flipped her hand over, and again Scott imitated the action. That was one of the clearest signs that Scott was beginning truly to imitate.

It was a bit more difficult to teach the chimpanzees to reproduce relatively unfamiliar actions independent of the demonstrator, such as "shake hand." Katrina rarely shook her hand. She learned after inadvertently putting her hand in a puddle of water, then shaking it vigorously. Immediately the demonstrator verbally rewarded her and showed her the action again, which Katrina repeated. The demonstrator rewarded her and continued to demonstrate 'shake hand' and reward her for repeating the action until Katrina got bored and moved away. From then on she was a little sporadic in her responses, but she did shake her hand more frequently than she had before.

Next, it was necessary for the chimpanzees to learn to take turns: to attend rather than act as the experimenter demonstrated the action, and only then to imitate the act. The demonstrator was careful, once the chimpanzees performed the actions regularly and well, to reward them only after the action had been demonstrated. It was important to ensure that a clear turn-taking routine had been established so that the chimpanzees knew they were to perform an action only after it had been demonstrated.

It took some time before Scott and Katrina engaged in turn-taking exchanges, which involved self-directed behaviors such as touching one's chin or patting one's stomach. The training procedure eventually adopted was to demonstrate the action and then, for example, to touch Scott's chin, which emphasized the desired action and that Scott was to perform it on his own body (e.g., Adamson & Bakeman, 1984). This is a familiar procedure, one that every human infant has probably experienced thousands of times as people play "touch your body part" games. This social game, however, had never been played with these nursery-reared chimpanzees. We feel that games like this one provide necessary experiences that can be considered tutorials in imitation (note that Mitchell, 1993, states a similar hypothesis).

We also used positive reinforcement for the combination of independent action with self-directed behavior, such as with the items "wipe hands," "pat stomach," and "wipe face." For "pat stomach," the demonstrator first demonstrated the action, then gently poked Scott in the stomach; he waved his

Figure 12.4. After a 3-month period of teaching in which he learns imitative skills, Scott readily imitates even self-directed actions, such as 'grasp wrist.'

hands over the area, presumably because it tickled. That was the best imitative action that he seemed able to do for many weeks. Then one day the demonstrator paused, and very slowly and deliberately patted by alternating hand pats to her stomach. Scott paused in his waving hands, watched and then clearly patted alternately with each hand. That was one of the first indications that he could truly imitate arbitrary actions.

Learning progressed during the 3-month teaching period; near the end of that time, Scott and Katrina learned new actions with minimal shaping and imitated one or two new actions within the first few demonstrations. For instance, "grasp wrist" was learned by Scott with minimal shaping, even though Scott consistently performed this action with his hands first extended through the fencing (Figure 12.4). Katrina did not require the same level of prolonged shaping throughout training, and by the time this action was learned, she was able to perform it within her own space.

After the 3-month teaching period, a series of 45 novel actions was interspersed with the training actions. Imitation tests were given to Scott and Katrina for 30-min sessions twice a day. Preliminary evidence suggested that Scott and Katrina were able to imitate novel actions (Custance et al., 1993).

In summary, at the beginning of the imitation testing, at $4\frac{1}{2}$ years of age, Scott and Katrina did not imitate. A period of learning and tutorials emphasizing social-imitative rules lasted 3 months. During this time, aspects of imitation were learned in steps that followed those typically taken in human development. The chimpanzees learned the social rules of imitation, that is, turn taking, matching the models' actions, and directing behavior toward oneself; then both Scott and Katrina understood the concept of "do as I do." Thereafter, they were capable of imitating many novel actions and always recognized the novel actions interspersed with the training actions.

Facial imitation model

Parker (1991) argued that the manner in which a subject discovers that the marked face in the mirror is its own, is by producing and monitoring facial actions. In the Facial Imitation Model, three factors are necessary for the development of mirror self-recognition:

1. voluntary control of the facial muscles;
2. the recognition that one's own facial actions are the cause of the identical actions in the mirror image (which, according to Parker, requires at least a Piagetian Stage 4 understanding of causality); and
3. at least Piagetian Stage 4 imitation in the facial modality.

We have already discussed the roles played by contingency-movement testing and an understanding of causality in the development of MSR. We have also considered the development of imitation and the necessary and sufficient conditions involved in imitation. Now we will briefly consider the possible role played by imitation in the development of self-recognition. Our basic plan, in this section, is to discuss the role of facial imitation and facial-contingent behavior in the development of MSR. Although not our original intent, many of the arguments could also be applied to Mitchell's (1993) theory that kinesthetic–visual matching is the ability that underlies both imitation and mirror self-recognition.

If imitation is a necessary condition for the development of self-recognition, then one would expect to find it among species that exhibit self-recognition. In fact, just as with self-recognition, the strongest evidence for true imitation does come from three species, namely humans (e.g., Meltzoff, 1988) and chimpanzees (Hayes & Hayes, 1952; Mathieu, 1982), as well as orangutans (Miles, 1990; Russon & Galdikas, 1993b). This lends support to Parker's (1991) suggestion that Stage 4 (or higher) facial imitation is a necessary condition for self-recognition.

However, as we mentioned earlier, there was very little evidence of facial-contingent actions in young chimpanzees either prior to or following the mark test. The youngest subjects did not engage in more facial-contingent behavior than the older subjects. We have not considered whether scaffolding is necessary for self-recognition as well as imitation; however, there is evidence to suggest that direct social contact *is* necessary for self-recognition (Hill et al., 1970). Social contact is sufficient for self-recognition and is not equivalent to developmental scaffolding from a competent adult. So, although scaffolding is required for the development of a robust imitative capacity, it does not appear necessary for the development of self-recognition.

Direct test

In the process of conducting longitudinal comparative developmental studies at the Yerkes Center we were able to test both self-recognition and imitation in Scott, a single chimpanzee subject. This opportunity allowed us to perform a preliminary test of our hypothesis that imitation is not necessary for mirror self-recognition, a topic related to the Facial Imitation Model (Parker, 1991) and to the kinesthetic–visual matching model of Mitchell (1993).

When Scott was 3 years old he was given a demonstration of how to use a stick as a tool to obtain a grape from a long tube (Bard, Fragaszy, & Visalberghi, 1993). Scott watched intently, but rather than imitate the task, he gave the stick to the demonstrator in order for her to solve the task. Eventually, he did learn to solve the task by himself in fewer trials than Katrina, who served as the matched-age control chimpanzee (i.e., without a demonstration). This task illustrates:

1. his ability to use trial-and-error strategies to solve a problem (Piagetian Stage 5),
2. Stage 5 understanding of others as casual agents,
3. some benefit of demonstration, but
4. a lack of imitative copying of actions.

In the months prior to the self-recognition test and the imitation of arbitrary actions, Scott was provided an opportunity to solve another task by imitating a model (Whiten, Custance, Gómez, Bard, & Texidor, 1993). Although he did benefit from observation of the model, again there was little overt evidence of an ability to imitate specific actions.

When first confronted with the mirror, Scott exhibited an ability to use the properties of the mirror to locate objects. Figure 12.5 illustrates his ability to locate the camera and photographer in the mirror. In addition, he was able to direct action on his own body when he saw the mark on the mirror image. Figure 12.1 illustrates mirror-guided, mark-directed behavior. Thus, at $4\frac{1}{2}$ years of age, Scott exhibited self-recognition.

In the descriptions of learning in the "do as I do" task, we have discussed the importance of the socialization and teaching processes involved in performing imitation. Although Scott had cognitive skills sufficient both to solve a tool task and to recognize himself in a mirror, he had not been exposed to the benefits of imitating a knowledgeable model. Thus, he showed little evidence of an ability to imitate arbitrary actions (facial or gestural) even if objects were not involved. During the course of the 3-month training period, we found that imitative skills developed through processes similar to those that occur in early human development; that is, turn-taking skills are reinforced in social games; in communicative exchanges, meaning is attributed; and the

Figure 12.5. Scott, a $4\frac{1}{2}$-year-old chimpanzee, has the ability to locate objects in the mirror.

concept of "do as I do" has to have attained meaningful consequences to the young chimpanzee before imitative performance is possible.

The hypothesis that imitative ability is not necessary for MSR is based on an extremely small sample. We would recommend to future researchers that a test of imitation of arbitrary actions be conducted without any training for the individual subjects prior to the teaching phase. We are currently proposing a longitudinal research project to chart concurrent development in self-recognition and imitation in chimpanzee subjects. Furthermore, more subjects are needed to clarify the development of cognitive processes and the issue of necessary and sufficient conditions for both self-recognition and imitation.

Conclusions

We set out to specify the relationship between imitation and self-recognition. They share many similar necessary conditions such as motivation and mental representation, but we believe neither's development is dependent on the development of the other. For example, most apes who have not received scaffolding from humans do not develop a robust imitative capacity, but they still develop self-recognition (but see Swartz & Evans, 1991). For that matter, children with autism who exhibit severe deficits in motor imitation neverthe-less develop self-recognition (Dawson & McKissick, 1984). Therefore, we suggest that imitation is not necessary for development of self-recognition (but see Boysen, Bryan, & Shreyer, *SAAH*13; Hart & Fegley *SAAH*9; and Gopnik & Meltzoff, *SAAH*10 for alternative perspectives). Neither is self-recognition necessary for the development of true imitation.

Through a consideration of the developmental conditions for self-recognition and imitation, we support previous suggestions that mental representation, characteristic of the sixth stage of the sensorimotor period (Piaget, 1952, 1954) is necessary for the expression of both behaviors. Moreover, mental representation, when developed in a background of direct social experience, is sufficient for the expression of self-recognition. However, we find that although mental representation may be necessary, it is not sufficient for the development of imitation. In order to develop the ability to perform delayed imitation and to imitate a complex sequence of actions on the first exposure, an additional condition is necessary. Imitative abilities develop in social interactions of a specific type: social interactions in which one partner *teaches* specific imitative skills.

Acknowledgments

Preparation of this chapter was supported, in part, by NIH grant RR-00165 to the Yerkes Regional Primate Research Center from the National Center of Research Resources, NIH Grant RR-03591 to R. B. Swenson of the Yerkes Center, and NIH Grant RR-06158 to K. A. Bard. The Yerkes Center is fully accredited by the American Association for Accreditation of Laboratory Animal Care. We are grateful to the veterinary staff, the animal care staff, Cindy Cross, Kathy Gardner, Albert Lin, and Dr. Andrew Whiten, for their assistance. We appreciate the comments made on preliminary versions of this chapter by Drs. William Hopkins, Daniel Hart, Sue T. Parker, and R. W. Mitchell.

References

Adamson, L., & Bakeman, R. (1984). Mother's communication acts: Changes during infancy. *Infant Behavior and Development, 7,* 467–487.

Amsterdam, B. (1972). Mirror self-image reactions before age two. *Developmental Psychobiology, 5,* 297–305.

Anderson, J. R. (1984). Development of self-recognition: A review. *Developmental Psychobiology, 17,* 35–49.

Antinucci, F. (1989). *Cognitive structures and development in nonhuman primates.* Hillsdale, NJ: Erlbaum.

Baldwin, J. M. (1903). *Mental development in the child and the race.* New York: Macmillan.

Bakeman, R., & Gottman, J. (1986). *Observing interaction: An introduction to sequential analysis.* Cambridge University Press.

Bard, K. A. (1993). A developmental theory requires developmental data. *Behavioral and Brain Sciences, 16,* 511–512.

(in press-a). Evolutionary roots of intuitive parenting: Maternal competence in chimpanzees. *Early Development and Parenting.*

(in press-b). Very early social learning: The effect of neonatal environment on chimpanzees' social responsiveness. *Proceedings of the 14th Congress of the International Primatological Society,* Strasbourg, France.

(in press-c). Parenting in primates: The special case of the ontogeny of maternal competence.

In M. Bornstein (Ed.), *Handbook of parenting, Vol. 2: Ecology and Biology of Parenting*. Hillsdale, NJ: Erlbaum.

(in press-d). Similarities and differences in the neonatal behavior of chimpanzee and human infants. In G. Eder (Ed.), *The role of the chimpanzee in research*. Basel: Karger-Verlag.

Bard, K. A., Hopkins, W. D., & Fort, C. L. (1990). Lateral bias in infant chimpanzees (*Pan troglodytes*). *Journal of Comparative Psychology, 104*, 309–321.

Bard, K. A., Fragaszy, D., & Visalberghi, E. (1993). Acquisition and comprehension of a tool-using behavior by young chimpanzees: Effects of age and modeling. (Submitted for publication.)

Bard, K. A., Platzman, K. A., & Gardner, K. (1991). Young chimpanzees score well on the Bayley scales of infant development. *American Journal of Primatology, 24*, 89.

Bard, K. A., Platzman, K. A., Lester, B. M., & Suomi, S. J. (1992). Orientation to social and nonsocial stimuli in neonatal chimpanzees and humans. *Infant Behavior and Development, 15*, 43–56.

Bigelow, A. F. (1981). The correspondence between self- and image movement as a cue to self-recognition for young children. *Journal of Genetic Psychology, 139*, 11–26.

Boesch, C. (1991). Teaching among wild chimpanzees. *Animal Behavior, 41*, 530–532.

Brazelton, T. B., Koslowski, B., & Main, M. (1974). The origins of reciprocity: The early mother–infant interaction. In M. Lewis & L. Rosenblum (Eds.), *The effect of the infant on its caregiver*. New York: Wiley.

Brooks-Gunn, J., & Lewis, M. (1982). The development of self-knowledge. In C. B. Kopp & J. B. Krakow (Eds.), *The child: Development in a social context* (pp. 332–387). Reading, MA: Addison-Wesley.

Bruner, J. (1970). The growth and structure of skill. In K. J. Connolly (Ed.), *Mechanisms of motor skill development* (pp. 63–94). New York: Academic Press.

Chapman, M. (1987). A longitudinal study of cognitive representation, symbolic play, self-recognition, and object permanence during the second year. *International Journal of Behavioral Development, 10*, 151–170.

Collinge, N. E. (1989). Mirror reactions in a zoo colony of *Cebus* monkeys. *Zoo Biology, 8*, 89–98.

Custance, D., Whiten, A., & Bard, K. A. (1993). Can young chimpanzees (*Pan troglodytes*) imitate arbitrary actions? Hayes & Hayes (1952) revisited. (Submitted for publication.)

(in press). The development of gestural imitation and self-recognition in chimpanzees (*Pan troglodytes*) and children. *Proceedings of the 14th Congress of the International Primatological Society*, Strasbourg, France.

Dawson, G., & McKissick, F. G. (1984). Self-recognition in autistic children. *Journal of Autism and Developmental Disorders, 14*(4), 383–394.

Galef, B. G., Jr. (1988). Imitation in animals: History, definitions and interpretations of data from the psychological laboratory. In T. R. Zentall & B. G. Galef, Jr. (Eds.), *Social learning: Psychological and biological perspectives* (pp. 3–28). Hillsdale, NJ: Erlbaum.

Gallup, G. G., Jr. (1970). Chimpanzees: Self-recognition. *Science, 167*, 86–87.

Gallup, G. G., Jr., McClure, M. K., Hill, S. D., & Bundy, R. A. (1971). Capacity for self-recognition in differentially reared chimpanzees. *The Psychological Record, 21*, 69–74.

Hayes, K. J., & Hayes, C. (1952). Imitation in a home-raised chimpanzee. *Journal of Comparative and Physiological Psychology, 45*, 450–459.

Hill, S. D., Bundy, R. A., Gallup, G. G., Jr., & McClure, M. K. (1970). Responsiveness of young nursery-reared chimpanzees to mirrors. *Proceedings of the Louisiana Academy of Sciences 33*, 77–82.

Hopkins, W. D., Bard, K. A., Gardner, K., & Bennett, A. (1993). Sleeping posture and hand preferences in infant chimpanzees (*Pan troglodytes*): Ontogenetic correlates. Manuscript submitted for publication.

Johnson, D. B. (1983). Self-recognition in infants. *Infant Behavior and Development, 6*, 211–222.

Kaye, K. (1982). *The mental and social life of babies: How parents create persons*. Chicago: University of Chicago Press.

Köhler, W. (1925). *The mentality of apes*. London: Routledge & Kegan Paul.

Kugiumutzakis, G. (1992). Self-recognition as precondition of the infant imitative communication. Paper presented at the 14th Congress of the International Primatological Society. Strasbourg, France.

Lewis, M., & Brooks, J. (1978). Self-knowledge and emotional developmental. In M. Lewis & L. A. Rosenblum (Eds.), *The development of affect* (pp. 205–226). New York: Plenum.

Lewis, M., & Brooks-Gunn, J. (1979). *Social cognition and the acquisition of self*. New York: Plenum.

Lin, A. C., Bard, K. A., & Anderson, J. R. (1992). Development of self-recognition in chimpanzees (*Pan troglodytes*). *Journal of Comparative Psychology, 106*, 120–127.

Mathieu, M. (1982). *Intelligence without language*. Paper presented at the joint meetings of the International Primatological Society and the American Society of Primatologists, Atlanta, GA.

Meltzoff, A. (1988). The human infant as *Homo imitans*. In T. R. Zentall & B. G. Galef, Jr. (Eds.), *Social learning: Psychological and biological perspectives* (pp. 319–341). Hillsdale, NJ: Erlbaum.

Meltzoff, A., & Moore, M. K. (1977). Imitation of facial manual gestures by human neonates. *Science, 198*, 75–78.

Miles, H. L. W. (1990). The cognitive foundations for reference in a signing orangutan. In S. T. Parker & K. R. Gibson (Eds.), *"Language" and intelligence in monkeys and apes: Comparative developmental perspectives* (pp. 511–539). Cambridge University Press.

Mitchell, R. W. (1987). A comparative-developmental approach to understanding imitation. *Perspectives in Ethology, 7*, 183–215.

(1993). Mental models of mirror self-recognition: Two theories. *New Ideas in Psychology, 11*, 295–325.

Natale, F., Antinucci, F., Spinozzi, G., & Poti', P. (1986). Stage 6 object concept in nonhuman primate cognition. *Journal of Comparative Psychology, 100*, 335–339.

Parker, S. T. (1990). Origins of comparative developmental evolutionary studies of primate mental abilities. In S. T. Parker & K. R. Gibson (Eds.), *"Language" and intelligence in monkeys and apes* (pp. 3–64). Cambridge University Press.

(1991). A developmental approach to the origins of self-recognition in great apes. *Human Evolution, 6*, 435–449.

(1993). Imitation and circular reactions as evolved mechanisms for cognitive construction. *Human Development, 36*, 309–323.

Parker, S. T., & Gibson, K. R. (1977). Object manipulation, tool use, and sensorimotor intelligence as feeding adaptations in cebus monkeys and great apes. *Journal of Human Evolution, 6*, 623–641.

Plooij, F. (1979). How wild chimpanzee babies trigger the onset of mother–infant play and what the mother makes of it. In M. Bullowa (Ed.), *Before speech: The beginnings of interpersonal communication* (pp. 223–243). Cambridge University Press.

Piaget, J. (1952). *The origins of intelligence in children*, New York: Norton.

(1954). *The construction of reality in the child*, New York: Basic.

(1962). *Play, dreams and imitation in childhood*. New York: Norton.

Rijt-Plooij, H., van de & Plooij, F. X. (1987). Growing independence, conflict and learning in mother–infant relations in free-ranging chimpanzees. *Behaviour, 101*, 1–86.

Russon, A., & Galdikas, B. M. F. (1993a). Imitation in free-ranging rehabilitant orangutans (*Pongo pygamaeus*). *Journal of Comparative Psychology, 107*, 147–161.

(1993b). Imitation in rehabilitant orangutans: Model and action selectivity. Submitted to the *Proceedings of the 14th Congress of the International Primatological Society*, Strasbourg, France.

Suarez, S. D., & Gallup, G. G., Jr. (1981). Self-recognition in chimpanzees and orangutans, but not gorillas. *Journal of Human Evolution, 10*, 175–188.

Swartz, K. B., & Evans, S. (1991). Not all chimpanzees show self-recognition. *Primates, 32*, 483–496.

Tomasello, M., Kruger, A., & Ratner, H. (1993). Cultural learning. *Behavioral and Brain Sciences, 16*, 495–552.

Trevarthen, C. (1979). Communication and cooperation in early infancy: A description of primary intersubjectivity. In M. Bullona (Ed.), *Before speech: The beginning of interpersonal communication* (pp. 321–347). Cambridge University Press.

Visalberghi, E., & Fragaszy, D. (1990). Do monkeys ape? In S. T. Parker & K. R. Gibson (Eds.), *Language and intelligence in monkeys and apes: Comparative development perspectives* (pp. 247–273). Cambridge University Press.

Vygotsky, L. S. (1978). *Mind in society: The development of higher psychological processes.* Cambridge, MA: Harvard University Press.

Whiten, A., & Byrne, R. (1988). Tactical deception in primates. *Behavioral and Brain Sciences, 11*, 233–273.

Whiten, A., Custance, D., Gómez, J. C., Bard, K. A., & Texidor, P. (1993). Observational learning of an artificial food-processing problem in child and chimpanzees. Manuscript in preparation.

Whiten, A., & Ham, R. (1992). On the nature and evolution of imitation in the animal kingdom: Reappraisal of a century of research. In P. J. B. Slater, J. S. Rosenblatt, C. Beer, & M. Milinski (Eds.), *Advances in the Study of Behavior* (Vol. 211, pp. 239–283). New York: Academic Press.

Wood, D., Bruner, J. S., & Ross, G. (1976). The role of tutoring in problem solving. *Journal of Child Psychology and Psychiatry, 17*, 89–100.

13 Shadows and mirrors: Alternative avenues to the development of self-recognition in chimpanzees

Sarah T. Boysen, Kirstin M. Bryan, and Traci A. Shreyer

Self-awareness has long been described as a capacity unique to, or perhaps defining, humans (Gallup, 1977b, 1983; Suarez & Gallup, 1981), and has been evaluated in numerous nonhuman species since the innovative studies of Gallup (1970). The ontogeny of self-recognition in human infants, that is, the acquisition of knowledge of one's physical appearance as a component of self-awareness, has also been explored by those with interests in human development, with most studies employing a variant of the Gallup (1970) mark test (Amsterdam, 1972; Bertenthal & Fischer, 1978; Brooks-Gunn & Lewis, 1984; Dixon, 1957; Papoušek & Papoušek, 1974). This test entails a surreptitiously applied mark (rouge placed on the child's nose) and attentional orientation to the mark as the dependent variable as the child views itself in a mirror. Similarly, self-recognition has been evaluated, principally through the use of the mark test, with a number of nonhuman primate species and elephants (see review by Anderson, 1984; Calhoun & Thompson, 1988; Gallup, 1970, 1977a,b, 1983, 1991, *SAAH*3; Gallup, Wallnau, & Suarez, 1980; Ledbetter & Basen, 1982; Lin, Bard, & Anderson, 1992; Povinelli, 1989; Robert, 1986; Suarez & Gallup, 1981; Swartz & Evans, 1991). Self-recognition, as one facet of self-awareness, has been demonstrated through the mirror mark test in chimpanzees (Gallup, 1970; Lin, Bard, & Anderson, 1992; Suarez & Gallup, 1981), and orangutans (Lethmate & Dücker, 1973), and the gorilla Koko (Patterson & Cohn, *SAAH*17). Prior to the testing of Koko, no definitive evidence for self-recognition in gorillas, as measured by the mark test, had been obtained (Ledbetter & Basen, 1982), nor has any monkey species tested to date shown similar recognition of its own reflected image (see review by Anderson, 1984; see Gallup, 1977a; Gallup, Wallnau, & Suarez, 1980). More recently, the appropriateness of the emphasis placed on evaluation of self-awareness and self-recognition by the mark test alone has been called into question, and the suggestion made that other approaches with nonhuman species might provide additional insights into the capacity for a concept of

227

self (Gallup, *SAAH*3; Mitchell, 1993, *SAAH*6; Parker, 1991; Parker & Milbrath, *SAAH*7; Swartz & Evans, 1991).

Other possible contributions to the emergence of self-recognition certainly exist outside reflected mirror information. One potential natural source is shadows. In an innovative attempt to explore other environmental contributions to self-awareness, Gallup and his colleagues (Cameron & Gallup, 1988) examined the developmental sequence of behavioral reactions to shadows in human infants, proposing that shadows may be more readily available than reflective surfaces in a natural setting for a number of species. They tested 75 children ranging in age from 13 to 42 months on three shadow-recognition tasks explicitly designed to be performed by young children. All three tasks required minimal attention and no verbal instruction. These tasks investigated three abilities hypothesized to be related to self-recognition:

1. the ability to locate an object by seeing only the object's shadow, termed *shadow permanence*;
2. the ability to manipulate one's own body to *create shadows* (shadow imitation of "puppets"); and
3. the ability to demonstrate recognition of one's own shadow, termed *shadow self-recognition* (Cameron & Gallup, 1988).

The investigators found an age-related pattern of responding to shadows similar to developmental patterns of self-recognition derived from tests using mirror information. Shadow permanence was the first to emerge, at approximately 13 months, whereas the production of shadow puppets emerged around 19 months. The third ability, shadow self-recognition, was observed only in infants older than 25 months, and the majority of children tested did not exhibit shadow self-recognition until around 40 months. The developmental delay in shadow self-recognition was hypothesized to be related to two factors:

1. the richer quality of the mirror image, relative to detail, color, etc., and
2. the previous exposure of the infants to mirrors being as great or greater than their exposure to their own shadows, based upon a survey of parents.

Shadow self-recognition in chimpanzees

In our laboratory, we were interested in exploring the question of shadow self-recognition with our two youngest chimpanzees (*Pan troglodytes*), Bobby, a male who was 2.8 years old at the beginning of the study period, and Sheba, an 8-year-old female. We expected that similar developmental differences, as noted by Cameron and Gallup (1988) for human infants, might be seen between the two animals tested with the same shadow recognition tasks. We anticipated that overall the older animal would exhibit more awareness of shadows than the younger one, particularly for the shadow self-recognition task, as abilities related to this facet of self-awareness were demonstrated only by older children in the sample tested by Cameron and Gallup (1988).

Figure 13.1. Sheba examines the inside of her mouth after stealing and drinking non-toxic red dye that had been left in the room several minutes earlier for a mirror mark-test session with Bobby.

Bobby and Sheba served as subjects. Both animals were captive-born and human cross-fostered, and thus had prior, but undocumented, experience with mirrors. Sheba had been raised in a human home as a zoo public relations animal from age 4 months to $2\frac{1}{2}$ years, at which time she joined the Primate Cognition Project at Ohio State. Bobby was raised from birth by human caregivers for the entertainment business and arrived at the project at 19 months of age. Both animals has also been tested in our laboratory for mirror recognition prior to the shadow study, using the methods described by Gallup (1970). Sheba clearly demonstrated self-directed behaviors at the marks on her brow ridge and opposite ear when tested at age 6 (see Figure 13.1). Bobby, however, did not exhibit similar responses when tested with the mark test just prior to the present study, but did show self-recognition approximately 1 year later. Neither animal had been formally trained or tested on shadow recognition, although each had access to natural lighting in their outdoor cage area, as well as exposure to directional indoor lighting during routine videotaping.

The experimental context was a quiet test room. The apparatus included a color video camcorder, a 250-watt spotlight, and large table. A soft sponge football (noiseless) about 20 cm in length, and a round sponge ball, approximately 14 cm in diameter, were used for the first test, the shadow permanence task. For Task 2, the experimenters demonstrated the creation of shadow puppet stimuli using their hands, and in the third task, a large cotton floppy-brimmed hat was used for the shadow self-recognition task.

Both animals were tested individually while seated on the table facing a blank white wall. The spotlight was placed about 3 m behind the subject, and illuminated the wall. The room was darkened at the beginning of each trial, with only the spotlight providing light, and thus it was possible for the experimenters and subjects to cast shadows on the wall that the chimps faced. The animals first received several minutes of orientation, during which the experimenters drew their attention to shadows on the wall by pointing to them or tapping them. Both animals then received repeated trials of all three of Cameron and Gallup's (1988) shadow recognition tasks in the same order, over a 2-week period. The three tasks included:

Task 1: Ball task (shadow permanence). Two experimenters, standing behind the subject, passed a noiseless sponge ball back and forth three or four times, in an arc above and behind the subject. This created a shadow of a moving ball in front of the subject. The experimenter last receiving the ball would then lower it so that the ball's shadow was no longer visible, and hide it behind her or within her clothing. Responses were then scored depending on the experimenter toward which the subject became oriented. Formal criterion required that the subjects orient their head and/or body toward the experimenter who had retained the ball, and in most cases, the animal began immediate searching behaviors directed toward that person (looking behind her back, attempting to explore her pockets, etc.; see Figure 13.2a–d).

Task 2: Puppet task (shadow imitation). In this task, the chimps were encouraged to imitate the experimenters, who were making shadow puppets with their hands held behind and above the subjects, so that shadow images reminiscent of animals appeared on the wall in front of the chimps. The chimps were verbally prompted and gesturally encouraged to imitate by the experimenters. Formal criterion for a correct response was any attempt at producing puppets through directed hand movements that produced moving shadow figures on the wall in front of the animals (Figure 13.3).

Task 3: Hat task (shadow self-recognition). In the hat task, which represented a test of shadow self-recognition, one experimenter raised a large brimmed hat behind the animal so that the chimp's shadow appeared to be "wearing" the hat. Formal criterion, as specified for human infants by Cameron & Gallup (1988), required that the subjects exhibit one of three possible responses:

1. some attempt to "disengage" his or her shadow from that of the hat, accomplished by a left, right, or downward movement of the head and/or body;
2. touching their own heads; or
3. reaching behind their heads (without turning around) and displacing the actual hat (which was positioned just a short distance from their heads).

Turning around to simply orient toward the hat was not scored as a correct response, just as it was not considered an adequate response for the children tested by Cameron & Gallup (1988) (see Figure 13.4a–c).

All trials on all tasks were videotaped for further analysis. On a given trial, a correct score was recorded if the animal performed the criterion response,

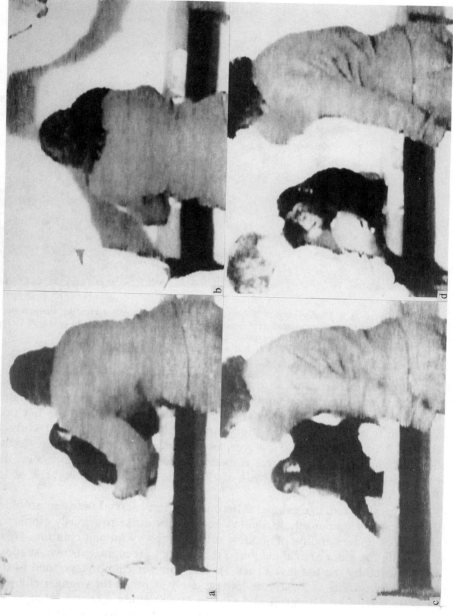

Figure 13.2. In this series of photographs taken from videotape, Bobby is seated on a table, facing the wall, and gently restrained by one experimenter. Two other experimenters, standing behind, pass a noiseless ball back and forth to one another, creating a shadow of the ball and their hands as they do so. This series depicts the final pass of the ball, back to the experimenter on the left. Bobby then orients toward her, indicating the final location of the ball and his recognition that the shadow of the ball represents an actual ball ("shadow permanence").

Figure 13.3. Three-year-old Bobby imitates the experimenter creating shadow puppets on the wall; a second experimenter gently restrains Bobby, who is seated facing
the wall, with illumination coming from a spotlight positioned behind them.

and all other responses were scored as incorrect. Tapes were scored by two
observers, and interrater reliability was +95%; disputed trials were evaluated
by a third observer. The older animal responded correctly on the hat task
significantly more times than the younger chimp (Sheba, 8 of 11 correct;
Bobby, 4 of 12 correct). However, no significant differences were found
between the number of correct responses made by the two animals on the
ball task (Sheba, 7 of 16; Bobby, 4 of 12) or the puppet task (Sheba, 9 of 15;
Bobby, 10 of 20).

   As hypothesized, age-related differences were observed between an older
and younger chimpanzee on shadow-recognition tasks previously employed
to test for understanding of shadow relationships by human children. However, only the older animal exhibited the ability to recognize its own shadow,
as measured by the hat task (Task 3). The younger chimp responded to the
hat task by turning to locate the hat, precisely as noted for younger children
(< 24 mos.) in the Cameron and Gallup (1988) study. On the other hand,
Sheba responded repeatedly during testing with the hat by attempting to
"detach" her herself (and thus her shadow), through movements of the head
and upper body. Her behavior was highly reminiscent of the responses described by Cameron & Gallup as the most compelling single index of shadow
self-recognition in children.

   No developmental differences were found between the number of correct
responses the chimps made on the other two tasks, which measured their
ability to locate an object using only shadow information, as a measure of

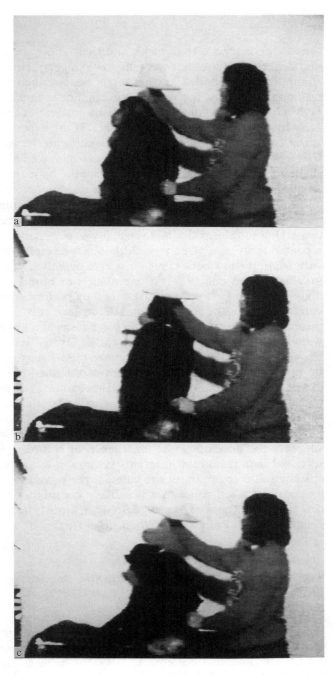

Figure 13.4. Following the positioning of the hat during Task 3 (hat task), which created the shadow of a hat on Sheba's shadow on the wall, Sheba (a) moves her head up slightly, exploring the possible contingencies between her actual head and changes in the shadowed hat position, then (b) reaches up and behind to touch the hat, and finally (c) shifts downward and "disengages" her herself (and her shadow) from the shadow of the hat.

shadow permanence (ball task), and their ability to manipulate their hands and fingers to create shadows (shadow puppets). However, Sheba, the older chimp, often responded in what might be described as a more sophisticated fashion than Bobby, the younger animal. For example, as Cameron & Gallup (1988) found with older children, Sheba sometimes did not respond immediately by orienting to the ball on Task 1. If the experimenters asked, "Where is the ball?" she would respond by pointing to the correct experimenter, as well as by orienting toward her. Sheba also created better approximations of shadow puppets on Task 2 than did Bobby, by manipulating her individual fingers when attempting to imitate the experimenters' behaviors.

In addition to developmental differences, attention variables may have also contributed to the results of the chimps' performance on the three shadow tasks. Due to the small sample size, it was necessary for each animal to have repeated trials on a given task. Although we employed a minimal number of such trials, the low correct response rate overall on the ball and puppet tasks may have been due to retest habituation or boredom, rather than the animals' ability to perform the tasks, as they performed promptly, reliably and correctly within the first test session for each task. Although it is generally assumed that human subjects who solve tasks at a given developmental level will also tend to solve items testing earlier levels (Bertenthal & Fischer, 1978; Brooks-Gunn & Lewis, 1984), Chapman (1987) suggested that children tested on tasks measuring mirror analogies of the shadow-recognition tasks (e.g., object permanence, imitation, or self-recognition) often failed tasks measuring earlier emerging abilities though passing later emerging-ability tasks, due to lack of attention or interest.

The appearance of shadow self-recognition in only the older animal supports the developmental emergence of self-recognition found in cross-sectional data from human infants (Cameron & Gallup, 1988). Evidence of shadow self-recognition in the chimpanzee provides further support for the existence of a self-concept, which likely contributes to a theory of mind in this great ape species, as suggested previously (Gallup & Suarez, 1986; Premack, 1988; Premack & Woodruff, 1978).

The emergence of self-recognition in two chimpanzees: Five minutes to awareness

There is little question that the issues of self-recognition, self-awareness, and the phenomena of mind and consciousness in nonhuman animals continues to pique the curiosity and interest of the scientific community (Gallup, 1991, *SAAH*3; Griffin, 1976; Premack, 1988; Premack & Woodruff, 1978; Ristau, 1991). To further explore this issue, we analyzed a videotaped session of the introduction of a mirror to two young male chimpanzees, Kermit and Darrell, ages 4 and $4\frac{1}{2}$. A portion of the footage had been presented previously as part of the workshop from which the present volume emerged (Boysen, 1991).

Figure 13.5. Laboratory setting for introduction of mirror to Kermit and Darrell.

This archival videotaped record captured the apparent rapid emergence of self-recognition in these two young chimpanzees.

Both animals were captive-born and peer-raised together from infancy in the nursery of Yerkes Regional Primate Research Center, Emory University, Atlanta, Georgia, and were selected by the nursery staff for transfer on permanent loan to the Primate Cognition Project at Ohio State in 1983. Darrell was $3\frac{1}{2}$ years old, and Kermit 3 years, when they arrived at OSU in May 1983. Since that time, and prior to the mirror introduction, they had been immersed in intensive social interaction with the first author and with periodic student volunteers throughout the day, 7 days a week. The chimps acquired facility through training with a match-to-sample strategy on color and shape matching, drawing, and a variety of similar conceptual tasks during two or three brief daily teaching sessions, which lasted 15–30 min. The balance of each day was spent in play bouts with their human teachers and/or caregivers, in walks outdoors, and in opportunities for vigorous gymnastic play with each other in the outdoor exercise areas adjacent to their indoor sleeping quarters. Prior to the test session with mirrors, to the best of our knowledge, the animals had no prior experience with mirrors or reflective surfaces of any kind.

A 60 cm × 60 cm square of mirrored acrylic was secured to a wooden frame, which permitted the mirror to stand on the floor at a 45° angle. The mirror was placed at the far end of the corridor adjacent to the animals' indoor cages (see Figure 13.5). The mirror was put in place, and the animals were released from their indoor cage so that they were free to interact directly with the plastic mirror. The entire testing session lasted a total of 22 min, at which time the animals showed a lessening of interest in the mirror, and it

Table 13.1. *Duration of time spent interacting at mirror, in 5-min time blocks*

Time block	Kermit (age 4)	Darrell (age 4.5)
Block 1	2:19 min	2:41 min
Block 2	2:42 min	0.40 sec
Block 3	0:53 sec	1:18 min
Block 4	3:28 min	1:29 min
Total	9:22 min	6:08 min

Table 13.2. *Response by behavior category*

	Kermit		Darrell	
Behavior	1st 10 min	Total	1st 10 min	Total
Mirror-oriented				
Look behind mirror	11	13	5	7
Hit/kick mirror	45	80	6	48
Lick mirror	11	27	2	6
Look from behind mirror	3	5	1	2
Reach behind mirror	1	1	20	20
Look down into mirror	8	12	0	4
Thrust against mirror	7	7	44	44
Touch mirror	27	35	15	19
Adjust mirror angle	0	34	0	3
Self-oriented				
Touch body part while looking in mirror	13	14	0	11
Make faces	20	29	14	16
Turn upside down	1	1	0	0
Touch image in mirror	40	44	0	2
Use mirror to inspect nonfacial body part	0	0	3	4
Shake hands while looking in mirror	10	10	0	0
Shake head while looking in mirror	0	0	0	1
Other				
Stomp while looking in mirror	0	0	0	18
Vocalize	0	0	2	2
Jump	76	105	12	19

was removed from the room. The total amount of time during which Kermit directed attention toward the mirror was 9 min, 36 sec, while Darrell spent a total of 6 min, 22 sec, responding to the mirror (see Table 13.1). The test session was analyzed in 5-min time blocks, and the frequency of response for each behavior category is shown in Table 13.2 for each animal.

Remarkably, one of the chimps (Kermit) displayed mirror-guided, self-directed behaviors within the first 5-min block (Table 13.2). Kermit exhibited more responses across all three broad behavioral categories, including mirror-oriented behaviors and self-oriented behaviors, which included numerous contingent behaviors that provided feedback of facial expression, body movements, and overall image. The greater number of occurrences of all behaviors by Kermit occurred in the first 10 min of mirror exposure, while similar categorical behaviors emerged in the second 10-min block of mirror exposure in Darrell. Relative to mirror-oriented behaviors such as exploring the physical dimensions of the mirror by hitting or kicking it, Kermit investigated these features of the mirror early in the test session, whereas Darrell exhibited a greater number of social responses to the mirror, particularly in the form of social/sexual thrusting against the mirror and its reflected image (see Table 13.2). In contrast, Kermit exhibited more self-oriented behaviors earlier in the test session, as well as more such behaviors overall, than Darrell. This category included touching the image in the mirror and touching body parts, particularly his nose, while looking in the mirror. The nature of Kermit's behavioral responses to his image in the mirror, as well as the rapidity with which these behaviors appeared when he was first exposed to a mirrored surface, suggest that he very rapidly recognized the reflected image as his own. Darrell's understanding of the physical dimensions of the mirror, and the relationship between his behaviors and those depicted in the mirror, were slightly slower to emerge, but were nevertheless apparent by the end of the 20-min exposure period. This was most dramatically exhibited by his immediate use of a hand-held mirror provided by the experimenter following the removal of the larger framed mirror from the test room. Darrell immediately began to vigorously groom the bottom of his foot, guided by the image of this otherwise visually inaccessible area, reflected in the hand-held mirror (Figure 13.6). This intense grooming bout was also accompanied by continuous grooming vocalizations. Had Darrell not come to recognize the significant features of a mirror during his recent experience with the larger mirror – that is, that he could see places on his body not usually visible – it is unlikely that he would have immediately been able to use the hand-held mirror, given the first opportunity to do so, in a clearly functional manner. These observations further document that chimpanzees given a period of mirror exposure, even a brief one, may readily assimilate mirror-contingent behavior into meaningful feedback about their physical appearance. More important, with experience (and in some cases, within minutes), they can integrate such feedback toward the ability to recognize themselves, thus contributing to an emergent sense of self. With the exception of humans and the great apes, such behaviors are a significant departure from those exhibited by other nonhuman primates. Monkeys have been reported to have consistently failed the mark test; they continue to respond to mirror images, even after many years of mirror exposure, as if the reflected image were a conspecific (Anderson, 1984; Anderson & Roeder, 1989; Gallup, Wallnau, & Suarez, 1980). However, other indications of self-recognition have been suggested in monkeys (Eglash & Snowdon, 1983) and gorillas (Parker, *SAAH*19). Hence, this measure of

Figure 13.6. Darrell immediately uses a hand-held mirror to groom the bottom of his foot, following the test session in which the larger mirror was introduced for the first time.

self-awareness may be a relatively unique endowment of the pongids and hominids. However, it is also possible, indeed plausible, that self-awareness is not an all-or-none phenomena, but instead emerges developmentally as well as phylogenetically. Thus, just as other human (and nonhuman) cognitive structures and capabilities progress through a number of developmental stages and may be seen in more primitive and advanced states cross-specifically, self-awareness may be represented by shades of awareness across the range of extant primate species. Indeed, it may be that the mark test taps some cognitive capacity to simultaneously integrate incoming sensory information demonstrable only among the great apes and humans, such as kinesthetic–visual matching, as proposed by Mitchell (1993). Nonetheless, the demands of group living, including complex reciprocal social interactions, alliance formation, and cooperation, have likely contributed in broader evolutionary strokes toward establishing a generalized sense of self that subserves social needs in a wide range of monkey species whose levels of self-awareness have not achieved comparable status as accessed by the mark test. In that case, the mark test, as the standard of comparison, would be insensitive to levels of social awareness that appear to support the range and depth of complex relationships observed among other species, including monkeys.

References

Amsterdam, B. (1972). Mirror self-image reactions before age two. *Developmental Psychology, 5*, 297–305.

Anderson, J. R. (1984). The development of self-recognition: A review. *Developmental Psychobiology, 17*, 35–49.

Anderson, J. R., & Roeder, J-J. (1989). Responses of capuchin monkeys (*Cebus apella*) to different conditions of mirror-image stimulation. *Primates, 30*, 581–587.

Bertenthal, B. I., & Fischer, K. W. (1978). Development of self-recognition in the infant. *Developmental Psychology, 14*, 44–55.

Boysen, S. T. (1991). Shadow recognition in the chimpanzee. Workshop presentation: Self-recognition in infants and nonhuman primates, Sonoma State University, Rohnert Park, CA.

Brooks-Gunn, J., & Lewis, M. (1984). The development of early visual self-recognition. *Developmental Review, 4*, 215–239.

Calhoun, S., & Thompson, R. L. (1988). Long-term retention of self-recognition by chimpanzees. *American Journal of Primatology, 15*, 361–365.

Cameron, P. A., & Gallup, G. G., Jr. (1988). Shadow self-recognition in human infants. *Infant Behavior and Development, 11*, 465–471.

Chapman, M. (1987). A longitudinal study of cognitive representation in symbolic play, self-recognition, and object permanence during the second year. *International Journal of Behavioral Development, 10*, 151–171.

Dixon, J. C. (1957). Development of self-recognition. *Journal of Genetic Psychology, 91*, 251–256.

Eglash, A. R., & Snowdon, C. T. (1983). Mirror-image responses in pygmy marmosets (*Cebuella pygmaea*). *American Journal of Primatology, 5*, 211–219.

Gallup, G. G., Jr. (1970). Chimpanzees: Self-recognition. *Science, 167*, 86–87.

(1977a). Absence of self-recognition in a monkey (*Macaca fascicularis*) following prolonged exposure to a mirror. *Developmental Psychobiology, 10*, 281–284.

(1977b). Self-recognition in primates: A comparative approach to the bidirectional properties of consciousness. *American Psychologist, 32*, 329–338.

(1983). Toward a comparative psychology of mind. In R. Mellgren (Ed.), *Animal cognition and behavior* (pp. 473–510). New York: North Holland.

(1991). Toward a comparative psychology of self-awareness: Species limitations and cognitive consequences. In G. R. Goethals & J. Strauss (Eds.), *The self: An interdisciplinary approach* (pp. 121–135). New York: Springer-Verlag.

Gallup, G. G., Jr., & Suarez, S. D. (1986). Self-awareness and the emergence of mind in humans and other primates. In J. Suls & A. Greenwald (Eds.), *Psychological perspectives on the self* (Vol. 3, pp. 3–26). Hillsdale, NJ: Erlbaum.

Gallup, G. G., Jr., Wallnau, L. B., & Suarez, S. D. (1980). Failure to find self-recognition in mother–infant and infant–infant rhesus monkey pairs. *Folia Primatologica, 33*, 210–219.

Griffin, D. R. (1976). *The question of animal awareness*. New York: Rockefeller University Press.

Ledbetter, D. H., & Basen, J. A. (1982). Failure to demonstrate self-recognition in gorillas. *American Journal of Primatology, 2*, 307–310.

Lethmate, J., & Dücker, G. (1973). Untersuchungen zum Selbsterkennen im Spiegel bei Orang-Utans und einigen anderen Affenarten. (Studies on self-recognition in a mirror by orangutans and some other primate species.) *Zeitschrift für Tierpsychologie, 33*, 248–269.

Lin, A. C., Bard, K. A., & Anderson, J. R. (1992). Development of self-recognition in chimpanzees (*Pan troglodytes*). *Journal of Comparative Psychology, 106*, 120–127.

Mitchell, R. W. (1993). Mental models of mirror self-recognition. *New Ideas in Psychology, 11*, 295–325.

Papoušek, H., & Papoušek, M. (1974). Mirror image and self-recognition in young human infants: I. A new method of experimental analysis. *Developmental Psychobiology*, 7, 149–157.

Parker, S. T. (1991). A developmental approach to the origins of self-recognition in the great apes. *Human Evolution*, 6, 435–449.

Povinelli, D. J. (1989). Failure to find self-recognition in Asian elephants (*Elephas maximus*) in contrast to their use of mirror cues to discover hidden food. *Journal of Comparative Psychology*, *103*, 122–131.

Premack, D. (1988). "Does the chimpanzee have a theory of mind" revisited. In R. W. Byrne & A. Whiten (Eds.), *Machiavellian intelligence: Social expertise and the evolution of intellect in monkeys, apes and humans* (pp. 160–179). Oxford University Press.

Premack, D., & Woodruff, G. (1978). Does the chimpanzee have a theory of mind? *Behavioral and Brain Sciences*, *1*, 515–526.

Ristau, C. A. (Ed.) (1991). *Cognitive ethology: The minds of other animals (Essays in honor of Donald R. Griffin)*. Hillsdale, NJ: Erlbaum.

Robert, S. (1986). Ontogeny of mirror behavior in two species of great apes. *American Journal of Primatology*, *10*, 109–117.

Suarez, S. D., & Gallup, G. G., Jr. (1981). Self-recognition in chimpanzees and orangutans, but not gorillas. *Journal of Human Evolution*, *10*, 175–188.

Swartz, K. B., & Evans, S. (1991). Not all chimpanzees (*Pan troglodytes*) show self-recognition. *Primates*, *32*, 483–496.

14 Symbolic representation of possession in a chimpanzee

Shoji Itakura

Introduction

Self-recognition in nonhuman primates is often equated with mirror self-recognition (Anderson, 1984; Gallup, 1970; Itakura, 1987a,b; Povinelli, 1987). Nevertheless, for humans, "it is the ability to recognize and respond to self independent of contingency which represents the important developmental milestone in self-recognition" (Lewis & Brooks-Gunn, 1979b, p. 218). Human children begin to recognize their contingent-independent image in photographs only after mirror self-recognition (MSR), with most children at 22 months of age correctly labeling a photograph of themselves among photographs of other infants, and some infants doing so as early as 16 months (Lewis & Brooks-Gunn, 1979a, p. 12). The recognition of a contingency-independent self in photographs is relatively unstudied in primates (but see Miles, *SAAH*16; Patterson & Cohn, *SAAH*17). The only previous instance described is open to interpretation: The sign-using chimpanzee Lucy, after observing an image of herself and her cat (which had died three months previously), signed "Lucy's cat, Lucy's cat" repeatedly for ten minutes as she stared at the picture (Temerlin, 1977).

This instance suggests not only that Lucy recognized her own image and her cat's, but also that she viewed the cat as her own – as a possession. Although possession in nonhuman primates has been studied in the context of control over objects (Kummer, 1973; Thierry, Wunderlich, & Gueth, 1989; Torii, 1975), there is no evidence that possession in these primates is mediated by a self-concept or a concept of the other, such that they view these objects as symbolically representing themselves or another. In this chapter I explore whether a symbol-trained chimpanzee can symbolically associate objects and individuals (including herself) after observing an arbitrarily created relationship between them. In particular, I show that a chimpanzee not only can recognize and name photographs of herself and others, but also can recognize the symbolic relation between these individuals and objects distinctly their own.

241

Method

Subject

The subject was a 13-year-old female chimpanzee (*Pan troglodytes*) named Ai. Before this study she had received extensive training in the use of visual symbols called lexigrams that represent objects and attributes of objects. She had also learned to name individuals, including chimpanzees, orangutans, and humans, by choosing a letter of the alphabet (Matsuzawa, 1990; Itakura, 1988). Moreover, Ai had learned to use personal pronouns, YOU, ME, HIM, and HER, in a way similar to humans' use (Itakura, 1991, 1992a,b; Itakura & Matsuzawa, in press). Such an ability of apes to use personal pronouns is referred to in the chapters by Miles (*SAAH*16) and Patterson and Cohn (*SAAH*17).

Procedure

Feeding training with fixed-color bowls. Ai received feeding training together with another male chimpanzee, Akira, and a male orangutan named Doudou (see Figure 14.1). Ai was dominant to Akira. Ai was the target subject and the two other individuals were stimulus subjects. Ai and Akira were in the same room; Doudou was alone in another room. The chimpanzees and the orangutan could see each other through wire netting separating the cages. Ai was always fed with a green bowl, Akira was fed with a red bowl, and Doudou with a yellow bowl. A trial ended after the three individuals were fed with these bowls, and a session consisted of 12 trials according to the combination of feeding order; for example, in some trials Ai was fed first, Akira was fed second, and Doudou was fed last. Each trial was followed by a 60-sec intertrial interval. This training was conducted for five sessions. During before-feeding time, Ai's food-demanding behavior was coded by a one–zero type sampling method. *Before-feeding time* was defined as the time from when the experimenter picked up one of the three bowls until the experimenter moved to the target individual. *Food-demanding behavior* was defined as reaching a hand to, opening the mouth to, or following the experimenter. Following feeding training Ai received three kinds of tests.

Feeding with different bowls. In the first test, each individual was fed with a different bowl; for example, Ai was fed with the red bowl, Doudou with the green bowl, and Akira with the yellow bowl. A session consisted of six trials according to the combination of three bowls and three individuals. Ai's responses were observed during this time.

Preference of bowls. In the second test, Ai's preference for bowls was tested by pairing two of three bowls in counterbalanced right–left position. Two bowls, each containing a piece of peanut, were presented to Ai. We then

Figure 14.1. Schematic representation of the feeding training situation in the apes' home room. The experimenter feeds each of them sequentially with bowls that are associated with each individual arbitrarily by the experimenter.

coded which bowl's peanut was taken first by Ai. Because a session consisted of 60 trials, the same trial appeared 10 times. This test was conducted for one session.

Symbolic association between individuals and bowls. In the third test, Ai sat in front of the TV monitor and 5 × 6 matrix keyboard on a console in the experimental room. The TV monitor displayed the still-photograph stimuli via a laser disc player (TEAC-LV200). Letters were mounted on the keyboard. There was a food tray by the TV monitor. A food reward, such as a peanut, was automatically delivered after a correct response. Ai would begin a trial by pressing a start key, which presented photographs of each individual (Ai, Akira, or Doudou) or each bowl (green, red, or yellow) on the monitor. After a 2-sec delay, one of the two rows of keys on the console was illuminated. Each row contained the name of Ai (L), Akira (A), and Doudou (U). Ai's task was to choose the correct name among the three letters in the illuminated row that corresponded to the photograph presented on the TV monitor. Before this test it had been confirmed that Ai could name each of three individuals with a letter of the alphabet and each of the three bowls by a lexigram. The correct responses were to associate L (Ai) with a photograph of the green bowl, A (Akira) with one of the red bowl, and U (Doudou) with one of the yellow bowl. Thirty photographs were used: Half were of the three individuals (five for each) and the remainder were of the three bowls (five for each). Stimuli were presented at random. Because a session consisted of 90 trials, the stimulus set was repeated three times. This test was conducted for two sessions, with a different stimulus set for each session.

Results

Feeding training with fixed-color bowls. In the first session Ai showed food-demanding behavior toward bowls of all three colors, and showed some aggressive behavior to Akira when he was the target individual (this behavior was not counted as a food-demanding behavior). Food-demanding behavior decreased, however, when the experimenter held the red and yellow bowls, used for Akira and Doudou, respectively. Figure 14.2 shows the number of trials during which Ai made food-demanding behavior.

Ai came to ignore the experimenter when he picked up the red or yellow bowls, and came to show food-demanding behavior only when the green bowl was picked up.

Feeding with different bowls. When Ai was fed with the red or yellow bowl, she accepted them with no hesitation; however, when the green bowl was used for Doudou, Ai followed the experimenter, then gave up quickly. When the green bowl was used for Akira, Ai showed violent, aggressive behavior to Akira as if to assert her privilege to be fed with the green bowl. The data are presented in Table 14.1.

Preference of bowls. Although when presented with two bowls Ai could take both peanuts at once, initially she took only one at a time. When the green bowl was paired with another bowl she took the peanut from the green bowl first and then the one from the other bowl. When the red and yellow bowls were paired, Ai preferred to take the one in the left position first, then the nut from the other bowl (see Table 14.2).

In the case of the green bowl, color preference was dominant to position preference. The preference of first selection for the green bowl disappeared soon because Ai took the peanuts from both bowls simultaneously. Position preference during the first selection of the bowls seemed to depend upon the position of Ai herself relative to the bowls.

Symbolic association between individuals and bowls. Ai showed high accuracy in associating individuals and bowls, as shown in Figure 14.3. She chose

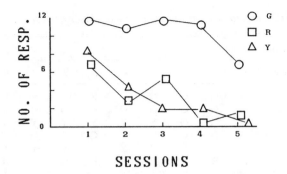

Figure 14.2. Frequency of occurrence of food-demanding behaviors to each bowl by Ai during before-feeding time.

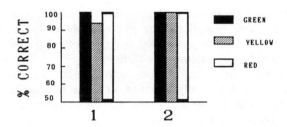

Figure 14.3. The percentage of correct responses in the individuals–objects association test.

Table 14.1. *Ai's responses during different-bowl feeding*

Trial	Combination of individual and bowl color	Responses
1	Doudou – Red	No responses
2	Akira – Yellow	A little emotional
3	Ai – Yellow	No responses
4	Doudou – Green	Follow the experimenter
5	Ai – Red	No responses
6	Akira – Green	Aggressive behavior to Akira

the correct letter from the first trial, when each bowl was presented on the TV monitor, making only one error, which occurred after 30 trials.

Discussion

These results clearly indicate that the chimpanzee Ai developed the representation of a symbolic relationship between objects and individuals, including herself, solely by participating in the feeding protocol described. These

Table 14.2. *First selection of bowl in bowl preference test*

Combination of bowls		Trial									
Left	Right	1	2	3	4	5	6	7	8	9	10
R	G	G	G	G	G	G	R	R	R	R	R
G	R	G	G	G	G	G	G	G	G	G	G
Y	G	G	G	G	Y	Y	Y	Y	Y	Y	Y
G	Y	G	G	G	G	G	G	G	G	G	G
Y	R	Y	Y	Y	Y	Y	Y	Y	Y	Y	Y
R	Y	R	R	R	R	R	R	R	R	R	R

Abbreviations: G, green bowl; R, red bowl; Y, yellow bowl.

results also confirm that the chimpanzee understood the letter as representing the name of an individual, and recognized the photographs of her companions, as well. Significantly, she also recognized her own photograph. These accomplishments were not based on stimulus transfer.

In conclusion, Ai showed emotional responses to objects interpreted as "possessions," when viewed behaviorally. Further, in terms of cognition, she also made symbolic associations between objects and individuals based solely on her observation of objects used in feeding.

The present study was aimed at clarifying whether a chimpanzee can associate objects and individuals symbolically by observing their arbitrary relationships.

Although Ai's behavior represents a simple level of possession, it is a prerequisite for representation of "ownership" in the context of the relationship between objects and individuals. True ownership requires stricter criteria. One such criterion is the respect of ownership by other individuals (Bachmann, 1980; Kummer, Gotz, & Angst, 1974; Sigg & Falett, 1985). Thus, when Ai was fed with the red or yellow bowls, she should have rejected them. In the future, data should be collected to test other such paradigms and procedures.

Acknowledgments

The author is grateful to the staff of the Psychology Department of the Primate Research Institute, Kyoto University. He would like to express special thanks to Prof. Kiyoko Murofushi, PhD, whose guidance made this work possible. He also thanks Yuko Itakura for drawing Figure 14.1. This work was supported in part by a grant from Monbusho (the Ministry of Education, Science, and Culture, Japan), No. 01790033.

References

Anderson, J. R. (1984). Monkeys with mirrors: Some questions for primate psychology. *International Journal of Primatology, 5*, 81–98.

Bachmann, G., & Kummer, H. (1980). Male assessment of female choice in hamadryas baboons. *Behavioral and Ecological Sociobiology, 6,* 315–321.

Gallup, G. G., Jr. (1970). Chimpanzees: Self-recognition. *Science, 167,* 86–87.

Itakura, S. (1987a). Mirror guided behavior in Japanese monkeys (*Macaca fuscata fuscata*). *Primates, 28,* 149–161.

(1987b). Use of a mirror to direct their responses in Japanese monkeys (*Macaca fuscata fuscata*). *Primates, 28,* 343–352.

(1988). *Individual recognition by a chimpanzee.* Paper presented at the 53th Congress of the Japanese Psychological Association, Hiroshima, Japan.

(1991). Use of personal pronouns by a chimpanzee. In A. Ehara, T. Kimura, O. Takenaka, & M. Iwamoto (Eds.), *Primatology today* (pp. 317–318). Amsterdam: Elsevier.

(1992a). Sex discrimination of photographs of humans by a chimpanzee. *Perceptual and Motor Skills, 74,* 475–478.

(1992b). A chimpanzee with the ability to learn the use of personal pronouns. *Psychological Record, 42,* 157–172.

Itakura, S., & Matsuzawa, T. (in press). Acquisition of personal pronouns by a chimpanzee. In H. Roitblat, L. Herman, P. Nachatigall (Eds.), *Language and communication: Comparative perspectives.* Hillsdale, NJ: Erlbaum.

Kummer, H. (1973). Dominance versus possession: An experiment on hamadryas baboon. In E. W. Menzel (Ed.), *Precultural primate behavior* (Vol. 1, pp. 226–231). Basel: Karger.

Kummer, H., Gotz, W., & Angst, W. (1974). Triadic differentiation: An inhibitory process protecting pair bonds in baboons. *Behavior, 49,* 62–87.

Lewis, M., & Brooks-Gunn, J. (1979a). Toward a theory of social cognition: The development of self. *New Directions for Child Development, 4,* 1–20.

(1979b). *Social cognition and the acquisition of self.* New York: Plenum.

Matsuzawa, T. (1990). Form perception and visual acuity in a chimpanzee. *Folia Primatologica, 55,* 24–32.

Povinelli, D. J. (1987). Monkeys, apes, mirrors and mind: The evolution of self-awareness in primates. *Human Evolution, 2,* 493–509.

Sigg, H., & Falett, J. (1985) Experiments on respect of possession and property in hamadryas baboons (*Papio hamadryas*). *Animal Behavior, 33,* 978–984.

Temerlin, M. K. (1977). *Lucy: Growing up human.* New York: Bantam.

Thierry, B., Wunderlich, D., & Gueth, G. (1989). Possession and transfer of objects in a group of brown capuchins (*Cebus apella*). *Behavior, 110,* 294–305.

Torii, M. (1975). Possession by non-human primates. In S. Kondo, M. Kawai, & A. Ehara (Eds.), *Contemporary primatology* (pp. 310–314). Basel: Karger.

15   Self-awareness in bonobos and chimpanzees: A comparative perspective

Charles W. Hyatt and William D. Hopkins

Bonobos are the only species of great ape for which there are no data concerning self-recognition. Although there is little evidence reported for self-recognition in New or Old World monkeys (see Anderson, 1984; Boccia, *SAAH*23; Thompson & Boatright-Horowitz, *SAAH*22), mirror image stimulation (MIS) has been established to elicit behaviors indicative of visual self-recognition in humans, chimpanzees, orangutans, and gorillas (see Gallup, 1987; Lewis & Brooks-Gunn, 1979, for reviews; Miles, *SAAH*16; Patterson & Cohn, *SAAH*17). In an effort to determine to what extent, if any, visual self-recognition exists in bonobos, a study was conducted at the Yerkes Regional Primate Research Center main station and at the Yerkes field station.

Bonobos have until recently been classified as pygmy chimpanzees, but closer examination has revealed significant differences between what are now recognized as two distinct species of the genus *Pan* (see reviews by Susman, 1984; de Waal, 1991). Bonobos are physically smaller, walk bipedally more often (Doran, 1992), and exist in larger and more sexually active social groups than do "common" chimpanzees (Kano, 1982). Bonobos also exist in smaller ranges and total numbers in the wild than do chimpanzees, and so their exposure to and interaction with humans has been limited. Less than fifty bonobos are currently in captivity in the United States. The language-trained bonobos Matata and her son Kanzi (see Savage-Rumbaugh, 1986) have been observed by the authors to engage in mirror-aided grooming of their teeth and heads; yet until now, no controlled studies have been conducted on self-awareness in bonobos.

Methods

Ten bonobos (*Pan paniscus*) aged 2.3–34 years in two social groups were selected for the study; eight chimpanzees (*Pan troglodytes*) aged 11–37 years in four social groups were randomly selected as a comparison sample (see Table 15.1). All subjects were naive to mirror experimentation and language training, but the chimpanzees had seen reflective metal that had been installed 3–5 months earlier as enrichment devices 1.6–2 m outside their cages.

Table 15.1. *Subject information*

Name	Sex	Social group	Birthdate	No. of mirror exposures	No. of self-directed behaviors	
					Mirror	Control
Pan paniscus						
Brian	M	1	01-08-89	3	43	49
Bosondjo	M	1	01-01-71[a]	3	18	5
Jill	F	1	07-15-85	5	65	24
Kidogo	M	2	12-06-74	2	no response	
Laura	F	1	08-27-67	3	79	8
Linda	F	1	01-01-54[a]	3	0	5
Lorel	F	1	04-17-69	3	4	16
Mabruki	M	1	01-23-83	2	27	2
Murphy	M	1	04-15-90	3	28	8
Zalia	F	2	03-26-84	2	24	10
Pan troglodytes						
Ada	F	1	01-01-55[a]	1	no response	
Columbus	M	2	10-10-77	1	3	7
Joseph	M	1	11-25-80	1	43	0
Leslie	F	3	01-01-70[a]	1	82	27
Lux	M	3	08-16-81	1	38	8
Mickie	F	4	05-17-72	1	34	32
Ossobaw	M	2	11-17-76	1	53	22
Su	F	1	10-15-58	1	21	9

[a] Ages estimated.

These surfaces were somewhat cloudy, however, and chimps had not been observed to interact with them.

The experimental treatment consisted of placing a 1.3 m × 0.3 m mirror just outside each of their cages on its horizontal axis. The mirror was close enough to be seen clearly, but far enough away so that hands reaching through the cage could not touch it. Subjects were exposed in their social groups to the mirror for 1 hour each session. For 30 min the reflective side of the mirror was exposed (Mirror "on" condition), and for 30 min the non reflective cardboard backing of the mirror was exposed (mirror "off" condition) as a baseline. On/off conditions were varied systematically. All of the bonobo subjects except Mabruki and Zalia received at least three hour-long sessions, while these two animals each received two hour-long sessions. Chimpanzee subjects received one hour-long session (see Table 15.1).

The sessions were videotaped and analyzed using a coding scheme similar to that of Lin, Bard, and Anderson (1992). We looked for self-directed touching of the eyes, face, head, or body, for gazing at the mirror, reaching for it, vocalizing, displaying, or any other activities observed in front of the mirror. Behaviors were coded in real time using a computer program developed in

QuickBasic by the authors, which simply tabulated length of time spent on each activity. A sample of the tapes were recoded by another observed and yielded significant interrater agreement ($r = .92$, $df = 8$, $p < .05$). Data concerning frequencies of behaviors for the bonobos and the chimpanzees were totaled and averaged for exposure time across conditions. (As there were no reactions to the mirror from the bonobo Kidogo and the chimpanzee Ada, these subjects were not included in our statistical analyses.) For the data concerning durations of behaviors, a number of species differences were noted in reaction to MIS. Individual differences were also noted.

Results and discussion

Species differences

Results indicated that both species engaged in higher frequencies of self-directed behaviors with the mirror on than with it off (Westergaard, Hyatt, & Hopkins, in press). More important, we observed several clear instances of mirror-aided self-grooming, which provided conclusive evidence of self-recognition in bonobos, a previously undocumented finding (see Figure 15.1).

On average the chimpanzees demonstrated significantly longer durations of looking into the mirror and of self-directed behaviors than did the bonobos. Results also indicated that the apes that spent more time in front of the mirror when the reflection was facing them spent less time in front of the mirror when the reflection was turned away from them. These animals were also more likely, in the "on" condition, to demonstrate the mirror-aided behaviors indicative of self-recognition.

Individual differences

The reactions of some of the chimpanzees to the mirror were more immediately and obviously indicative of self-recognition than those of the bonobos. The mated chimp pair Leslie and Lux showed great interest in their reflections as soon as the mirror was presented to them. Leslie examined her eyes and teeth while Lux watched closely. Cagemates Columbus and Ossobaw also showed interest in the mirror. Cagemates Joseph and Su also examined their faces with aid of the mirror, but Ada, another chimp in their cage, showed no interest at all. This should not be surprising: Swartz and Evans (1991, *SAAH*11) have shown that not all chimps demonstrate self-recognition.

Clear interest in the mirror accompanied by instances of mirror-aided grooming was seen in only some of the bonobos. The subadult female Jill in particular exhibited the most interest in and activity toward the mirror, intently gazing, walking away, and then returning to it many times after introduction. She repeatedly tested the angles of sight in which her reflection was visible, and she bounced, turned, and contorted herself to examine her back, mouth, and sexual swellings. At one point, she turned backward to look at her swellings upside down through her legs; she then proceeded to climb

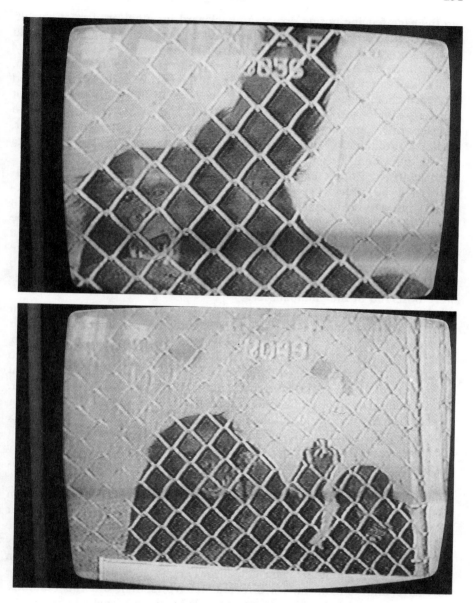

Figure 15.1. Bo "knows" Bo: The bonobo Bosondjo exploring his mouth with the aid of a mirror.

up the cage fencing upside down and backward for a better view of herself. She also spent several minutes tossing a ball into the air and rolling it around on the floor as she observed it and herself in the mirror.

The three adult females of the bonobo group at the field station, Laura, Linda, and Lorel, each spent time looking at the mirror, and they were noted to touch themselves while doing this. In general, however, they did not appear as interested in the mirror as did Jill. The subadult female bonobo housed at the main station, Zalia, engaged in extensive gazing at the mirror, including examination of her eyelid; she was also observed to roll a large plastic tub around behind her while watching her reflection.

In contrast, Zalia's cagemate Kidogo showed no indications of self-awareness; he pushed his tub around the cage in a display pattern he frequently used toward humans. When not displaying, he showed no interest in the mirror, nor any awareness of his reflection, and for much of the time he went to a part of his cage where he could not see it. The adult male Mabruki demonstrated more pronounced displays toward the mirror: He would charge and kick his foot at the wall. He did this three times but demonstrated no behaviors indicative of self-recognition. No displays or object manipulations were observed in the chimpanzees.

Although the two infant bonobos showed great fascination in gazing at their reflections in the mirror, including extensive reaching behaviors, their brief episodes of self-touching while looking in the mirror were inconclusive. Murphy, however, did poke at Jill's sexual swellings while watching her and himself in the mirror. The two infants also spent much of their time in front of the mirror playing together or near adults who were looking at the mirror.

The adult male bonobo Bosondjo showed very little interest in the mirror overall, but the one instance in which he did show interest was conclusive. After generally ignoring the reflective material and the others' interest in it for all of the first exposure period and most of the second, Bosondjo finally sat down and spent more than 2 min examining in detail the inside of his mouth. Apparently Bo knows Bo.

In conclusion, the bonobos in this study clearly demonstrated self-recognition (see Figure 15.1). Although the term "self-awareness" may have deeper implications than were observable here, the examination of otherwise unobservable body parts such as teeth and sexual swellings indicate undeniably that the image seen in the mirror was interpreted by the subjects as a reflection and not as a foreign other. The chimpanzees in this study demonstrated more extensive and reliable mirror-directed behaviors than did the bonobos. However, the considerable overlap of individual differences between the bonobo and chimpanzee subjects prohibits making any quantitative inferences concerning differences between the two species.

Acknowledgments

Support for this research was provided by NIH grant RR-00615 to the Yerkes Regional Primate Research Center. Additional support was provided by NINDS grant 29574 to

WDH. APA guidelines for the ethical treatment of animals were adhered to during all phases of this project. Yerkes Regional Primate Research Center is fully accredited by the American Association for Accreditation in Laboratory Animal Care.

References

Anderson, J. R. (1984). Monkeys with mirrors: Some questions for primate psychology. *International Journal of Primatology*, *5*, 81–98.

Doran, D. M. (1992). The ontogeny of chimpanzee and pygmy chimpanzee locomotor behavior: A case study of paedomorphism and its behavioral correlates. *Journal of Human Evolution*, *23*, 139–157.

Gallup, G. G., Jr. (1987). Self-awareness. In G. Mitchell & J. Erwin (Eds.), *Comparative primate biology, Vol. 2B. Behavior, cognition, and motivation* (pp. 3–16). New York: Liss.

Kano, T. (1982). The social group of pygmy chimpanzees (*Pan paniscus*) at Wamba. *Primates*, *23*, 171–188.

Lewis, M., & Brooks-Gunn, J. (1979). *Social cognition and acquisition of self*. New York: Plenum.

Lin, A. C., Bard, K. A., & Anderson, J. R. (1992). Development of self-recognition in chimpanzees (*Pan troglodytes*). *Journal of Comparative Psychology*, *106*(2), 120–127.

Savage-Rumbaugh, S. (1986). *Ape language: From conditioned response to symbol*. New York: Columbia University Press.

Susman, R. L. (1984). *The pygmy chimpanzee: Evolutionary biology and behavior*. New York: Plenum.

Swartz, K. B., & Evans, S. (1991). Not all chimpanzees (*Pan troglodytes*) show self-recognition. *Primates*, *32*, 483–496.

de Waal, F. (1991). Behavioral contrasts between bonobo and chimpanzee. In P. G. Heltne & L. A. Marquardt (Eds.), *Understanding chimpanzees* (pp. 145–175). Cambridge, MA: Harvard University Press.

Westergaard, G. C., Hyatt, C. W., & Hopkins, W. D. (in press). The responses of bonobos (*Pan paniscus*) to mirror-image stimulation. *International Journal of Primatology*.

16　ME CHANTEK: **The development of self-awareness in a signing orangutan**

H. Lyn White Miles

Introduction

The investigation of the linguistic and cognitive abilities of great apes offers insights into self-awareness by providing the additional modality of language with which self-awareness can be investigated in animals (see Patterson & Cohn, *SAAH*17). In ape language experiments, chimpanzees, gorillas, and an orangutan have been enculturated in human settings and have learned to communicate according to human linguistic and cultural conventions using a set of gestural signs, computer lexigrams, or plastic tokens (Fouts, 1973; Gardner & Gardner, 1969; Miles, 1980, 1986; Patterson, 1978; Premack, 1972; Rumbaugh, Gill, & von Glaserfeld, 1973; Savage-Rumbaugh, 1986; Terrace, Petitto, Sanders, & Bever, 1979).

These studies provide several major advantages for investigating self-awareness in animals. First, they permit us to study self-awareness in our closest biological relatives, the great apes, with whom we share a large number of biological and behavioral similarities. A relatively small number of genetic alterations are required to extrapolate humans and all of the great apes from a theoretical common ancestor most like the orangutan (Yunis & Prakash, 1982). The orangutan karyotype is the most conservative (primitive), with humans and the African apes displaying more derived features (Mai, 1983; Schwartz, 1987; Stanyon & Chiarelli, 1982; Weiss, 1987). This interpretation is also supported by fossil data (Pilbeam, 1982). Thus, the orangutan, having the most pleisiomorphic traits of the great apes, is the most primitive and has been termed "a living fossil" (Lewin, 1983). Orangutans have a high intelligence similar to that of the other great apes, and are not as solitary and asocial as has been previously assumed (Bard, 1988, 1990; Chevalier-Skolnikoff, 1983; Furness, 1916; Galdikas, 1982; Gómez, *SAAH*5; Harrison, 1963; Laidler, 1980; Lethmate, 1982; Maple, 1980; Miles, 1980; Milton, 1988; C. Parker, 1969; S. Parker, 1991; Povinelli & Cant, 1992; Rumbaugh & Gill, 1973; Rumbaugh & Pate, 1984; Russon & Galdikas, 1993; Shapiro, 1982; Suarez & Gallup, 1981; Tuttle, 1986; Wright, 1972). The cognitive abilities of orangutans (and gorillas) have often been underemphasized in contrast with those of chimpanzees (Maple, 1980; Miles, 1990; Russon & Galdikas, 1993).

254

Second, these studies permit us to relate self-awareness to associated abilities, as well as to the social and physical context in which self-awareness emerges. This relatedness allows us to trace several lines of cognitive evidence, determine their degree of association, and map their developmental sequence. Third, such studies include a linguistic component that provides an additional mode in which cognitive abilities are expressed and can be tested. It is possible to actually make inquiries of the animal, however rudimentary, that provide supporting evidence. For example, in mirror self-recognition (MSR) studies we can ask the animal in gestural signs, WHO YOU? Because the animal's responses to these inquiries can be compared to data collected from behavioral observations, we need not presume that the response is meaningful. With full contextual data, we are able to determine whether the animal's response is likely a self-identification, an error, or a random response.

Fourth, these studies are particularly well suited for reconstructing the evolution of self-awareness in early hominids, notably in determining the importance of the role of human culture (Miles & Harper, in press). Finally, because most of these experiments attempt to replicate a human child-rearing environment, they constitute an animal model that permits closer comparisons with human children, including children with various linguistic or cognitive impairments. These comparisons can point to domains that are critical for the development of self-awareness, suggest areas for therapeutic intervention, or confirm hypotheses developed in studies of human children.

Project Chantek

Project Chantek was established in 1978 to study the linguistic and cognitive abilities of a male orangutan named Chantek. Chantek was born at the Yerkes Regional Primate Research Center and from the age of nine months was cross-fostered at the University of Tennessee at Chattanooga. Cross-fostering is a technique of raising an ape as a child, not as a pet, in a humanlike physical and psychological setting to test how much human culture can be absorbed (Gardner & Gardner, 1989; Kellogg, 1968). Cross-fostering, mediated by the two-way use of language to communicate, results in at least a partial enculturation of the animal.

Enculturation is the anthropological term for the cross-generational transmission of human culture. Culture was first defined by Tylor (1877, p. 1) as "that complex whole which includes knowledge, belief, art, morals, law, custom, and any other capabilities and habits acquired by man [*sic*] as a member of society." It came to mean a society's patterns of language, behavior, and beliefs (Linton, 1936). However, structuralists and materialists expanded the definition to refer to all human behavior (Harris, 1964), including the "holistic, integrated totality . . . of human activity" (Rosman & Rubel, 1981), whereas cognitive and symbolic anthropologists restricted its meaning to refer only to the *rules* for behavior, but not the behavior itself (Goodenough, 1957, 1981). Using the more narrow definition, Geertz (1965, pp. 106–107) argued

that culture is composed of symbolic rules, "a set of control mechanisms –
plans, recipes, rules, instructions (what computer engineers call 'programs')
– for the governing of behavior."

We are following a psychological definition of culture as "the complex of
shared concepts and patterns of learned behavior that are handed down from
one generation to the next through the means of language and imitation"
(Barnouw, 1985). There is growing support for regarding cross-fostered apes
as enculturated because, whereas anthropologists once emphasized the absolute uniqueness of human culture and communication, we are now seeing
more clearly the continuities between humans and the great apes (Nanda,
1984, p. 39). Animal learning is often the result of conditioned associations
produced through repeated training, whereas enculturation refers to learning
the *patterns* for behavior based on language and conceptual thought, which
we feel occurs in cross-fostered language studies. In fact, ape language experiments provide a unique opportunity to study enculturation as symbolic
rule acquisition in animal studies (Haviland, 1991, p. 286).

For self-awareness to flourish, the individual must be provided with a spatial, social, and behavioral context in which he or she can learn about a world
of objects and individuals whose differentiation from the self is aided by
symbols (Rogoff, 1990). The individual is an apprentice who learns his or her
culture and language through a guided process of participation and reinvention
(Lock, 1980), aided by scaffolding, or the creation of helping structures to
manage and segment what is learned (Bruner, 1983; Vygotsky, 1978; Wertsch,
1978; Wood, Bruner, & Ross, 1976). That is, the objective environment is
given intersubjective meaning, organized culturally, and mediated symbolically primarily through language (Rogoff, 1990; Shore, 1991). Self-awareness
is expressed in mirror self-recognition and extends to identifying oneself as
an object, reacting to oneself, assuming a variety of social roles, and apprais-
ing or evaluating oneself in terms of others (Lewis & Brooks-Gunn, 1979).

Chantek's enculturation was carried out by a small group of human
caregivers on the campus of the University of Tennessee at Chattanooga,
who communicated with him using gestural signs based on the American Sign
Language for the deaf (Miles, 1986, 1990). He lived in a house trailer, took
frequent trips in a car, and interacted with the people, animals, and objects
in his human surroundings. Signs were taught first by molding his hands into
the proper shape, and later by imitation. Emphasis was not place on vocabulary acquisition or an effort to demonstrate that an orangutan had learned
adult human language, but on the development of Chantek's linguistic and
cognitive skills, and on his use of signs within a cultural context to meet his
own needs (Miles, 1990). By teaching signs to Chantek through guided participation, we were not only transmitting an encoded linguistic system, we
were also creating a cultural framework within which we developed shared
meanings.

We collected data over an 8-year period with a variety of methods, including diary-form behavioral observations, videotaped interactions with caregivers,

and cognitive experiments. We gathered information on Chantek's daily activities, social behavior, and interactions with his environment; the context or
situation in which Chantek signed; and interactions between Chantek and his
caregivers. Of particular importance was the contextual information routinely
collected for Chantek's signing, which allowed us to have a full picture of his
use of signs and not rely on anecdotal examples. Chantek's signing interactions were videotaped and translated into English glosses based on the procedures of Hoffmeister, Moores, and Ellenberger (1975) and Klima and Bellugi
(1979). In addition, experimental testing procedures were carried out to
monitor Chantek's cognitive and linguistic development, and to test specific
abilities such as MSR, imitation, and other cognitive skills. After several
years, Chantek developed a vocabulary of approximately 150 signs, including
at least five he invented on his own. He used his signs regularly in multisign
combinations to meet his needs, comment on his environment, and interact
with his caregivers. He showed evidence of a symbolic capacity and learned
to imitate, invent signs of his own, and engage in both linguistic and
nonlinguistic deception (Miles, 1986, 1990; Miles, Mitchell, & Harper, 1992).

Mirror self-recognition

One of our goals was to study the development of Chantek's self-awareness
through his ability to recognize himself in a mirror and use the mirror for
grooming and other self-directed behaviors, following techniques derived from
Gallup (1970, 1977) and Lewis and Brooks-Gunn (1979). Lewis and Brooks-
Gunn pointed out that self-recognition involves the discrimination of one's
body and body parts, and also knowing that the self cannot exist in two places
at once, that is, that there is a continuity of identity through time and space.
Mirror self-recognition has been shown in humans and in great apes (Gallup,
1970; Patterson & Cohn, *SAAH*17; Suarez & Gallup, 1981), including
unenculturated orangutans (Lethmate & Dücker, 1973; Suarez & Gallup,
1981). When presented with a mirror, orangutans and other apes showed self-
directed behaviors such as grooming themselves, rather than mirror-directed
behaviors, indicating that they knew they were looking at their own image.

Because the ability of orangutans to recognize themselves in mirrors had
already been established, our goal with Chantek was to focus on the effects
of enculturation and to relate self-recognition to the emergence of other
cognitive and linguistic abilities. Because Chantek was enculturated in a human
setting, mirrors of several varieties were a part of his environment, although
no mirror training was specifically carried out nor was special attention drawn
to them. In our collection of observational data in the daily record, we noted
his reaction to mirrors. As soon as Chantek showed contingent behaviors
while looking at himself in the mirror at 24 months of age, we conducted
several mirror mark tests. We modified the Gallup mark test as adapted by
Anderson (1983). In most mirror tests, the animal is anesthetized, a dot is
placed on the animal's ear, forehead or other location on the head, and the

animal is observed upon awakening to determine if it touches or attempts to groom the spot while looking into a mirror presented by the researcher. When conducting our mirror tests, we did not anesthetize Chantek, because MSR had already been demonstrated in orangutans, and because it would not have been appropriate to the larger goals of our enculturation study to anesthetize him repeatedly. Instead, a dot of room-temperature tempera (children's nontoxic) paint or ink was placed, as unobtrusively as possible, on his forehead during relaxed interaction or play. The motion of placing the mark was done in the course of other touching motions and activities so as not to draw special attention to the mark placement. A few minutes later, a mirror was presented to him (we used several different types of mirrors) and his reactions were observed. Chantek also had opportunities to see mirrors when no marks were placed on his face, as mentioned above. On none of these occasions (with or without a mirror) did he touch the mark prior to seeing himself in the mirror or appear to sense that a mark had been made. During the study period, 28 instances of Chantek's mirror use were recorded from 21 months of age through nearly 8 years of age. In addition, six mark tests were conducted from 24 to 41 months of age.

One significant pattern that emerged was the relationship between Chantek's referential use of signs and the development of his self-awareness through mirror self-recognition. Evidence suggested that Chantek passed through three stages of linguistic and cognitive development: Stage I, Instrumental Association; Stage II, Subjective Representation; and Stage III, Objective Representation (Miles, 1990). These stages are useful points of reference in the interpretation of the development of his MSR and other indicators of self-awareness, and reflect milestones that may be, wholly or in part, homologous with those found in human children. Further, it is likely that similar cognitive processes underlie both referential linguistic ability (including deception, imitation, and displaced reference) and MSR, including the objectification of self through self-awareness.

Stage I: Instrumental Association

Our preliminary analysis suggests that Stage I of Chantek's language acquisition is characterized by associative learning and pragmatic communication, and lasted from 9 to 24 months of age (Miles, 1990). In Stage I, Chantek performed signs that he associated with people, objects, and actions, but there is no clear evidence that his use of signs was referential or symbolic; Chantek's goal was simply to meet his immediate needs. After 1 month of training at 12 months of age, Chantek performed his first signs and sign combinations. Soon he began to overextend the meaning of his signs in ways similar to those of human children, for example signing DRINK for juicy grapes, and TICKLE for chase. During this stage, he learned to use a vocabulary of 20 signs of actions (such as EAT, MORE, and GO) and objects (such as FLOWER, KEY, and BRUSH).

Table 16.1. *Chantek's Stage I reactions to mirror presentation and mark test (9–24 months of age)*

No.[a]	Age (mo)	Reaction
1	21	Looked at self; reached behind mirror
2	22	Looked persistently at self
3	22	Looked at self; reached behind mirror
4	22	Looked persistently at self; touched mirror
5	22	Looked at self; reached behind mirror
6	23	Looked persistently at self for 15 minutes; engaged in contingent behaviors
7*	24	Looked at self; touched mirror; scraped back of mirror & air in front of mirror
8*	24	Looked at self; reached behind mirror; touched & licked mirror

[a] Asterisk indicates a mark test.

Here, Chantek completed Piagetian sensorimotor Stage 5 (Piaget, 1952). For example, Chantek experimented with objects by vacuuming himself with the vacuum cleaner. At 10 months of age, he engaged in simple imitations, and by the time he reached his first birthday, he began to respond to games of pretense. At 13 months of age, he spontaneously made signs with his feet, not just his hands, which suggests he understood that a sign is spatial and relational, and not just a motor movement. His first behavioral deceptions occurred at 14 months of age. Not all deception requires perspective taking (Miles, 1986; Mitchell, 1986), and Chantek's early deceptions do not exhibit perspective taking or objectification and were likely unintentional associations (Miles, 1986). At 15 months of age, he signed to himself, his toys, and other animals for the first time, and made his first sign imitation (WIPER). Two months later, he showed rudimentary evidence of sympathetic identification by "protecting" his caregiver from an attacking toy animal. By 20 months of age, he elaborated this play and engaged in active pretense and animism by feeding his toy animals.

Toward the end of Stage I Chantek's casual attention to mirrors changed to genuine interest. Mirrors were a frequent element in his environment and, just like human children, he was constantly exposed to his reflection in the bathroom mirror or had opportunities to observe cosmetic mirrors and other reflecting surfaces. He began to engage in contingent behaviors while observing himself in the mirror. We tested for his mirror self-recognition, which normally occurs in human children $1\frac{1}{2}$–2 years of age (Lewis & Brooks-Gunn, 1979). Chantek's Stage I reactions to mirrors, both in the environment and during the mark test, are presented in Table 16.1.

Table 16.1 shows that in Stage I, neither during casual encounters with household mirrors nor during two mark tests we conducted, did we observe that Chantek recognized his image. However, his interest in mirrors and contingent behaviors increased during this period. At 21 months of age, he looked at his reflection and reached his hand in back of the mirror as if to see if the image had substance. At 22 months of age, when presented with a mirror on four different occasions, he experimented with the mirror and its surface. He showed keen interest and played with the mirror while looking at his image. On one of these occasions, he took the mirror into his bedroom and sat on his bed and held the mirror in front of him while he reached and grabbed behind it as if to capture the image he saw.

The use of name signs are also indicators of awareness of self and others. At 20 months of age, Chantek began to show even greater interest in and identifications with others. Interestingly, shortly after this point in his development, he first named himself by signing CHANTEK spontaneously. For the first 2 years of his development Chantek had used 20 different signs, but never performed the signs ME or CHANTEK or otherwise referred to himself, despite his caregivers' frequent use of these signs (spontaneous usage required that he initiate the sign independently without imitating his caregiver's immediate signing). At 22 months of age, he used the sign CHANTEK in response to the question WHO YOU? and then rapidly used it in requests for different actions such as to be fed, chased, or tickled. He used CHANTEK immediately in combination with other signs, such as CHASE CHANTEK, and it became an active sign in his vocabulary.

The rather sudden onset of his use of his name sign, after a year of exposure, suggests that it might be related to the development of pointing and reference, which occurred around the same time. At 10 months of age, Chantek had developed the first step of protodeclarative pointing, which begins in human children at the same age and is an important precursor to referential language (Bates, 1979). Children first draw attention to themselves, then to objects, and finally begin to point. Chantek followed a similar developmental sequence. At 10 months of age, he began to exhibit himself and make noises to attract attention. Next, he began to maintain eye contact while he first showed us objects and then gave them to us. Later he gestured toward various directions, and by 21 months of age, could imitate pointing to objects. At this time, he also learned to point to his body parts in imitation games when asked, WHERE NOSE? and WHERE EYES?

Both Mitchell (1993) and Parker (1991) have suggested that imitation games that utilize pointing are significant actions that encourage self-recognition. These games may have helped Chantek objectify his body parts and make associations between his caregiver's body parts and his own, the implicit purpose of the game. He may also have made associations between the label CHANTEK and those body parts. Although he was reliably and validly using the sign CHANTEK for himself, it is interesting to note his failure during this period to recognize himself in the mirror. This suggests that he used the sign CHANTEK as a performative. His name sign was either a marker for a request or an indication that he wished to be a recipient of an action. Thus, his use of the sign CHANTEK in this stage is not referential, that is, not yet a name.

At 23 months of age in addition to touching and reaching behind the mirror presumably for the source of the image, he began to engage in more elaborate mirror-directed behavior, such as scraping and licking the mirror surface and air; and making movements, faces, and other contingent behaviors. These explorations and experimentations prompted us to examine his self-recognition ability with the mirror mark test. We administered two mark tests at the close of Stage I, when Chantek was 24 months of age. In the first test, we placed a dot of yellow paint on his forehead and allowed him to see himself in a mirror. He looked at his image and seemed puzzled while he scraped at the mirror's front and back surface, and at the air between himself and the image in the mirror. In the second test, he repeated these actions, and also licked the mirror's surface.

At the end of Stage I at 24 months of age, Chantek first began to use name signs for his caregivers. He learned the name signs LYN and ANN for his primary caregivers and frequently generalized and overextended these to refer to other caregivers. That he had not used these name signs previously, and developed them shortly after he learned CHANTEK, supports our conclusion that his earlier use of his name was not a self-identification, but a self-association. Thus, although his use of his name sign may have been aided by pointing to body parts, and he was clearly experimenting with mirrors and engaging in contingent behaviors, he was not signing referentially and did not recognize himself in Stage I.

Stage II: Subjective Representation

Stage II, Subjective Representation, began at approximately 2 years of age and was differentiated from Stage I based on the development of linguistic reference. The protodeclarative phases of Stage I led to spontaneous, proximal pointing to objects, and finally to the development of linguistic reference in Stage II. In their study of a deaf child, Hoffmeister and Moores (1973, p. 5) showed that "Within the early stages of sign language development . . . proximal pointing indicates specific reference." Although it occurred several months later than it does in human children, the development of Chantek's spontaneous pointing paralleled that of hearing and deaf children. However, Stage II was characterized by a period of inconsistent mirror self-recognition.

Chantek's responses to mirrors during Stage II is shown in Table 16.2. At the start of Stage II at 25 months of age, Chantek had one failure to show mirror self-recognition. Because he was able to respond, CHANTEK, to our general question, WHO YOU?, we pointed to his image in a hand mirror and asked, WHO THAT? Chantek did not name himself, but looked intently at his image and continued to reach behind the mirror.

A week later, still 25 months old, Chantek first recognized himself in a mirror by passing the mark test. We again placed a yellow dot of paint on his forehead, but this time instead of reaching behind the mirror, he immediately

Table 16.2. *Chantek's mirror self-recognition in Stage II*
(2–4.5 years of age)

No.[a]	Age (yr; mo)	Reaction
1	2; 1	Looked at self; reached behind mirror
2*	2; 1	Used mirror to groom mark (first passing of mark test)
3*	2; 1	Used mirror to groom mark
4*	2; 2	Used mirror to groom mark
5*	2; 4	Looked at self; reached behind mirror
6*	3; 0	Looked at self; reached behind mirror; scraped back of mirror
7	3; 2	Looked at self in reflecting window; displayed at image
8*	3; 5	Used mirror to groom mark
9	3; 8	Used mirror to groom swollen eye; opened and closed his mouth
10	4; 0	Sought out bathroom mirror; groomed lip and inspected face
11	4; 1	Sought out bathroom mirror; long period of grooming lip
12	4; 1	Used mirror to groom cut lip
13	4; 1	Used mirror to groom both lips
14	4; 3	Used mirror to groom paint off face

[a] Asterisk indicates a mark test.

began to adjust his body position to see his image in the mirror and use it to groom the mark off of his forehead with his finger. Chantek's age is comparable to the age at which children normally first pass the mark test: 18–24 months of age (Lewis & Brooks-Gunn, 1979). Several days later we repeated the test, and once again he used the mirror to guide his finger in grooming off the painted mark.

The next month, at 26 months of age, Chantek began to point spontaneously to himself and use the POINT sign to respond to questions such as WHERE YOU WANT TICKLE? By now Chantek was pointing to objects to bring his caregiver's attention to the object or simply to acknowledge its presence, and also pointing in response to questions such as WHERE HAT?, WHICH DIFFERENT?, and WHAT WANT? As he began to point to aspects of the environment and map it representationally through signs, as well as name and point to himself and others, he discovered the power of language and began to develop a theory of mind (Premack & Woodruff, 1978). Once he was able to point to and name himself, others, and objects, and identify himself in a mirror, his rate of vocabulary growth more than tripled, and his vocabulary expanded to 89 signs.

In referential use of language, a sign must also be used independently of context to convey meaning intentionally (Miles, 1990). In Stage II, Chantek began to use his signs when their referents were not present, and he was able to generalize their meanings to new referents. This first evidence of displaced reference (signing about objects not present) indicates that he had mental representations. He showed evidence of planning through mental representations and signed to himself about objects not present. For example, he signed to himself IN MILK RAISIN before going into his trailer and asking for milk.

Cognitive testing showed that Chantek was in sensorimotor Stage 6 (Piaget, 1952). We observed Chantek engaging in delayed imitation and using tools in sequence to solve problems. For example, he used a screwdriver to pry two boards apart, attempted to "cook" his oatmeal in the kitchen, and attempted to wash "dishes," including a camera (Miles, 1990). He could complete tool tasks using more than 20 different sequential steps; for example, he used a key to open a box, where he obtained wire cutters to release a screwdriver, which he used to unscrew a screw that held two pieces of wood together.

Chantek manipulated objects in novel relations to one another to create new meanings. For example, at 26 months of age, he gave his caregiver a knife and bottle nipple, two objects needed to prepare his milk formula, and fixed his stare at the cabinet where the remaining ingredient was located. He used tools to explore his environment, and he manipulated his caregivers as social tools to obtain attention. His deceptions became more elaborate, and he began to use signs in his deceptive activity (Miles, 1986). Chantek also engaged in rudimentary symbolic play, and identified and imitated two-dimensional representations such as a picture of the signing gorilla, Koko, pointing to her nose.

However, Table 16.2 shows that after several early successes, Chantek failed to recognize himself in the mirror from approximately $2\frac{1}{2}$ to $3\frac{1}{2}$ years of age, and specifically failed the mark test twice, at the ages of 28 and 36 months. In both mark tests, he looked at himself in the mirror but reached behind it, sometimes touching the mirror or scraping the back of it as he had done earlier. However, in both tests we used a new, small, compact mirror, and it is possible that the novel mirror called for more experimentation. During this stage, at 38 months of age, he also played in a tree next to a building and noticed his image reflected in the large windows. He stared at it and made aggressive displays, so we surmise that he saw the image in the window but did not realize that it was himself.

A similar interrupted or inconsistent pattern of MSR has also been reported for human children (Zazzo, 1982). Zazzo described the reactions of children and reported that after successfully completing the test, they too sometimes reached behind the mirror, experimented, and seemed fearful of it. It is possible that Chantek maintained his self-recognition during this period, but was confused by extremes of the small new mirror and large reflective windows. Also, he may have been engaging in pretend play when seeing his reflection in the large windows, given that he had been engaging in pretense from the time he was 20 months old. It is difficult to choose between these alternatives, but the most conservative conclusion is that, similar to human children, Chantek showed variable self-recognition during this period.

However, from the age of approximately $3\frac{1}{2}$ years onward, Chantek gave continuous and unambiguous evidence of self-recognition. We conducted a final mark text at 41 months of age (three months after the window image reflection incident) and he immediately used the mirror to guide his hand to groom off a dot of ink we placed on his forehead. We discontinued the mark test at this point because he began to spontaneously use the large bathroom mirror for self-grooming. In six other instances in the second half of Stage II, he used the bathroom mirror to not only groom paint off of his face, but also for self-care to groom his swollen eye and cut lip. He would make requests to go to the bathroom to look in the mirror and also seek the mirror out on his own. He would sit for long periods in front of the mirror inspecting his face, and opening and closing his mouth to look inside. In fact, by $3\frac{1}{2}$ years of age, he used the mirror for grooming nearly every time he entered the bathroom. Thus, we began to note only lengthy or unusual instances of his grooming before the bathroom mirror in our daily record.

We surmise that following his first mirror self-recognition, coupled with his linguistic use of displaced reference, Chantek used his name sign referentially, that is, as a true name. During the later period of inconsistent MSR, Chantek began to sign ME at 32 months of age. His first use was in the combination FOOD–EAT ME when asked WHO EAT? He later initiated communications such as ME CHANTEK, ME GO CHANTEK GO, and ME KEY CHASE during games, and could reply ME when asked who was requesting or engaging in a specific activity. For example, if Chantek was sipping a soda and his caregiver asked WHO DRINK?, Chantek would reply ME or CHANTEK.

However, Chantek also signed ME when caregivers asked WHO ME? referring and pointing to themselves. ME is a confusing sign since it was the first sign that had a varying referent depending upon who was sending the message. This led to numerous confusing sign exchanges that resembled the comedy routine, "Who's on first?" Chantek cleverly began to answer such questions as WHO GO, YOU OR ME? with the unambiguous answer CHANTEK while he continued to use ME when he initiated communications himself, such as ME DIRTY, ME GO, and other combinations. Chantek seemed to understand that both CHANTEK and ME (when he initiated communications) were the same, but he was reluctant to sign ME if that sign had already been used in the caregiver's previous communication and instead substituted CHANTEK. This more unambiguous use of CHANTEK is a lexical grammatical device invented by Chantek that served the purpose of a metacommunication, as he either sought to avoid the more confusing ME, or chose to add more linguistic clarity by replying with CHANTEK.

The sign YOU also entered Chantek's vocabulary at 34 months of age, and added potentially more confusion since the sign YOU had the same gestural shape as the sign ME but was pointed in the opposite direction. Chantek first signed YOU when his caregiver was tickling him and asked WHO (tickle)? Chantek responded to similar questions such as TICKLE WHO? with the answer YOU (the agent of the action) rather than ME (the object or recipient). (In American Sign Language the agent and object would be encoded in the articulation of the sign, but we often used citation forms of ASL signs in pidgin Sign English, which relied on English word order to convey grammatical distinctions, as well as encoded signs). Chantek's understanding of personal pronouns was affected by his subjective orientation. His use of YOU may have been in transition from a performative in which he reasoned: When asked WHO?, coupled with an action sign such as TICKLE, always respond YOU to become the recipient of the action. He may also have been affected by cognitive or linguistic limitations, in that he may have maintained a semantic primacy of agency in his signing by always emphasizing the agent of an action, not the recipient, even if it were himself. Eventual analysis of his total sign corpus and the context of his communications will allow us to reconstruct this development more precisely.

Thus, in Stage II, Chantek recognized himself in the mirror, used his signs as symbolic representations, and distinguished himself from others, whom he could perceive as agents. However, his perspective remained somewhat subjective and self-focused, with an emphasis on his needs, and was characterized by his reluctance to linguistically encode objective descriptions of actions, especially when he was the recipient. Thus, he was not yet able to totally frame a mental representation or linguistic reference from the perspective of the other.

Stage III: Objective Representation

The third stage, Objective Representation, introduced perspective taking – the ability to recognize and utilize the point of view of the other. Stage III ranged from about $4\frac{1}{2}$ to more than 8 years of age, during which Chantek's vocabulary increased to approximately 150 signs. In Stage III, Chantek developed a less egocentric point of view, reflected in both his cognitive behavior and his signing.

His cognitive development extended somewhat into the preoperational stage, which included more complex imitations, representation through symbolic play, and mastery of rudimentary categorical conceptualization (Inhelder & Piaget, 1964; Piaget, 1945/1962). He developed an understanding of a number of contrasting concepts, such as hot/cold and same/different, and completed complex sorting tasks that required an understanding of classification, classes, and categories.

In Stage II he had discovered the power of language; in Stage III he began to harness that power for himself. Chantek not only recognized himself and others, but he could represent the world to himself from the other's perspective and base his actions accordingly. He now understood that the sign ME could refer both to himself and to others, depending upon one's perspective and the context. Signs were not just referential, but they formed a contextually based shared cultural code founded on relationships and roles. For example, the sign ME refers to the speaker in a communication dyad that can vary from situation to situation, but not to a specific person. Chantek was also able to take the perspective of the other by getting the caregiver's attention and directing her eye gaze before starting to sign (see Gómez, *SAAH5*). His passive understanding that signs were representations became an active understanding that signs were quite separate from their referents. He could request the names of things by offering his hands to be molded, sometimes with a questioning intonation and facial expression. He understood meta-communication about signing itself, and that signs had components that could be altered toward a shared ideal form. He was able to respond to his caregiver's requests to SIGN BETTER by improving the manner in which he performed a sign. He could even carefully deconstruct a sign's articulatory elements by looking at his hands, slowly isolating each motion, and maintaining eye contact with his caregiver.

It was at this point that he invented symbols of his own. These sign inventions were not erratic, performative gestures associated with an object, but symbols that he used referentially in combination with other signs. His sign inventions were iconic, but they were based on the linguistic parameters of signing articulation. For example, he invented NO-TEETH to indicate that he would not use his teeth during rough play, EYE-DRINK for contact lens solution used by his caregivers, DAVE-MISSING-FINGER for a favorite person who had a hand injury. VIEWMASTER for a slide viewer, and BALLOON.

In Stage III, Chantek's MSR became more elaborated as his sense of self became more objectified and his perspective taking grew. His use of the bathroom mirror for grooming and inspection had became so commonplace during the later part of Stage II that in Stage III we recorded only more extensive or unusual mirror uses. He also began to routinely name his image in the mirror, doing so spontaneously in conversations, during self-signing, and in response to a caregiver's question. These also became commonplace and were not recorded.

Chantek's more lengthy or unusual reactions to mirrors during Stage III are shown in Table 16.3. As these examples illustrate, enculturated apes are the only animals to show any evidence of an ideal image of themselves (Miles, 1993; Mitchell, 1993, in press, *SAAH6*). Table 16.3 shows that he continued to use the mirror for grooming, such as inspecting inside his mouth and checking a cut lip. On these occasions, he spent long periods of time before the mirror, inspecting his face and upper body and turning his face from side to side in self-approving motions as if to study his image. He would watch himself intently as he made faces, looked inside his mouth, or simply leaned back to study his appearance. At 5 years of age, as he began to experience hormonal changes moving him toward sexual maturity, he began to use the mirror to watch himself while he stimulated his nipples.

Chantek also used objects in his self-explorations before the mirror and began to coordinate the use of the mirror with other tools. At 6 years of age, he received a pair of sunglasses and went to the bathroom mirror and put them on. He did not move the glasses around on his face or rub them on his skin as if to experiment with the feel of the glasses. Instead, he looked intently at his image from different angles and seemed to be experimenting

Table 16.3. *Chantek's mirror self-recognition in Stage III (4.5–8 years of age)*

No.	Age (yr; mo)	Reaction
1	4; 9	Used mirror to groom several areas of face
2	4; 9	Leaned back to study self in mirror; made faces; distorted lips; inspected inside mouth; put finger in mouth during inspection
3	5; 0	Groomed powdered milk off face
4	5; 0	Rubbed nipples while looking in mirror
5	5; 0	Groomed face, chest and shoulders; examined image for lengthy period
6	6; 2	Put on sunglasses and looked at self in mirror
7	6; 4	Spit on mirror; licked and touched nipple while looking in mirror
8	6; 9	Groomed mouth, teeth, and eyes; played with angle of mirror to observe others
9	6; 11	Groomed face
10	7; 1	Groomed eyes and nipple
11	7; 2	Groomed eyes; attempted to lick nipples in mirror
12	7; 10	Groomed eye; used mirror to curl his lashes with eyelash curler

with both his appearance and the effect the glasses had on his perceptions. He also played with small mirrors and incorporated them into his actions with other objects. At nearly 7 years of age, he used a mirror to observe others surreptitiously. By nearly 8 years of age, in an elaborate imitation, he attempted to curl his eyelashes with an eyelash curler in front of the bathroom mirror, as he had observed his caregiver do. Tomasello (1990) as well as Visalberghi and Fragaszy (1990) have argued that the great apes are poor imitators in the areas of communication and technology, but for enculturated cross-fostered apes this is not the case.

It is tempting to consider some of these instances as adornment or development of a social self based on others' evaluations, because he gave other evidence at this time of internalizing simple values and ideals. For example, Chantek signed BAD to himself when he misbehaved, including when he was alone. He also signed BAD to noisy birds and other animals when they were loud, and to his caregivers when they displeased him. This suggests that he had begun to internalize some sense of images and socially appropriate behavior, was able to make simple evaluations, and could attribute intent and responsibility. Thus, it is possible that some of his mirror behavior in this stage constituted the beginnings of adornment and self-evaluation. However, like very young human children (Guillaume, 1926/1971), Chantek seemed to create the adornment for his own appreciation and did not attempt to deliberately modify his image for our approval, as older human children and adults do (Mitchell, 1993).

During Stage III, he showed additional evidence of simulative perspective taking through imitating his caregiver's role during play and engaging in rudimentary role reversal. Imitation and pretense are important indicators of cognitive development, including the development of mirror self-recognition (Bretherton, 1984; Gopnik & Meltzoff, *SAAH*10; Guillaume, 1926/1971; Hart & Fegley, *SAAH*9; Lewis & Brooks-Gunn, 1979; Mitchell, 1987, 1990, 1993, in press; Parker, 1991; Parker & Milbrath, *SAAH*7; Piaget, 1945/1962). For example, at 7 years of age, Chantek spontaneously reversed roles with his caregiver and gave him an instruction to DO SAME followed by a slapping action. This command was normally addressed to him accompanied by an action that he was to imitate. He did this during a wrestling game in order to deceive his caregiver into imitating the action; this brought the caregiver closer so that Chantek could grab him. Bretherton (1984) has presented several categories of pretend play in human children, and apes have been thought to engage in only three of these categories (Parker & Milbrath, *SAAH*7). In this incident, Chantek may have achieved a more advanced category of exchanging roles with another during play, as the manner in which he commanded the caregiver is evidence that the role was being exchanged, not just the signs. This role exchange suggests that Chantek's uses of the mirror for adornment and self-evaluation may have incorporated just the beginnings of a social self based on the internalization of the perspective of the other. Thus, Chantek went from instrumental self, to egocentric self, to perspective-taking self, to possibly a nascent, socially evaluative self. Further research will help us to determine if the perspective of others played a major role in his formation of self. For example, an examination of a large sample of videotapes of Chantek's interactions with his caregivers may shed light on theories of the genesis of MSR, including the facial imitation model and kinesthetic–visual-matching model (Mitchell, 1993; Parker, 1991).

Conclusion

Chantek showed first evidence of self-recognition at 2 years of age and consistent self-recognition at age $3\frac{1}{2}$. This conclusion is based on mirror uses and mirror mark tests, and is supported by related linguistic and cognitive abilities that emerged in three stages in his development and included use of names and pronouns, the emergence of pointing, perspective taking, simulation, and pretense. Interestingly, the same pattern of early inconsistent MSR (followed by later consistent self-recognition) has been reported for children (Zazzo, 1982). It is possible that the inconsistent response to the mirror through Stage II is due to gradually learning the properties of a reflecting surface (Loveland, 1986).

It is significant that Chantek could clearly use his name sign CHANTEK and the pronoun ME before he consistently recognized himself and used a mirror for self-grooming. Additional evidence regarding perspective taking suggests that Chantek's early use of these signs was subjective, and became more objective when he could self-objectify through his grooming efforts. By the last stage of development, Chantek was able to use mirrors for self-grooming, experimentation, and possibly adornment, suggesting the formation of a social self based on the perspective of the other. Support for this perspective taking comes from Chantek's internalization of signed concepts such as BAD and SIGN BETTER, and from role reversal during games. Evidence from sign language research with orangutans and other apes can shed light on the evolution of self-recognition and awareness and provide clues for the relationship between these abilities and other cognitive areas, including imitation, pretense, mental state attribution, and a symbolic capacity.

Acknowledgments

This research was supported by the National Institutes of Child Health and Human Development grant NICHD 14918, National Science Foundation grant BNS 8022260, the Yerkes Regional Primate Research Center, supported by National Institutes of Health grant RR 0165, and grants from the UC Foundation. I would also like to thank Philip Lieberman for his initial support; and Stephen Harper, Robert Mitchell, Ann Southcombe, and the members of Project Chantek for their assistance in this research.

References

Anderson, J. (1983). Responses to mirror image stimulation and assessment of self-recognition in mirror- and peer-reared stumptailed macaques. *Quarterly Journal of Experimental Psychology, 35*, 201–212.

Bard, K. A. (1988). Behavioral development in young orangutans: Ontogeny of object manipulation, arboreal behavior, and food sharing. *Dissertation Abstracts International B49*, 4, 1407 (University Microfilms No. 8810704).

 (1990). "Social tool use" by free-ranging orangutans: A Piagetian and developmental perspective on the manipulation of an animate object. In S. T. Parker & K. R. Gibson (Eds.), *"Language" and intelligence in monkeys and apes: Comparative developmental perspectives* (pp. 356–378). Cambridge University Press.

Barnouw, V. (1985). *Culture and personality* (4th ed.). Homewood, IL: Dorsey.

Bates, E. (1979). *The emergence of symbols: Cognition and communication in infancy.* New York: Academic Press.

Bretherton, I. (1984). *Symbolic play: The development of social understanding.* New York: Academic Press.

Bruner, J. (1983). *Child's talk: Learning to use language.* New York: Norton.

Chevalier-Skolnikoff, S. (1983). Sensorimotor development in orang-utans and other primates. *Journal of Human Evolution, 12*, 545–561.

Fouts, R. S. (1973). Acquisition and testing of gestural signs in four young chimpanzees. *Science, 180*, 978–980.

Furness, W. (1916). Observations on the mentality of chimpanzees and orangutans. *Proceedings of the American Philosophical Society, 55*, 281–290.

Galdikas, B. (1982). Orang-utan tool-use at Tanjung Puting Reserve, Central Indonesian Borneo (Kalimantan Tengah). *Journal of Human Evolution, 10*, 19–33.

Gallup, G. G., Jr. (1970). Chimpanzees: Self-recognition. *Science, 167*, 86–87.

 (1977). Self-recognition in primates: A comparative approach to the bidirectional properties of consciousness. *American Psychologist, 32*, 329–338.

Gardner, R. A., & Gardner, B. T. (1969). Teaching sign language to a chimpanzee. *Science, 165*, 664–672.

(1989). *Teaching sign language to chimpanzees*. Albany, NY: SUNY Press.

Geertz, C. (1965). The impact of the concept of culture on the concept of man. In J. Platt (Ed.), *New views on the nature of man* (pp. 93–118). Chicago: University of Chicago Press.

Goodenough, W. (1957). Cultural anthropology and linguistics. In P. Garvin (Ed.), *Report of the 7th annual roundtable meeting on linguistics and language study* (Monograph Series on Languages and Linguistics, No. 9). Washington, DC: Georgetown University Institute of Languages and Linguistics.

(1981). *Culture, language, and society* (2nd ed.). Menlo Park, CA: Benjamin/Cummings.

Guillaume, P. (1926/1971). *Imitation in children*, 2nd ed. Chicago: University of Chicago Press.

Harris, M. (1964). *The nature of cultural things*. New York: Random House.

Harrison, B. (1963). *Orang-utan*. New York: Doubleday.

Haviland, W. (1991). *Anthropology*. New York: Holt, Rinehart & Winston.

Hoffmeister, R. J., & Moores, D. F. (1973). *The acquisition of specific reference in the linguistic system of a deaf child of deaf parents* (Research Report No. 53, Project No. 332189). Washington, DC: U.S. Office of Education.

Hoffmeister, R. J., Moores, D. F., & Ellenberger, R. L. (1975). Some procedural guidelines for the study of the acquisition of sign language. *Sign Language Studies, 7*, 121.

Inhelder, B., & Piaget, J. (1964). *The early growth of logic in the child*. New York: Harper & Row.

Kellogg, W. (1968). Communication and language in the home-raised chimpanzee. *Science, 162*, 423–427.

Klima, E., & Bellugi, U. (1979). *The signs of language*. Cambridge, MA: Harvard University Press.

Laidler, K. (1980). *The talking ape*. New York: Stein & Day.

Lethmate, J. (1982). Tool-using skills of orang-utans. *Journal of Human Evolution, 11*, 49–64.

Lethmate, J., & Dücker, G. (1973). Untersuchungen zum Selbsterkennen im Spiegel bei Orang-Utans und einigen anderen Affenarten. *Zeitschrift für Tierpsychologie, 33*, 248–269.

Lewin, R. (1983). Is the orangutan a living fossil? *Science, 222*, 1222–1223.

Lewis, M., & Brooks-Gunn, J. (1979). *Social cognition and the acquisition of self*. New York: Plenum.

Linton, R. (1936). *The study of man*. New York: Appleton-Century-Crofts.

Lock, A. (1980). *The guided reinvention of language*. New York: Academic Press.

Loveland, K. (1986). Discovering the affordances of a reflecting surface. *Developmental Review 6*, 1–14.

Mai, L. (1983). A model of chromosome evolution and its bearing on cladogenesis in the Hominoidea. In R. Ciochon & R. Corruccini (Eds.), *New interpretations of ape and human ancestry* (pp. 87–114). New York: Plenum.

Maple, T. (1980). *Orang-utan behavior*. New York: Van Nostrand Reinhold.

Miles, H. L. (1980). Acquisition of gestural signs by an infant orang-utan (*Pongo pygmaeus*). *American Journal of Physical Anthropology, 52*, 256–257.

(1986). How can I tell a lie?: Apes, language and the problem of deception. In R. W. Mitchell & N. S. Thompson (Eds.), *Deception: Perspectives on human and nonhuman deceit* (pp. 245–266). Albany, NY: SUNY Press.

(1990). The cognitive foundations for reference in a signing orangutan. In S. T. Parker & K. R. Gibson (Eds.), *"Language" and intelligence in monkeys and apes: Comparative developmental perspectives* (pp. 451–468). Cambridge University Press.

(1993). Language and the orangutan: The old person of the forest. In P. Singer & P. Cavalieri (Eds.), *A new equality: The Great Ape Project* (pp. 42–57). London: Fourth Estate.

Miles, H. L., & Harper, S. (in press). "Ape language" studies and the study of human language origins. In D. Quiatt & J. Itani (Eds.), *Hominid culture in primate perspective*. Denver: University Press of Colorado.

Miles, H. L., Mitchell, R. W., & Harper, S. (1992). Imitation and self-awareness in a signing orangutan. In K. Bard, S. Parker, & S. Boysen (Cochairs), *Comparative developmental approaches to the study of self-recognition and imitation*. Symposium conducted at the 14th Congress of the International Primatological Society, Strasbourg, France.

Milton, K. (1988). Foraging behaviour and the evolution of primate intelligence: Social expertise and the evolution of intellect in monkeys, apes, and humans. In R. Byrne & A. Whiten (Eds.), *Machiavellian intelligence* (pp. 285–305). Oxford: Oxford University Press (Clarendon Press).

Mitchell, R. W. (1986). A framework for discussing deception. In R. W. Mitchell & N. S. Thompson (Eds.), *Deception: Perspectives on human and nonhuman deceit* (pp. 3–40). Albany, NY: SUNY Press.

(1987). A comparative-developmental approach to understanding imitation. In P. G. Bateson & P. H. Klopfer (Eds.), *Perspectives in ethology: Vol. 7. Alternatives* (pp. 183–215). New York: Plenum.

(1990). A theory of play. In M. Bekoff & D. Jamieson (Eds.), *Interpretation and explanation in the study of animal behavior: Comparative perspectives* (pp. 197–227). Boulder, CO: Westview.

(1993). Mental models of mirror self-recognition: Two theories. *New Ideas in Psychology, 11,* 295–325.

(in press). The evolution of primate cognition: Simulation, self-knowledge, and knowledge of other minds. In D. Quiatt & J. Itani (Eds.), *Hominid culture in primate perspective.* Denver: University Press of Colorado.

Nanda, S. (1984). *Cultural anthropology* (2nd ed.). Belmont, CA: Wadsworth.

Parker, C. E. (1969). Responsiveness, manipulation and implementation behavior in chimpanzees, gorillas and orangutans. In C. R. Carpenter (Ed.), *Proceedings of the Second International Congress of Primatology, Vol. 1: Behavior* (pp. 160–166). New York: Karger Basel.

Parker, S. T. (1991). A developmental approach to the origins of self-recognition in great apes. *Human Evolution, 6,* 435–449.

Patterson, F. G. (1978). Linguistic capabilities of a lowland gorilla. In F. C. Peng (Ed.), *Sign language and language acquisition in man and ape: New dimensions in comparative pedolinguistics* (pp. 161–201). Boulder, CO: Westview.

Piaget, J. (1945/1962). *Play, dreams, and imitation in childhood.* New York: Norton.

(1952). *Origins of intelligence in children.* New York: International Universities Press.

Pilbeam, D. (1982). New hominoid skull material from the Miocene of Pakistan. *Nature, 295,* 232–234.

Povinelli, D. J., & Cant, G. H. (1992). *Orangutan clambering and the evolutionary origins of self-conception.* Paper presented at the 14th Congress of the International Primatological Society, Strasbourg, France.

Premack, D. (1972). Language in chimpanzees. *Science, 172,* 808–822.

Premack, D., & Woodruff, G. (1978). Does the chimpanzee have a theory of mind? *Behavioral and Brain Sciences, 1,* 515–526.

Rogoff, B. (1990). *Apprenticeship in thinking: Cognitive development in social context.* New York: Oxford University Press.

Rosman, A., & Rubel, P. (1981). *The tapestry of culture: An introduction to cultural anthropology.* Glenview, IL: Scott, Foresman.

Rumbaugh, D., & Gill, T. (1973). The learning skills of great apes. *Journal of Human Evolution, 10,* 175–188.

Rumbaugh, D., Gill, T., & von Glaserfeld, E. (1973). Reading and sentence completion by a chimpanzee *(Pan). Science, 182,* 731–733.

Rumbaugh, D., & Pate, J. (1984). Primates' learning by level. In G. Greenberg & E. Tobach (Eds.), *Behavioral evolution and integrative levels* (pp. 569–587). Hillsdale, NJ: Erlbaum.

Russon, A., & Galdikas, B. (1993). Imitation in free-ranging rehabilitant orangutans. *Journal of Comparative Psychology, 107*(2), 147–160.

Savage-Rumbaugh, E. S. (1986). *Ape language: From conditioned response to symbol.* New York: Columbia University Press.

Schwartz, J. H. (1987). *The red ape: Orang-utans and human origins.* Boston: Houghton Mifflin.

Shapiro, G. (1982). Sign acquisition in a home-reared/free-ranging orangutan: Comparisons with other signing apes. *American Journal of Primatology, 3,* 121–129.

Shore, B. (1991). Twice-born, once conceived: Meaning construction and cultural cognition. *American Anthropologist, 93*, 9–27.

Stanyon, R., & Chiarelli, B. (1982). Phylogeny of the Hominoidea: The chromosome evidence. *Journal of Human Evolution, 11*, 493–504.

Suarez, S., & Gallup, G. (1981). Self-recognition in chimpanzees and orangutans, but not gorillas. *Journal of Human Evolution, 10*, 175–188.

Terrace, H. S., Petitto, L., Sanders, R., & Bever, T. (1979). Can an ape create a sentence? *Science, 206*, 809–902.

Tomasello, M. (1990). Cultural transmission in the tool use and communicatory signaling of chimpanzees? In S. T. Parker & K. R. Gibson (Eds.), *"Language" and intelligence in monkeys and apes: Comparative developmental perspectives* (pp. 274–311). Cambridge University Press.

Tuttle, R. (1986). *Apes of the world: Their social behavior, communication, mentality, and ecology.* Park Ridge, NJ: Noyes.

Tylor, E. B. (1877). *Primitive culture: Researches into the development of mythology, philosophy, religion, language, art, and customs* (Vol. 1). New York: Holt.

Visalberghi, E., & Fragaszy, D. (1990). Do monkeys ape? In S. T. Parker & K. R. Gibson (Eds.), *"Language" and intelligence in monkeys and apes: Comparative developmental perspectives* (pp. 247–272). Cambridge University Press.

Vygotsky, L. S. (1978). *Mind in society: The development of higher psychological processes.* Cambridge, MA: Harvard University Press.

Weiss, M. (1987). Nucleic acid evidence bearing on hominoid relationships. *Yearbook of the American Journal of Physical Anthropology, 30*, 41–74.

Wertsch, J. V. (1978). Adult–child interaction and the roots of metacognition. *Quarterly Newsletter of the Institute for Comparative Human Development, 2*, 15–18.

Wood, D., Bruner, J., & Ross, G. (1976). The role of tutoring in problem-solving. *Journal of Child Psychology and Psychiatry, 17*, 89–100.

Wright, R. V. S. (1972). Imitative learning of a flaked tool technology – the case of an orangutan. *Mankind, 8*, 296–306.

Yunis, J., & Prakash, O. (1982). The origin of man: A chromosomal pictorial legacy. *Science, 215*, 1525–1530.

Zazzo, R. (1982). The person: Objective approaches. *Review of Child Development Research, 6*, 47–290.

17   Self-recognition and self-awareness in lowland gorillas

Francine G. P. Patterson and Ronald H. Cohn

Self-recognition in mirrors is considered an indicator of self-awareness, a capacity once assumed to be present only in human beings. Gordon Gallup's initial studies using the face-marking test, and the work they stimulated, have clearly demonstrated mirror self-recognition (MSR) in only three species: humans, chimpanzees, and orangutans (Anderson, 1983; Gallup, 1970; Platt & Thompson, 1985; Suarez & Gallup, 1981). Nevertheless field studies and laboratory tests, including simple discrimination learning, learning set, discrimination-reversal training, and oddity concept formation, have demonstrated that the cognitive abilities of all three genera of great apes are closely comparable. Gorillas and orangutans were ranked slightly above chimpanzees with respect to intelligence as measured by the transfer index, a refined measurement of learning-set ability in which species and individual differences in motivation and perceptuomotor skills are controlled (Rumbaugh & Gill, 1973). Piagetian studies also indicate that gorillas, chimpanzees, and orangutans undergo similar cognitive development (Chevalier-Skolnikoff, 1977, 1983; Mathieu & Bergeron, 1983; Redshaw, 1978). Nevertheless, the apparent inability of six gorilla subjects to recognize their mirrored images has led researchers to conclude that gorillas are the only great apes to lack the capacity for self-awareness (Ledbetter & Basen, 1982; Suarez & Gallup, 1981). Our current study with the female lowland gorilla Koko, which considers linguistic evidence as well as that provided by self-recognition tests, challenges this assertion.

Koko became the subject of an ongoing language study in July 1972, when she was 1 year old (Patterson, 1978b). She was taught sign language and continuously exposed to spoken English. Her rearing environment was similar to that of a human child. Other gorillas were not a part of this environment until a 3-year-old gorilla named Michael joined the language study when Koko was 5 years old. Mirrors, however, were a part of Koko's environment from the start of the study, although no training of any kind was associated with them. When she was about $3\frac{1}{2}$ years old (January 1975), Koko began consistently to exhibit mirror-guided self-directed behaviors. Similar behaviors are exhibited consistently by human children before 2 years of age (Lewis &

Brooks-Gunn, 1979), by chimpanzees at $2\frac{1}{2}$ (Custance & Bard, *SAAH*12; Gallup, *SAAH*3), and by an orangutan at $3\frac{1}{2}$ years (Miles, *SAAH*16). Koko would groom her face and underarms, pick at her teeth, and examine her tongue while studying her reflection. She would also comb her hair, make faces, and adorn herself with hats, wigs, and makeup in front of the mirror. Although, in contrast to Koko, the gorilla Michael has had only limited and sporadic exposure to mirrors, we have also documented him exhibiting such behaviors on videotape.

When Koko was 19 years old, a variant of the mirror mark test used by Gallup was administered to her to provide data strictly comparable to self-recognition studies done with other great apes. For Koko, this experiment was modified according to a procedure devised by Anderson (1983) so that she would not have to be anesthetized.

A mirror (0.6 m × 0.76 m) mounted on a plywood panel of the same size was used. Koko had direct access to this mobile mirror. As a control, to ensure that the actual marking procedure would be unobtrusive, Koko's brow was wiped with a damp washcloth at the beginning of each of three preliminary sessions. For the marking session, her face was marked with a mixture of red and white clown paint that approximated the color of the washcloth. Zaunders® clown paint (containing stearic acid, P.P.G. 76, water, triethanolamine, and color) was chosen because it is nontoxic, unscented, and water soluble.

Koko's exposure to the test mirror was restricted to six 10-min videotaped sessions recorded during 4 days (July 1–3 and 6, 1990). The mirror was placed in her room and propped in a upright position on the floor, providing a full-length view of her body when she was sitting on the floor. Koko was able to touch and move the mirror during all sessions. She became adapted to the new mirror during the first session. At the beginning of the second, third, and fourth sessions, Koko's brow, the target area, was wiped with a clean, damp washcloth that was approximately body temperature. At the beginning of the fifth session, her brow was wiped with an identical damp washcloth that had been dipped in the clown paint. Three days later (July 6, 1990), Koko was given a sixth, control session in which her brow was wiped with a clean, damp washcloth. Two experimenters, one who wiped Koko's forehead and one who videotaped the sessions, were present in the room with Koko during all six sessions of the test. Both experimenters had worked with Koko since she was 1 year old, and had similar direct contact sessions with her every day. Such sessions were frequently videotaped, and Koko was quite familiar with the camera.

The videotape of the mirror sessions was analyzed by two independent observers. They scored the number of times during each session that Koko touched the target area (and had 97.9% agreement). During the marking procedure, Koko turned her head quickly, with the result that the actual marked area included not only the brow, but also a spot over the right ear and the top of the head. For scoring purposes, this complete area that was

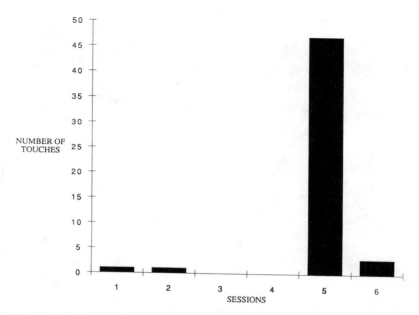

Figure 17.1. Mirror test target touches.

marked in the fifth session was considered the target area for all unmarked sessions as well. For each session the observers also scored the number of seconds Koko spent viewing her reflection (98.4% agreement) and the number of her self-directed mirror behaviors (95.0% agreement). Self-directed mirror behaviors are defined as "actions directed toward one's own body" (Brooks-Gunn & Lewis, 1982, pp. 353–354).

In the four sessions in which she was unmarked, Koko touched the target area an average of once per session, but in the fifth session, in which she was marked, she touched it 47 times (Figure 17.1). The time Koko spent viewing her reflection in each session adds support to the interpretation that she recognized that the paint spot she saw in the mirror was actually on her own body. She spent an average of 48% of the time viewing her reflection in the sessions in which she was unmarked. During the fifth (marking) session Koko's mirror viewing time increased to 88% of the session (Figure 17.2). Although this increase in viewing time during the marking session is characteristic of chimpanzees tested for self-recognition, previously no gorillas had shown an increase in attention to the mirror when marked (Gallup, 1987).

Her self-directed behavior in front of the mirror provides further support for self-recognition by Koko. Gallup (1987, p. 8) has argued that "the mark test serves merely as a means of validating impressions that arise out of seeing animals use mirrors spontaneously for purposes of self-inspection." That Koko exhibited any self-directed behavior at all is in itself significant,

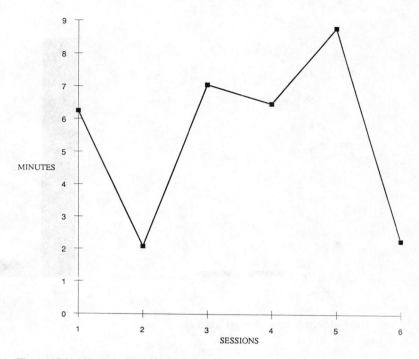

Figure 17.2. Mirror test viewing time.

as no gorilla in previous formal studies has clearly exhibited any such behavior (Ledbetter & Basen, 1982; Suarez & Gallup, 1981). There have, however, been informal observations of gorillas engaging in self-directed mirror behaviors. Riopelle, Nos, and Jonch (1971, p. 88) reported that the gorilla Muni, while sitting in front of a mirror, "lifted one leg and looked at his reflection, inspecting the parts of him that he ordinarily could not see." Lethmate (1974; also personal commun., Jan. 24, 1992) found indications of self-recognition in some of the gorillas he tested informally. Of six gorillas observed for mirror-guided, self-directed behaviors, one groomed his teeth and another groomed his head while watching the mirror. Four gorillas were then marked on the forehead with white paint while their keeper distracted them with tickling. The marked gorillas were subsequently confronted with their mirror images. Of the four, two touched the mark, one while viewing the mirror and the other after turning away from the mirror.

During every mirror session, Koko frequently showed self-directed behaviors such as grooming her underarms, making faces, and picking at her teeth with the aid of the mirror (see Figure 17.3). As one would expect, the incidence of these behaviors, including mark-directed behaviors, was highest during the fifth session, the one in which she had been marked. Whereas in the other

Figure 17.3. With the aid of a mirror, Koko examines parts of her body that she otherwise cannot see.

sessions she exhibited an average of 88 self-directed behaviors per session, in the fifth session she exhibited 202 (Figure 17.4), supporting the conclusion that she recognized the altered image as her own.

In each session, at least 80% of Koko's self-directed behaviors could be classified as self-grooming. In the unmarked sessions she groomed her underarms and teeth most frequently, with some attention paid to her eyes and chin. During the marked session, the majority of Koko's initial self-grooming behaviors involved removing the paint from her brow. Once most of the paint had been removed she concentrated her grooming efforts on her teeth and other parts of her face.

It might be argued that Koko sensed the mark's presence because she had not been anesthetized. However, on September 6, 1982, when Koko was 11 years old, she passed a "mark test" designed by nature in which this possibility was completely eliminated. The incident was captured on videotape during an hour-long sample of Koko's signing recorded by the same two longtime companions. Forty-seven min into the hour, Koko began to inspect her reflection in a Plexiglas mirror (0.53 m × 1.22 m) that was a fixture in her room. A continuous 10-min segment of the hour-long sample (comparable to the 10-min sessions of the formal test described above), starting at this point and encompassing all of this serendipitous mark test, was analyzed by two independent observers. They scored the number of Koko's self-directed behaviors (95.8% agreement) and target touches (94.1% agreement). During the $4\frac{1}{2}$ min of the segment that Koko was in front of the mirror, she exhibited 136 self-directed behaviors. Three min and 55 sec into this period, Koko discovered a spot of black pigment on her otherwise pink upper right gum. Before this discovery she had touched this target area just twice; after the discovery she touched it 14 times. She could not possibly have felt, smelled, or tasted this spot, and it was not visible even in the mirror unless her lip was held up. She picked at the pigment with the handle of a toothbrush and with her index and small fingers as if she were trying to remove it. She inspected both the toothbrush handle and her small finger after using them to pick at the spot. After studying her reflection, she signed COME to one of the experimenters. Koko opened her mouth while raising her arm toward the experimenter's mouth, as if trying to get the experimenter to open her mouth as well. When the experimenter opened her mouth, Koko inspected the very same area on the left side of the experimenter's mouth where she had seen the pigment on her own mirror image. The experimenter had not watched the mirror behaviors and therefore did not realize what Koko's intent was at the time.

We took advantage of Koko's linguistic abilities in this study to address more directly the question of self-awareness, as well as to cross-validate the evidence provided by her responses to mirrors. Two "Who are you?" questions were asked during each of the first four 10-min sessions while Koko was away from the mirror. After the fifth (marking) session, Koko was asked "Who are you?" once more while away from the mirror, and also "Who is that?" of her mirror reflection. During the sixth (posttest) session she was asked "Who are you?" one more time. Although correct answers to these questions can simply be learned responses, the data argue otherwise in Koko's case. Her responses (see Table 17.1) were all different, multisign utterances (mean length of utterances: 4.4 signs), but had one consistent theme. Each reply

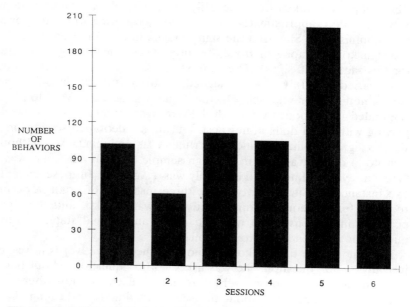

Figure 17.4. Self-directed behaviors during the mirror test.

Table 17.1. *Koko's responses to self-identity questions*[a]

Test session	First response	Second response
"Who are you?"		
1	ME GORILLA NIPPLES[b] TICKLE.	NIPPLES GORILLA.
2	POLITE-KOKO KOKO NUT NUT POLITE.	POLITE DEVIL HAIR HEAD. KOKO SWEET BAD.
3	KOKO POLITE ME THIRSTY.	GORILLA ME.
4	POLITE ME THIRSTY FEEL KOKO-LOVE.	POLITE POLITE-KOKO. KOKO FEEL THIRSTY SORRY.
5	KOKO POLITE SORRY GOOD.	KOKO; PLEASE GIMME BROW-WIPER.[c]
"Who is that?" (to mirror)		
5	ME THERE KOKO GOOD TEETH GOOD.	

[a] All responses were recorded on videotape as part of the mirror test July 1–3, 1990.

[b] Koko often uses the sign NIPPLE to mean female.

[c] Words joined with hyphens here indicate simultaneous signs. BROW-WIPER is an iconic sign invented by Koko, using the handshape and motion of wiper, but on the brow rather than the mouth.

to the questions contained one or more of the following signs: KOKO, ME, GORILLA.

A sample of 200 of Koko's multisign utterances was analyzed to determine the frequency with which Koko uses these three signs in everyday situations.

This sample was taken from the daily log of Koko's sign combinations from July 1–11, 1991, approximately the same period during which the formal test was administered. She used the sign GORILLA in 3%, the sign ME in 2%, and the sign KOKO in none of these 200 utterances (mean length of utterances for this sample: 3.0 signs). The contexts of Koko's uses of these three signs were varied. On July 4, Koko signed, GORILLA THIRSTY LIPSTICK to ask her companion for some lipstick. On July 6, she was asked, "Want to play?" and responded, CHASE GORILLA. On July 9, Koko signed GORILLA NUT to herself while playing with small dolls in her nest. During a videotaped session on July 8, Koko's sign combinations included THIRSTY ME and TICKLE ME. Although the sign KOKO did not appear in the sign sample, on July 23 Koko was asked, "How are you feeling?" and her reply was, FINE KOKO. In these examples, the two instances of ME and two of the three instances of GORILLA occurred in requests for someone to direct actions toward Koko, and the sign KOKO occurred in response to a question about an internal state, in which the questioner referred to Koko as "you."

To demonstrate that Koko does not use these three words in response to just any "who" question, she was asked "who" questions about her closest companions. Her responses were different for each individual but consistent over repeated questions about the same individuals, indicating that Koko distinguishes others from herself and from each other on a symbolic level. Koko's responses to "Who are you?" and "Who is Penny?" were similar in that they included the sign NIPPLE, which Koko uses frequently and consistently to mean *female*. ("Penny" is the name by which Koko knows the senior author.) Her responses to "Who is Ron?" and "Who is Mike?" included signs with negative connotations: DEVIL, TROUBLE, and BAD. Koko's descriptions of Ron, the disciplinarian in her life, as DEVIL TROUBLE, and of the gorilla Mike as DEVIL HEAD and BAD were consistent with her past comments about these two individuals (Patterson & Linden, 1981).

Language behavior is often cited as the quintessential activity that implicates mind. Through her signing, Koko has shown a number of cognitive correlates of self-awareness in both testing and nontesting situations. These include the acquisition and use of personal pronouns and proper names, reference to her own internal and emotional states, attribution of mental states to others, self-conscious behaviors, value judgments, self-talk, humor, symbolic play, expressions of intentionality, deception, and embarrassment. Many of these emerged at about the same time, when Koko was between 3 and 4 years old.

The use and comprehension of personal pronouns, proper names, and possessives is indicative of self-awareness in that it implies the ability to distinguish symbolically between self and other. Analysis of vocabulary acquisition data from the language project's longitudinal records (Patterson & Cohn, 1990) indicates that Koko first acquired reliable use of the signs for KOKO, ME, MINE, PENNY, and YOU all within a 5-month period when she was between $3\frac{1}{2}$ and 4 years old (see Table 17.2). This is the same period during

Table 17.2. *Koko's acquisition of personal pronouns, proper names, and possessives*

Sign	Date acquired[a]	Age in months
ME	5/74	34
KOKO	6/74	35
MINE	6/74	35
YOU	9/74	38
PENNY	11/74	40

[a] For a sign to qualify as part of Koko's vocabulary it must be used spontaneously at least half the days of a month, and its use must be recorded by two independent observers.

which Koko began to exhibit self-directed behavior in mirrors and first responded appropriately to questions about her mirror image. In human children, whether they are learning ASL or spoken language, the use of personal pronouns seems to emerge at the end of the second year, also the same time as the ability for self-recognition (Petitto & Bellugi, 1988).

Koko also shows evidence that she is able to identify herself and Michael in photographs (Patterson, 1978a), although she has not been formally tested for this ability. Koko has been asked specifically about differences and similarities between humans and gorillas (Patterson & Linden, 1981). On January 26, 1991, Koko identified the reason the senior author is not a gorilla as CLOTHING GOOD. On November 11, 1978, she had the following conversation with a companion, Maureen:

M: What's the difference between you and me?
K: HEAD.
M: And how are our heads different?
   (Koko beats on her head with her open hands quite hard, harder than a human would ever do.)
M: What else is different between us?
   (Koko moves her hands up on her stomach, a gesture resembling BLANKET.)
M: Do you mean something about your stomach?
K: STOMACH GOOD THAT.
M: Oh, but what were you saying about blanket, different?
   (Koko moves her hands up and down her torso, then pulls the hair on her belly. Maureen interprets these new gestures as meaning "body hair.")
M: Now can you name something the same?
K: EYE.
M: Yes, that's right, we both have eyes.

Spontaneous verbal expression of emotion also appeared when Koko was approximately 4 years old. Lewis (1986) has argued that only the self-aware are capable of reflecting on their emotional states. Emotional states are internal, private conditions. Animals cannot reflect on such internal states unless they are capable of reflecting on themselves, that is, unless they are self-aware.

Table 17.3. *Internal state words in Koko's early vocabulary*

Age acquired[a] (months)	Internal state word	Category
31	LISTEN	Perceptual
31	SORRY	Emotional & affective
33	LOOK	Perceptual
37	HUG-LOVE	Emotional & affective
39	GOOD	Social evaluation/Emotional & affective
40	TASTE	Perceptual
43	SMELL	Perceptual
43	SLEEP	Physiological
44	HURT	Perceptual
44	THIRSTY	Physiological
45	KISS	Emotional & affective
45	DIRTY	Emotional & affective
47	SMILE	Emotional & affective
48	FROWN	Emotional & affective
49	STUPID	Emotional & affective
51	BAD	Social evaluation/Emotional & affective
52	HUNGRY	Physiological
53	THINK	Cognitive

[a] For a sign to qualify as part of Koko's vocabulary it must be used spontaneously on at least half the days of a month, and its use must be recorded by two independent observers.

We categorized words in Koko's early vocabulary according to a scheme by Bretherton and her colleagues (Bretherton & Beeghly, 1982; Bretherton, McNew, & Beeghly, 1981) who found that in human children, internal-state language emerges late in the 2nd year and undergoes a rapid spurt in the 3rd. Koko began using such signs when she was just over age $2\frac{1}{2}$, with eighteen internal state signs (see Table 17.3) qualifying as established vocabulary items between this age and $4\frac{1}{2}$. Reflection on her own emotional state is now a daily affair for Koko. When asked, "How do you feel?" Koko will usually respond FINE, THIRSTY, HUNGRY, or SAD. She will often sign FROWN, SAD, when asked why she is crying.

On July 7, 1976, 5-year-old Koko referred to her past emotional state of anger, 3 days after the event:

> P: What did you do to Penny?
> K: BITE.
> P: You admit it? (Koko had earlier called the bite a SCRATCH.)
> K: SORRY BITE SCRATCH.
>   (Penny shows the mark on her hand; it does resemble a scratch.)
> K: WRONG BITE.
> P: Why bite?
> K: BECAUSE MAD.
> P: Why mad?
> K: DON'T KNOW.

In a formal study parallel to one with human children 5–13 years old by Wolman, Lewis, and King (1971), Koko (age 6) was asked a series of questions about her feelings, with these frames: (1) Do you ever feel __? (2) When do you feel __? The target feeling states were fear, hunger, sadness, anger, sleepiness, happiness, thirst, and nervousness. Koko's responses demonstrated that she was capable of reflecting on her feelings. Like the younger human subjects mentioned by Wolman et al., Koko most frequently reported external events as conditions of emotional arousal (Patterson, 1980). For example, her response to "When do you feel hungry?" was FEEL TIME. A possible explanation of this reply is that Koko feels hungry when it is time to eat. Koko's most frequent and emphatic use of the sign TIME is to tell her human companion that it is time to deliver the next meal. Some of Koko's replies to the test questions seemed to be directly related to the events of the preceding months. During this period she had been directing an unusual amount of aggressive behavior toward a new caretaker named Marjie. Koko's responses to the question, "When do you feel mad?" included KOKO MAD GIRL and KOKO LOVE MARJIE BYE.

A hypothetical question relating to both feelings and self-perception was posed to Koko when she was 14 years old. The second author asked her, "How would you feel if someone said 'you have dirt all over your head'"? Koko's signed response was, GORILLA GOOD HAIR, BRUSH-OFF-TOP-AND-BACK-OF-HEAD.

Koko frequently deceives, using deceptions much like those used by humans, chimpanzees, orangutans, and other gorillas (Miles, 1990; Mitchell, 1991; de Waal, 1986; Whiten & Byrne, 1988). At age 1;9 Koko began to engage her companions in behaviors that would distract them so she could pursue a forbidden activity or stop undesirable behavior on the part of her companion (Patterson, 1980). Koko's use of deception blossomed, so to speak, at about age 3–3½, as did MSR. She has used techniques such as concealment (hiding a contraband toy under her arm to sneak it outside), and creating an image (applying a crayon to her lips like lipstick when her companion asked if she was eating it). More recently, at age 19, Koko was asked to throw one of her small plastic toys (which Michael tends to eat) back into her room before going outside. Koko made a convincing throwing motion toward her room, but kept the toy hidden and did not let go of it.

Koko uses verbal as well as nonverbal tactics for deception. This ability began to reveal itself in a few isolated incidents at about age 3 (Patterson, 1980). At age 3½, during a scolding for stealing a spoonful of butter, Koko signed, TIME ME TOILET GO. She didn't use the toilet, however, after being excused. In May 1974 Koko broke a toy cat while with caretaker Kate. The next day she was asked "Who broke this cat?" Koko's reply was KATE CAT. At about age 5, more frequent and convincing evidence of her use of the lie to get herself out of trouble appeared. When questioned about a sink that she had broken by sitting on it, Koko asserted, KATE THERE BAD. Caught in the act of trying to pry a window screen with a stolen chopstick, and asked what she was doing, Koko replied SMOKE MOUTH, and placed the stick in her mouth as if smoking it. The "lipstick" incident mentioned above was recorded on videotape in January 1978. While the experimenter was busy writing notes, Koko snatched a red crayon and began chewing on it. In response to, "You're not eating that crayon, are you?" Koko signed LIP and began moving the crayon first across her upper then her lower lip as if applying lipstick. When asked what she was really doing she signed BITE. Asked, "Why?" she replied, HUNGRY.

Michael, too, lies to avoid the consequences of his misbehavior. The first recorded instance was on April 22, 1978, when he was approximately 5 years old. Michael had ripped a gaping hole in a volunteer's lab coat. When she asked him, "Who did this?" Michael responded, PENNY. He was told that was wrong and asked again, but his response was KOKO. Finally he admitted, MIKE.

One must be conscious of having experienced an emotional state oneself before attributing such a state to others (Gallup, 1987). Statements in which Koko appears to attribute emotional states to others have been documented frequently. On December 27, 1977, Koko heard Mike crying because he couldn't come out of his room. When asked how Mike felt she signed, FEEL SORRY OUT (Patterson, 1984a). Another such incident occurred November 3, 1977, when 6-year-old Koko saw a picture of a horse with a bit in its mouth. She signed, HORSE SAD. When her companion asked her why the horse was sad she signed, TEETH (Patterson, 1979a). On September 24, 1977, when shown a picture of the albino gorilla Snowflake struggling against being bathed, Koko, who also hates getting wet, signed ME CRY THERE, indicating the picture (Patterson, 1980).

The ability to take the perspective of another, which human children develop at about age 4, also requires knowledge of self and other (Astington & Gopnik, 1991). Koko demonstrated this ability in a videotaped sequence on June 22, 1991, in which she spontaneously answered a question first from the perspective of her "baby," a stuffed orangutan doll, and then from her own perspective. Koko asked to have the doll, cradled it like a baby and kissed it, and then signed DRINK. Penny asked, "Where does the baby drink?" In response, Koko molded the doll's hands to form the signs DRINK MOUTH, with the doll's hand indicating the doll's mouth. Then Koko pointed to the doll's mouth with her own finger and held the doll out toward Penny. From the doll's perspective, drinking occurs in its mouth. Penny then said, "The baby said, 'Drink mouth.' That's right, but where on you?" Koko put the doll to her breast. Penny commented, "Right, on your chest. Your baby signs!" As Koko continued to hold the doll to her breast she signed TOILET NIPPLE DRINK and DRINK NIPPLE. From Koko's perspective, a baby drinks from her nipples.

Koko has displayed evidence of her ability to take another's perspective in her humor as well (Patterson, 1986). While nesting with white towels, Koko pointed to one of the towels and signed, THAT RED. Her companion, Barbara, corrected her, telling her that it was white. Koko repeated her statement with additional emphasis: THAT RED. Barbara stated again that the towel was white.

After several more exchanges, Koko picked up a tiny piece of red lint, held it out to Barbara, grinning, and signed, THAT RED. Koko's typical incongruity-based humor is similar to that of children between the ages of 3 and 6. In this earliest form of humor, children may laughingly apply different names to familiar objects or use one object as if it were another (McGhee, 1980). Asked outright what she thought was funny, Koko signed THAT NOSE, indicating a toy bird's mouth, and HAT, referring to a rubber key she had placed on her head.

Koko has also made verbal jokes based on her self-identity. On October 30, 1982 Barbara showed Koko a picture of a bird feeding its young.

> K: THAT ME. (Pointing to the adult bird.)
> B: Is that really you?
> K: KOKO GOOD BIRD.
> B: I thought you were a gorilla.
> K: KOKO BIRD.
> . . .
> B: Can you fly?
> K: GOOD. ("Good" can mean "yes".)
> B: Show me.
> K: FAKE BIRD, CLOWN. (Koko laughs.)
> B: You're teasing me. (Koko laughs.) What are you really?
>   (Koko laughs again and after a minute signs:)
> K: GORILLA KOKO.

It has been suggested that there are certain emotional states that only the self-aware can have (Lewis, 1986). One such state is embarrassment. In order to be embarrassed, animals must be capable of reflecting on their own behavior and comparing it to standards set by society or by themselves. Koko seems embarrassed when her companions note that she is signing to herself, especially when the signing involves her dolls and animal toys: She will abruptly stop the activity, often turning away from the observer. Although clear instances of embarrassment have been observed only rarely, one incident recorded when Koko was 5 years old provides an example. A companion observed her create what appeared to be an imaginary social situation between two gorilla dolls. She signed, BAD, BAD, while looking at one gorilla and KISS, while looking at the other. Next, she signed, CHASE TICKLE, hit the two dolls together, and then wrestled with them. When she was through she signed GOOD GORILLA, GOOD GOOD. At this point, Koko noticed that her companion was watching the play session. She immediately put the dolls down (Patterson, 1979b).

The evidence provided by both the formal and serendipitous mirror tests leaves little doubt that Koko is capable of self-recognition. The verbal measures provide evidence that goes beyond this to give us glimpses of a gorilla's self-concept. Previous observations of self-inspection indicate that other gorillas also have self-recognition.

Why, then, have others failed to find convincing evidence of self-recognition in gorillas through formal testing? There are a number of possible explanations, including the rearing histories, lack of motivation, and sensitivity to anesthesia of the few gorillas tested (see Patterson, 1984b, and Povinelli, 1987, for in-depth discussions of these possibilities). However, a likely explanation is that the gorillas' behavior was inhibited by the presence of unfamiliar experimenters. It has been our experience that the presence of strangers profoundly affects gorilla behavior. We have found that it can take from several months to a full year for Koko and Michael to habituate to the presence of a new caretaker. Hence, when we performed the mirror study on Koko we chose to have only two experimenters present who had known Koko for 18 years. In each of the previous formal self-recognition studies with gorillas, experimenters who were not the gorillas' caretakers were in the room with them in very close proximity to the mirror (Ledbetter & Basen, 1982; Suarez & Gallup, 1981). Averting their gaze from strangers is a common behavior in gorillas. Observed social responses to the mirror may have been elicited by the experimenters, whereas mirror gazing and self-directed behaviors may have been inhibited by their presence. It is interesting to note that Ledbetter and Basen (1982, p. 308) state that they had been informed by the zoo staff, who were presumably familiar to the gorilla subjects, "that the animals had been exposed to small mirrors on several occasions in the past and it was their opinion that both animals 'recognized themselves.'" Ironically, it may have been the gorillas' very capacity for self-consciousness that prevented them from exhibiting behaviors indicative of self-recognition in the testing situations.

The lack of appreciation of these factors has led to the assumption that gorillas failed tests of self-recognition because they lacked the requisite cognitive capacity (Gallup, 1987; Ledbetter & Basen, 1982; Suarez & Gallup, 1981). A method that works with one species may be totally inappropriate for another species. It also may be that animals who are self-aware will not always exhibit self-recognition behavior in mirrors, even under ideal circumstances.

In July–August 1991 we tried a mark test with the 18-year-old gorilla Michael, whose mirror-guided, self-directed behaviors had already been documented on film. Although it seemed to the experimenters at the time, and to independent observers reviewing the videotape later, that Michael indeed recognized himself in the mirror during the test sessions, the results were inconclusive. This failure illustrates some of the special problems inherent in testing gorillas for mirror self-recognition.

The procedure used to test Michael was similar to that used with Koko, but circumstances required several significant differences. Due to his destructive tendencies, Michael could not be given free access to the test mirror, which was propped or held up outside his room approximately 2 ft from the mesh fence. The experimenters also remained separated from Michael by the mesh. Although both experimenters had cared for Michael since he was $3\frac{1}{2}$, their presence together in his quarters with a video camera was not a part of his normal daily routine. Perhaps the most important difference, however, was Michael's prior experience with the marking procedure. On June 5, 1990,

Michael's brow had been marked as a pretest before Koko's test to ensure that the clown paint would not be tactilely detectable, but would be clearly visible on black gorilla skin. During the pretest Michael did not respond to the marking until he saw his marked brow in a pocket mirror. He then turned his head sharply away from the mirror with a startled look. Next, he removed some of the paint with a finger and licked his finger. He repeated this several times. At the time of his own mirror test, more than a year later, whenever one of the experimenters touched Michael's brow with anything he immediately brushed his hand across his brow. Others have demonstrated that chimpanzees who have passed the mark test once may lose interest and fail if retested (Thompson & Boatright-Horowitz, *SAAH*22). This observation and Michael's behavior suggest that the subject's memory of previous mark tests is also a factor to consider.

During each of the first four sessions in which his brow was wiped but unmarked, Michael spent a few casual seconds making faces, displaying his teeth and waggling his tongue while watching himself in the mirror. He also watched himself eating, drinking, and signing in front of the mirror. He spent relatively little time viewing the mirror, however, preferring to interact with the experimenters, attempting to obtain treats or to be let outside. In the fifth and final sessions, Michael's right brow was marked with white clown paint while he was peering into a white cardboard roll containing a kaleidoscopic pattern. (This procedure was used because he proved to be too suspicious of the washcloth wipe method.) Michael appeared to be unaware of the mark at first, but less than a minute after the marking he brushed his finger over his brow where the tube had touched it and discovered the paint before the mirror was presented. The marking process also got a dot of paint in the center of Michael's nose. While he continued to remove the paint from his brow with his finger and then a cloth, the conspicuous white dot on his nose remained undisturbed and apparently undetected.

When the mirror was presented, Michael approached casually as before and began signing to the experimenter holding the mirror. He glanced at the mirror, began to rub at the paint on his brow, then suddenly "froze" and leaned forward, staring intently at his reflection. This time he did not display the same playful teeth-baring and tongue-waggling faces as in the previous session, but only stared with a serious expression. He rubbed the fading mark on his brow again with his finger, then sniffed and licked the finger, but did not touch the more obvious spot on his nose. He leaned close to the mirror again and turned his head from side to side, looking at his face from different angles. Then he leaned back from the mirror and asked in sign for the experimenters to turn the lights off. During the remainder of the session he asked several more times for the lights to be turned off and the drapes to be closed, but neither request was granted. He wiped his brow again with his finger and with a cloth, but still did not touch his nose. After moving around the room for a few minutes he returned to the mirror, looked closely, looked away and picked at his left eye, brushed his right brow, then looked back at the mirror. Finally, he moved into a corner of the room and rubbed his nose on the wall. He repeated this in the opposite corner. Perhaps if Michael could have been tested while alone in his room, with a hidden camera, the results would have been different. The presence of even very familiar observers obviously distracted him and may also have severely inhibited his mirror behavior.

Koko did not begin to exhibit MSR and other behaviors indicative of self-awareness until age $3\frac{1}{2}$–4 years. This may or may not be typical of her species. The age of acquisition of a self concept is undoubtedly affected by such factors as early social experience and general health. Koko's natal group at the San Francisco zoo included her mother, her father, two other adult females, and a male infant (her half-brother). She saw these gorillas continuously from birth to age 6 months, when she and her half-brother were removed for hand-rearing because of serious illness. (Her half-brother died.) From age 6 months until 5 years, when Michael arrived, Koko had some very limited contact with other gorillas. Before leaving the zoo at age 3 she was taken occasionally to see her parents, and between the ages of 3 and 4 she had two or three visits with a young male gorilla. She was also exposed to films and photos of gorillas. But from 6 months to 5 years her social companions were almost exclusively adult humans. Between the ages of 1 and 2 years she was exposed to a wide variety of human faces while on display in the zoo nursery, where zoo visitors observed her through glass windows. Thus the face she saw in the mirror was quite different from the faces with which she was most familiar, and this may have caused delayed recognition her own gorilla features. Another consideration is that a life-threatening illness at the age of 6 months may have caused some delay in her overall cognitive development.

Our studies with Koko and Michael indicate that we must be wary of making assumptions about an animal's cognitive abilities based solely on the results of the mark test. There are clearly many problems with the administration and interpretation of this test, especially when it is applied to apparently self-conscious animals such as gorillas. Nor should self-recognition in mirrors be considered synonymous with self-awareness. The many behaviors other than mirror self-recognition discussed earlier in this chapter, such as deception, symbolic play, and naming of internal states, would indicate self-awareness even if the gorillas failed to show MSR. There are other possible behavioral correlates of self-awareness not discussed here, including games, teasing, showing off, cheating, grief, gratitude, grudging, sympathy, empathy, reconciliation, sorrow, saving face, denial, argument, definition, representational art and dance, and concern about one's image. Other types of tests, such as those that make use of interspecies communication, are potentially more informative for exploring self-awareness than tests of MSR in that they can provide more direct information about an individual's self-concept. Further studies will be necessary, concentrating on the behavioral correlates to self-recognition and self-awareness, before we can claim to know which individuals and which species are self-aware.

Acknowledgments

We would especially like to thank Wendy Gordon for her ideas and her invaluable assistance in the preparation of this manuscript. We would also like to thank Lara Mendel for her hard work during this study. This work was funded by the Gorilla Foundation.

References

Anderson, J. R. (1983). Responses to mirror image stimulation and assessment of self-recognition in mirror and peer-reared stumptailed macaques. *Quarterly Journal of Experimental Psychology, 35,* 201–212.
Astington, J. W., & Gopnik, A. (1991). Theoretical explanations of children's understanding of the mind. *British Journal of Developmental Psychology, 9,* 7–31.
Bretherton, I., & Beeghly, M. (1982). Talking about internal states: The acquisition of an explicit theory of mind. *Developmental Psychology, 18,* 906–921.
Bretherton, I., McNew, S., & Beeghly-Smith, M. (1981). Early person knowledge as expressed in gestural and verbal communication: When do infants acquire a "theory of mind"? In M. E. Lamb & L. R. Sherrod (Eds.), *Infant social cognition* (pp. 333–373). Hillsdale, NJ: Erlbaum.
Brooks-Gunn, J., & Lewis, M. (1982). The development of self-knowledge. In C. B. Kopp & J. B. Krakow (Eds.), *The child: Development in a social context* (pp. 334–387). Reading, MA: Addison-Wesley.
Chevalier-Skolnikoff, S. (1977). A Piagetian model for describing and comparing socialization in monkey, ape, and human infants. In S. Chevalier-Skolnikoff & F. E. Poirier (Eds.), *Primate bio-social development: Biological, social, and ecological determinants* (pp. 159–187). New York: Garland.
(1983). Sensorimotor development in orang-utans and other primates. *Journal of Human Evolution, 12,* 545–61.
Gallup, G. G., Jr. (1970). Chimpanzees: Self-recognition. *Science, 167,* 86–87.
(1987). Self-awareness. In G. Mitchell & J. Erwin (Eds.), *Comparative primate biology, Vol. 2, Part B: Behavior, cognition, and motivation* (pp. 3–16). New York: Liss.
Ledbetter, D. H., & Basen, J. A. (1982). Failure to demonstrate self-recognition in gorillas. *American Journal of Primatology, 2,* 307–310.
Lethmate, J. (1974). Selbst-Kenntnis bei Menschenaffen. *Umschau, 15,* 486–487.
Lewis, M. (1986). Origins of self-knowledge and individual differences in early self-recognition. In A. Greenwald & J. Suls (Eds.), *Psychological perspectives on the self* (Vol. 3, pp. 55–78). Hillsdale, NJ: Erlbaum.
Lewis, M., & Brooks-Gunn, J. (1979). Toward a theory of social cognition: The development of self. In I. Uzgiris (Ed.), *New directions for child development: Social interaction and communication during infancy* (pp. 1–20). San Francisco: Jossey-Bass.
McGhee, P. E. (1980). Development of the creative aspects of humor. In P. E. McGhee & A. J. Chapman (Eds.), *Children's humour* (pp. 119–139). Chichester: John Wiley.
Mathieu, M., & Bergeron, G. (1983). Piagetian assessment on cognitive development in chimpanzee (*Pan troglodytes*). In A. B. Chiarelli & R. Corruccini (Eds.), *Primate behavior and sociobiology* (pp. 142–147). New York: Springer-Verlag.
Miles, H. L. W. (1990). The cognitive foundations for reference in a signing orangutan. In S. T. Parker & K. R. Gibson (Eds.), *"Language" and intelligence in monkeys and apes: Comparative developmental perspectives* (pp. 511–539). Cambridge University Press.
Mitchell, R. W. (1991). Deception and hiding in captive lowland gorillas (*Gorilla gorilla gorilla*). *Primates, 32,* 523–527.
Patterson, F. G. (1978a). Conversations with a gorilla. *National Geographic,* pp. 438–465.
(1978b). The gestures of a gorilla: Language acquisition in another pongid. *Brain and Language, 5,* 72–97.

(1979a). Linguistic capabilities of a lowland gorilla. *Dissertation Abstracts International, 40,* 2B.

(1979b). Linguistic capabilities of a young lowland gorilla. In R. L. Schiefelbusch & J. H. Hollis (Eds.), *Language intervention from ape to child* (pp. 327–356). Baltimore: University Park.

(1980). Creative and innovative uses of language by a gorilla: A case study. In K. E. Nelson (Ed.), *Children's language* (Vol. 2, pp. 497–561). New York: Gardner.

(1984a). Gorilla language acquisition. *National Geographic Society Research Reports, 17,* 677–699.

(1984b). Self-recognition by *Gorilla gorilla gorilla. Gorilla, 7,* 2–3.

(1986). The mind of the gorilla: Conversation and conservation. In K. Benirschke (Ed.), *Primates: The road to self-sustaining populations* (pp. 933–947). New York: Springer-Verlag.

Patterson, F. G., & Cohn, R. H. (1990). Language acquisition in a lowland gorilla: Koko's first ten years of vocabulary development. *Word, 41,* 97–143.

Patterson, F. G., & Linden, E. (1981). *The Education of Koko.* New York: Holt, Rinehart and Winston.

Petitto, L. A., & Bellugi, U. (1988). Spatial cognition and brain organization: Clues from the acquisition of a language in space. In J. Stiles-Davis, M. Kritchevsky, & U. Bellugi (Eds.), *Spatial cognition: Brain bases and development* (pp. 299–325). Hillsdale, NJ: Erlbaum.

Platt, M. M., & Thompson, R. L. (1985). Mirror responses in a Japanese macaque troop (Arashiyama West). *Primates, 26,* 300–314.

Povinelli, D. J. (1987). Monkeys, apes, mirrors and minds: The evolution of self-awareness in primates. *Human Evolution, 2,* 493–509.

Redshaw, M. (1978). Cognitive development in human and gorilla infants. *Journal of Human Evolution, 7,* 133–41.

Riopelle, A. J., Nos, R., & Jonch, A. (1971). Situational determinants of dominance in captive young gorillas. In *Proceedings of the 3rd International Congress on Primatology,* Zurich, 1970 (Vol. 3, pp. 86–91). Basel: Karger.

Rumbaugh, D. M., & Gill, T. V. (1973). The learning skills of great apes. *Journal of Human Evolution, 2,* 171–179.

Suarez, S. D., & Gallup, G. G., Jr. (1981). Self-recognition in chimpanzees and orangutans, but not gorillas. *Journal of Human Evolution, 10,* 175–188.

de Waal, F. (1986). Deception in the natural communication of chimpanzees. In R. W. Mitchell & N. S. Thompson (Eds.), *Perspectives on human and nonhuman deceit* (pp. 221–244). Albany, NY: SUNY Press.

Whiten, A., & Byrne, R. W. (1988). Tactical deception in primates. *Behavioral and Brain Sciences, 11,* 233–244.

Wolman, R. W., Lewis, W. C., & King, M. (1971). The development of the language of emotions: Conditions of emotional arousal. *Child Development, 42,* 1288–1293.

18   How to create self-recognizing gorillas (but don't try it on macaques)

Daniel J. Povinelli

Several years ago, I used the phrase "the anomalous gorillas" while reviewing the nonhuman primate self-recognition literature (Povinelli, 1987, p. 496). At that time, there had been a number of careful, unsuccessful attempts to produce a convincing demonstration of self-recognition in this species (Ledbetter & Basen 1982; Lethmate, cited in Gallup, 1982; Suarez & Gallup, 1981). However, there was one gorilla (Koko) who, based upon uncontrolled (at the time) observations, appeared to show clear signs of self-recognition. The descriptions of Koko's behaviors offered by Patterson (1984) were compelling and comparable to that obtained by Gallup and his colleagues, as well as by other independent researchers working under more controlled conditions with chimpanzees (Calhoun & Thompson, 1988; Gallup, 1970; Gallup, McClure, Hill, & Bundy, 1971; Lethmate & Dücker, 1973; Suarez & Gallup, 1981). Despite this rather remarkable behavior on the part of one member of the species, gorillas as a whole appeared to fall into a different category than chimpanzees, orangutans, and humans with regard to self-recognition. This seemed difficult to reconcile with the evolutionary relatedness of the species within the great ape–human clade; hence my invocation of the phrase "the anomalous gorillas" (Povinelli, 1987).

Several potential explanations for why most gorillas did not appear like chimpanzees on mirror self-recognition tasks suggested themselves, ranging from the hypothesis that they are differentially sensitive to the effects of being reared in captivity, to the possibility that they are simply not motivated to solve the learning problem imposed by the introduction of mirrors (Povinelli, 1987). I also addressed the possibility that gorillas have undergone a character state reversal in the underlying psychological capacity necessary for self-recognition, but I dismissed it on the grounds that was "extremely difficult to envision an organism evolving so unique a mental faculty [as self-awareness] and then losing it through the forces of evolution" (p. 499). Furthermore, I argued that such an explanation was "*ad hoc* and inconsistent with much of what we know about gorilla behavior" (p. 499). In retrospect, it is unclear to what data I was referring; at the time no such data existed for gorillas, and even today no one has yet attempted to provide an experimental demonstration

291

of mental state attribution in gorillas. Thus, I am left in the embarrassing position of admitting that I discounted the possibility that gorillas had secondarily lost the capacity for self-recognition on the basis of the assumption that complex mental characteristics should be perfectly correlated with phylogeny. In other words, I viewed complex psychological characteristics as somehow fundamentally different than complex morphological characteristics. However, after exposing a captive group of socially housed gorillas to a large mirror for nearly 4 years now without any evidence of self-recognition (Shumaker & Povinelli, unpublished observations), I appear to have undergone a character state reversal of sorts myself. Upon rethinking the issue, I now believe that the gorillas only seemed anomalous because I allowed certain preconceptions to guide my interpretation of the data.

Self-recognition and gorillas

I believe that the best available interpretation of the status of self-recognition in gorillas is that the ontogeny of most gorillas does not include the construction of the cognitive structures that ultimately support self-exploratory behaviors in front of mirrors. However, although most gorillas tested to date for self-recognition have not shown compelling evidence (Ledbetter & Basen, 1982; Parker, 1990; Shumaker & Povinelli, unpublished observations; Suarez & Gallup, 1981), one home-reared gorilla has shown the full suite of behavioral indices of mirror self-recognition (Patterson, 1984). Some may feel that this description does not do justice to the current state of affairs for self-recognition research in gorillas, citing either facial movements or isolated reports of what appear to be self-directed responses on the part of one gorilla or another (Parker, 1990, 1991; Patterson & Cohn, *SAAH*17). However, in our laboratory, as part of a long-term study of the development of self-recognition in chimpanzees, my colleagues and I have thus far collected extensive systematic observations of 105 mature and immature chimpanzees (Povinelli, Rulf, Landau, & Bierschwale, in press). The chimpanzees that do not display compelling evidence of self-recognition do occasionally show instances of what could arguably be called contingent facial or bodily movements, face making at the mirror, or other instances of self-directed behaviors while the subjects are near the mirror, but not looking into it. But these ambiguous behaviors are not correlated with compelling bouts of contingent facial and body movements or self-exploratory movements, nor do they predict group differences in success of Gallup's mark test. Thus, in no way can they be used as unique evidence of self-recognition.

In clear contrast, subjects described as capable of self-recognition typically show clear, sustained, and unambiguous instances of self-exploration while carefully monitoring their reflections, and this predicts dramatically higher levels of contingent body and facial movements than subjects not so classified. In addition, this classification scheme accurately predicts differences between the groups in terms of their ability to pass controlled mark tests (see

Povinelli et al., in press). We conclude that ambiguous instances of self-directed behaviors should not be used as evidence for self-recognition.

Thus, in direct contrast to my earlier, subjective impressions, I now believe that it is reasonable to predict that only *abnormally* reared gorillas, such as Koko, will ever show convincing signs of self-recognition. In this chapter, I outline my reasons for making this prediction in the form of a hypothesis about the relationship between heterochronic shifts in development and psychological evolution.

Reappearance of ancestral traits

A variety of lines of evidence strongly suggest that the great apes and humans all descended from a common ancestor living in the Miocene (Figure 18.1). Furthermore, since the 1960s most researchers in the field have been convinced that chimpanzees, gorillas, and humans form a monophyletic clade, with orangutans representing the nearest living outgroup (Andrews & Martin, 1987; Goodman, 1962; Marks, 1988). Nonetheless, orangutans show obvious evidence of self-recognition, whereas to date there is convincing evidence for only one gorilla. It is precisely this state of affairs that has made the failure to find self-recognition in gorillas so troublesome to many comparative psychologists. My interpretation of this situation, on the basis of cladistic parsimony, is that the trait of self-recognition was present in the population of the common ancestors of the entire great ape–human clade (see Figure 18.1). If the failure to find self-recognition in gorillas is reliable, then the most parsimonious interpretation is that the species has undergone a secondary loss of the capacity (see Gallup, 1985; Povinelli, 1987). Although I believe this to be the case, I also believe that a much richer understanding of this issue can be obtained by considering the effect of environmental perturbations upon early development.

The reappearance of ancestral characteristics within an individual following experimental interventions during ontogeny are well known to embryologists and developmental biologists in general (Buss, 1987, pp. 112–115; see Frazzetta, 1975). Hampé (1960), for instance, in a widely cited study, demonstrated that by allowing the fibula of chicken embryos to touch the metatarsal bones during development, a series of cascading changes occurred that resulted in the development of the ancestral condition found in *Archaeopteryx*, the first bird (see Figure 18.2). Alberch, Gould, Oster, and Wake (1979) have commented on Hampé's experiments, noting that "it is probable that the genetic capacity for producing these ancestral traits was never lost during more than 200 million years of avian evolution, but merely 'turned off' by failure of the inducing fibula to establish contact with the metatarsals" (p. 311). Numerous other examples are available that emphasize the same point: Perturbations in development can result in the secondary appearance of ancestral traits not typically found in descendant species (Kollar, 1972; Raikow, 1975; Raup & Kaufman, 1983). Perhaps one of the most dramatic instances has come from

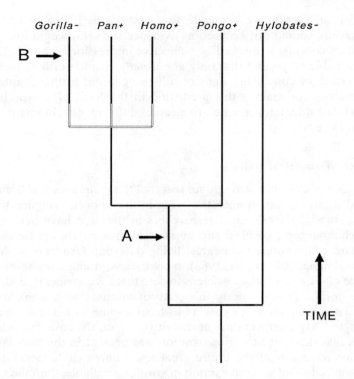

Figure 18.1. Evolutionary relationships of the living apes and humans and the distribution of self-recognition (indicated as + or −). The most parsimonious explanation is obtained by assuming that the underlying capacity for self-recognition evolved once in the ancestor of the great ape grouping (event *A*) and that is has been secondarily lost during the 5–10 million years of independent evolution in the gorilla lineage (event *B*). Note that the African ape–human clade is represented as an uncertain trichotomy due to the continuing controversey that now surrounds the exact relationships of these species (see Marks, 1991).

an experiment by Kollar and Fisher (1980), who demonstrated that the jaw epithelium of chicken embryos will develop teeth when simply placed on the jaw mesoderm of mice. This represents the retention of a capacity to respond to developmental signals that have not been in operation for at least 80 million years. Alberch and his colleagues note, "Latent capacities for the generation of ancestral structures probably exist in the genomes of all specialized animals" (p. 311).

But what of the loss of such characteristics in the first place? There is ample evidence to suggest that many (if not most) evolutionary innovations (losses or additions of particular traits) are achieved through *heterochrony*, changes in the timing and onset of developmental pathways of specific cell lineages (Buss, 1987; Gould, 1977; Raup & Kaufman, 1983). Thus, natural selection may often favor specific variants that are ultimately produced through heterochronic shifts in development. There are several important implications

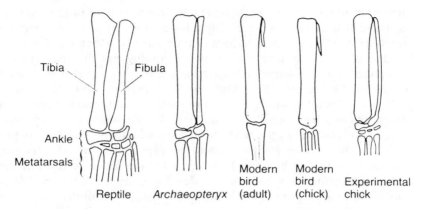

Figure 18.2. Hampé (1960) discovered that if the fibula of the developing chick embryo is preventing from touching the tibia, the fibula grows to its full length, inducing the tarsal bones to develop independently from the tibia, thus resulting in a pattern strikingly remincent of the ancestral condition found in the first bird, *Archaeopteryx*. These experimentally produced metatarsal bones are presumed to be homologous to the ancestral metatarsals of *Archaeopteryx*. (Figure from Hampé, 1960).

of this mode of evolutionary change. One is that mutations that reverse a heterochronic shift that originally resulted in the elimination of an ancestral trait should ultimately result in the reemergence of the trait that is presumed to have been present in the phyletic ancestor of the lineage in question (Buss, 1987). In more general terms, the work of embryologists such as Hampé (1960) reveals that experimental perturbations that mimic the emergence of natural variants can result in the reappearance of ancestral traits as well.

The "loss" of self-recognition in gorillas:
The heterochrony hypothesis

Understanding the atavistic implications of heterochrony places in an entirely different light recent evidence that gorillas may have undergone a heterochronic shift in the onset of locomotion relative to their acquisition of particular cognitive structures (Antinucci, 1989, Chap. 6). This finding is consistent with several lines of evidence suggesting that the developmental schedules of gorillas are accelerated relative to chimpanzees and orangutans in several domains such as age at weaning, age at first labial swelling, age at menarche, and age at first birth (see Watts & Pusey, 1993). In terms of physical growth in body size, gorillas are sequentially accelerated relative to chimpanzees, whereas humans are sequentially hypermorphic relative to chimpanzees (for an overview see McKinney & McNamara, 1991). That is, gorillas grow faster in each maturation phase than do chimpanzees; humans, in contrast, grow at the same rate as chimpanzees, but stay in each growth phase for a longer period by delaying the termination of each growth phase. It is possible that the gorilla's accelerated sexual and physical maturation (especially in the domain of independent locomotion) may interfere with future maturation in cognitive structures that were only allowed for in the great apes and their ancestors due to extended periods of immature development. This seems all the more likely because sensorimotor intelligence in apes (at least, in chimpanzees and orangutans) appears not to close until 6–8 years of age (Chevalier-Skolnikoff, 1983; Mathieu & Bergeron, 1981; Mignault, 1985), precisely the point at which our own research has detected the developmental emergence of self-recognition in chimpanzees (Povinelli et al., in press). Thus, dramatic alterations in physical maturation schedules prior to this point (as is clearly the case in gorillas), could be expected to affect later cognitive growth. Note that in humans, however, selection seems to have favored the early maturation of sensorimotor development. In humans, Stage 6 of sensorimotor development is complete by 18–24 months, but sequential hypermorphosis does not begin to become evident until well after that time (McKinney & McNamara, 1991).

Of course, this specific explanation need not be correct for the general point to be valid: Specific changes in the onset and timing (heterochronic shifts) of particular cognitive pathways may have led to the secondary loss of self-recognition in gorillas. If true, then Patterson's (1984) observations of self-recognition on the part of her home-reared gorilla that had been instructed in a variety of cognitive tasks since 1 year of age, are potentially understandable as an example of the reappearance of ancestral characteristics following perturbations in development. In this case, the experimental perturbation has taken the form of a potentially profound disturbance involving intensive tutoring on a variety of cognitive tasks not found in normal gorilla ontogeny (Patterson & Linden, 1981). It should be emphasized that I do not believe that this is language training per se, but rather aspects of cognitive stimulation related to the construction of a sense of personal agency.

In short, the general hypothesis I am advancing is that psychological ontogeny need not, and indeed, should not, be thought of in fundamentally different terms than morphological ontogeny (see also McKinney & McNamara, 1991). Because the development of the brain (like other morphological structures) is guided by epigenetic interactions, strong environmental inputs during ontogeny may affect psychological ontogeny in the same manner as morphological ontogeny, and thus result in the reappearance of psychological traits that have apparently been "lost" during evolution. To accept the hypothesis as having potential merit, several assumptions need to be made. These assumptions are widespread, but they ought to made explicit in order for the strengths and weaknesses of the hypothesis to be fully appreciated. First, I assume that phyletic differences in cognitive capacities are the result of the evolution of distinctive neural structures. Second, I assume that the ontogeny of psychological capacities are marked by maturation of brain structures. Of course, these latter developmental changes may range from the dramatic (such as increases in the size and complexity of the frontal cortex) to the subtle (complex readjustments of the physiological strength of neural networks underpinning particular representations of reality). The final assumption is that normal epigenetic interactions guide the development of brain structures. All of these assumptions seem reasonable, and many are supported by a variety of lines of evidence.

Summary and predictions

After 15 years of debate concerning the self-recognition capacity of gorillas; it may turn out that many gorillas may still possess the basic underlying instructions for the construction of cognitive structures necessary for self-recognition, but that these instructions have been "turned off" due to heterochronic shifts in other aspects of their development. Experimental interventions, such as those received by Patterson's gorilla, may be capable of reinstating the developmental pathways implicated in the expression of self-recognition. Suarez and Gallup (1981), therefore, might have been correct in arguing that gorillas may "lack some as of yet unspecified neuroanatomical features" implicated in the expression of self-recognition (p. 186). However, it may have been somewhat premature to speculate that the neurobiological explanation for the absence of self-recognition is the same in gorillas and other species that show no evidence for self-recognition (Suarez & Gallup, 1981, p. 186). In the account offered here, the relevant neurobiological differences separating gorillas and these other species from chimpanzees, orangutans, and humans may be quite different given that the nature of the secondary loss in gorillas remains unknown. In addition, Gallup (1985) may have been correct in arguing for the secondary loss of the underlying capacity for self-recognition in the gorilla lineage as well (see also Povinelli, 1987). On the other hand, Patterson (1984), too, may have been correct in observing, "Whatever the explanation for the negative findings of other researchers, we believe that we have definitive proof that self-recognition is not beyond the mental capacity of the gorilla" (p. 3). The hypothesis that I have outlined here bridges the gap between the two sides of this apparent deadlock (see Povinelli, 1991).

A final caveat seems warranted in light of mounting evidence that some chimpanzees may not be capable of self-recognition. Swartz and Evans (1991) have recently published the results of a study in which only 1–3 of 11 chimpanzees showed evidence of self-recognition. As noted previously, my colleagues and I have recently tested 105 chimpanzees ranging in age from 10 months to 39 years of age, and hence have had the opportunity to dissect this issue (and others) in far greater detail. We have found clear developmental evidence for at least two transitions in the chimpanzee's capacity for self-recognition. First, there appears to be a clear transition at around 6–8 years of age: about 5% of young chimpanzees 6 years and younger recognize themselves, as compared to nearly 80% of adolescent/young adults 8–16 years of age. The second transition appears to be a decline with age, so that only about 35% of chimpanzees older than 20 years show evidence of self-recognition. These subjects have been socially reared and housed, they are wild or captive-born, and mother or nursery-reared. None of these factors appear to have any systematic influence on the probability that an animal will recognize itself (Povinelli et al., in press). Thus, it may turn out that the capacity for self-recognition is polymorphic in chimpanzees; hence, the atavism hypothesis may apply to their species as well, although the heterochronic shift is less pronounced, hence the widespread intraspecies variation.

Elsewhere, we have argued that orangutan populations may be expected to show much higher rates (as well as earlier developmental onsets) of self-recognition than both chimpanzees and gorillas (Povinelli & Cant, 1992). We have hypothesized that evolution of an awareness of personal agency (primitive self-conception) was the result of the evolution of large body size in the arboreal ancestors of the great ape–human clade. In particular, we have argued that a sense of personal agency emerged as a necessary conceptual system for directing slow, cautious clambering through a habitat made fragile by excessive body weights (Povinelli & Cant, 1992). If our assumption is correct that orangutans more closely approximate the locomotor condition of the Miocene ancestor of the great apes, then both gorillas and chimpanzees represent derived conditions, both being far more terrestrial than orangutans, gorillas especially so.

At any rate, the possibility that self-recognition in chimpanzees is more polymorphic than was once assumed in no way invalidates the hypothesis offered here. Indeed, it only adds to the potential ways in which the atavism hypothesis might be falsified. In addition, it provides an opportunity for testing Gallup's (1985) model of the relationship between self-recognition and mental-state attribution using an intraspecies comparison.

Of course, any hypothesis is only as good as the set of falsifiable predictions that it makes. The atavism hypothesis advanced here makes at least two explicit predictions in terms of self-recognition research. First, it predicts that most infant gorillas provided with the cognitive stimulation that Patterson's gorilla received should show evidence of self-recognition. A clear weakness of the hypothesis is that it cannot specify the exact nature of the stimulation necessary, nor can it predict the critical windows of opportunity during ontogeny in which such interventions would produce the ancestral trait. Mitchell (1992, 1993) could argue that the type of tutoring necessary has to do with "body part objectification." However, although the hypothesis outlined here remains silent on what the necessary cognitive structures are that underpin and allow for self-recognition, elsewhere we have argued that it is the development of an awareness of the self as causal agent (a sense of personal agency) (Povinelli & Cant, 1992). Thus, if the general atavism hypothesis is correct, it provides an ideal case for testing these competing explanations of the cause of mirror self-recognition.

The second and potentially more important prediction from the hypothesis is that no amount of similar cognitive training should produce self-recognition in individual members of species outside the great ape–human clade. This phyletic difference is predicted because these species descended from lineages in which the underlying pathways necessary for the development of the relevant cognitive structures had not yet evolved. For those species (gibbons, Old and New World monkeys, prosimians – perhaps the remaining diversity of life itself) there exists no ancestral trait to reappear. An exception to this prediction would exist if self-recognition itself were produced by a heterochronic shift that could successfully be mimicked by experimental perturbations in, for example, macaque ontogeny, thus resulting in the appearance of "descendant characteristics."

References

Alberch, P., Gould, S. J., Oster, G. F., & Wake, D. B. (1979). Size and shape in ontogeny and phylogeny. *Paleobiology, 5,* 296–317.

Andrews, P., & Martin, L. (1987). Cladistic relationships of extant and fossil hominoids. *Journal of Human Evolution, 16,* 101–118.

Antinucci, F. (1989). *Cognitive structures and development in nonhuman primates.* Hillsdale, NJ: Erlbaum.

Buss, L. W. (1987). *The evolution of individuality.* Princeton, NJ: Princeton University Press.

Calhoun, S., & Thompson, R. L. (1988). Long-term retention of self-recognition by chimpanzees. *American Journal of Primatology, 15,* 361–365.

Chevalier-Skolnikoff, S. (1983). Sensorimotor development in orang-utans and other primates. *Journal of Human Evolution, 12,* 545–561.

Frazzetta, T. H. (1975). *Complex adaptations in evolving populations.* Sunderland, MA: Sinauer.

Gallup, G. G., Jr. (1970). Chimpanzees: Self-recognition. *Science, 167,* 86–87.

(1982). Self-awareness and the emergence of mind in primates. *American Journal of Primatology, 2,* 237–248.

(1985). Do minds exist in species other than our own? *Neuroscience and Biobehavioral Reviews, 9,* 631–641.

Gallup, G. G., Jr., McClure, M. K., Hill, S. D., & Bundy, R. A. (1971). Capacity for self-recognition in differentially reared chimpanzees. *Psychological Record, 21,* 69–74.

Goodman, M. (1962). Evolution of the immunologic species specifity of human serum proteins. *Human Biology, 34,* 101–150.

Gould, S. J. (1977). *Ontogeny and phylogeny.* Cambridge, MA: Harvard University Press.

Hampé, A. (1960). La compétition entre les éléments osseux du zeugopode de poulet. *Journal of Embryology and Experimental Morphology, 8,* 241–245.

Kollar, E. J. (1972). The development of the integument: Spatial, temporal, and phylogenetic factors. *American Zoologist, 12,* 125–135.

Kollar, E. J., & Fisher, C. (1980). Tooth induction in chick epithelium: Expression of quiescent genes for enamel synthesis. *Science, 207,* 993–995.

Ledbetter, D. H., & Basen, J. D. (1982). Failure to demonstrate self-recognition in gorillas. *American Journal of Primatology, 2,* 307–310.

Lethmate, J., & Dücker, G. (1973). Untersuchungen am Sebsterkennen im Spiegel bei Orang-Utans einigen anderen Affenarten. *Zeitschrift für Tierpsychologie, 33,* 248–269.

McKinney, M., & McNamara, K. J. (1991). *Heterochrony: The evolution of ontogeny.* New York: Plenum.

Marks, J. (1988). The phylogenetic status of orang-utans from a genetic perspective. In J. H. Schwartz (Ed.), *Orang-Utan biology* (pp. 53–67). New York: Oxford University Press.

(1991). What's old and new in molecular phylogenetics. *American Journal of Physical Anthropology, 85,* 207–219.

Mathieu, M., & Bergeron, G. (1981). Piagetian assessment on cognitive development in chimpanzees (*Pan troglodytes*). In A. B. Chiarelli & R. S. Corruccini (Eds.), *Primate behavior and sociobiology* (pp. 142–147). New York: Springer-Verlag.

Mignault, C. (1985). Transition between sensorimotor and symbolic activities in nursery-reared chimpanzees (*Pan troglodytes*). *Journal of Human Evolution, 14,* 747–758.

Mitchell, R. W. (1992). Developing concepts in infancy: Animals, self-perception, and two theories of mirror self-recognition. *Psychological Inquiry, 3,* 127–130.

(1993). Mental models of mirror self-recognition: Two theories. *New Ideas in Psychology, 11,* 295–325.

Parker, S. T. (1990). *Gorillas and chimpanzees engage in contingent facial behaviors in front of mirrors that are like those of human infants judged to recognize themselves.* Paper presented at the 13th annual meeting of the American Society of Primatologists, Davis, CA.

(1991). A developmental approach to the origins of self-recognition in great apes. *Human Evolution, 6,* 435–449.

Patterson, F. (1984). Self-recognition by *Gorilla gorilla gorilla. Gorilla* (Newsletter of the Gorilla Foundation), *7,* 2–3.

Patterson, F., & Linden, E. (1981). *The education of Koko.* New York: Holt, Rinehart & Winston.

Povinelli, D. J. (1987). Monkeys, apes, mirrors and minds: The evolution of self-awareness in primates. *Human Evolution, 2,* 493–507.

(1991). *Social intelligence in monkeys and apes.* Unpublished doctoral dissertation, Yale University.

Povinelli, D. J., & Cant, J. G. H. (1992). *Orangutan clambering and the evolutionary origins of self-conception.* Paper presented at the 14th Congress of the International Primatological Society, Strasbourg, France.

Povinelli, D. J., & Godfrey, L. R. (in press). The chimpanzee's mind: How noble in reason? How absent of ethics? In M. Nitecki (Ed.), *Evolutionary ethics.* Albany, NY: SUNY Press.

Povinelli, D. J., Rulf, A. B., Landau, K., & Bierschwale, D. (in press). Self-recognition in chimpanzees: Distribution, ontogeny, and patterns of emergence. *Journal of Comparative Psychology.*

Raikow, R. J. (1975). The evolutionary reappearance of ancestral muscles as developmental anomalies in two species of birds. *Condor, 77,* 514–517.

Raup, R. A., & Kaufman, T. C. (1983). *Embryos, genes, and evolution.* New York: Macmillan.

Suarez, S. D., & Gallup, G. G., Jr. (1981). Self-recognition in chimpanzees and orangutans, but not gorillas. *Journal of Human Evolution, 10,* 175–188.

Swartz, K. B., & Evans, S. (1991). Not all chimpanzees show self-recognition. *Primates, 32,* 483–496.

Watts, D. P., & Pusey, A. E. (1993). Behavior of juvenile and adolescent great apes. In M. E. Pereira & L. Fairbanks (Eds.), *Socioecology of juvenile primates* (pp. 148–171). Oxford University Press.

19   Incipient mirror self-recognition in zoo gorillas and chimpanzees

Sue Taylor Parker

Introduction

Laboratory studies of mirror self-recognition (MSR) using Gallup's controlled method of face marking under anesthesia (Gallup, 1970) have yielded positive results for chimpanzees and orangutans, but negative results for gorillas (Ledbetter & Basen, 1982; Suarez & Gallup, 1980). In contrast, both observational and informally controlled studies of the language-trained gorilla Koko have yielded positive results (Patterson & Cohn, *SAAH*17). In this chapter, I report observations of mirror-stimulated behaviors in a group of gorillas at the San Francisco Zoo and in a group of chimpanzees at the Oakland Zoo, which support the hypothesis that gorillas in general are capable of MSR.

Observations on gorillas

At the time of my first observations (in 1978) the gorilla group at the San Francisco Zoo was composed of an adult silverback male, Bwana; his 5-year-old son, Sunshine; his 3-year-old son, Mkumbwa; Sunshine's mother, Mrs.; Mkumbwa's mother, Jacqueline; and an adult nulliparous female, Pogo. During these observations, the animals were moving freely in and out of their night cages where they could see a 15 × 60–inch mirror I had set up outside the cage. During the hour that they had access to the mirror, I took ad-lib notes on any animal that was looking in the mirror.

On this occasion, the female Pogo inadvertently marked her own face with fingerpaint when she put her face down on the cardboard she had been marking. She subsequently looked in the mirror and wiped the paint off with her hand. The adult silverback, Bwana, whose face I had marked with fingerpaint, used a piece of paper to wipe the paint off his face while he looked in the mirror. Upon first contact with the mirror Bwana reached back behind himself, without turning, to grab Sunshine and repeated this a minute later. He also put cardboard on the floor and sat on it and looked at his reflection and jumped up and down while watching himself, and lay on his side and looked at his image.

At the time of my later observations (in 1989), the gorilla group was

301

composed of Pogo, Bwana, and Mkumbwa (now an adult), a subadult female, Zura, and a young adult female and her female infant (these two animals were not exposed to the mirror). These observations were made while the gorillas were caged in their night quarters. Each animal was observed individually in front of a 15×60–inch mirror set up outside the cages for periods of 17–41 min: Bwana for 41 min; Mkumbwa for 32 min; Pogo for 17 min; and Zura for 23 min. After an initial period of mirror exposure, each animal was marked on the face through the mesh with poster paint (without a control) by the keeper, Mary Kerr. During each mirror exposure, I took focal animal notes while Ms. Kerr videotaped the behaviors (Parker, 1990).

During the individual presentations in 1989 all of the gorillas looked, averted gaze, looked, left and returned, and repeated. Individuals differed in the percentage of time they spent looking at and testing the contingency between their actions and those of their images. While watching themselves in the mirror, the gorillas showed a variety of contingency-exploring behaviors, which I have classified into the following categories:

1. touching one's own face or head;
2. wiping paint off one's face;
3. making exploratory facial movements (mouth movements, tongue protrusion, lip eversion against mesh, raspberries [blowing air through loose lips], spitting, eye movements, blinking eyes);
4. making exploratory head movements (moving the head up and down and from side to side);
5. looking at oneself from different perspectives by changing one's own position and looking at the effect on the mirror image (from near and far, from front and side, and over the top of the head);
6. watching oneself eating food; and
7. watching oneself moving and arranging objects or wearing objects (e.g., burlap bags).

Frequencies of bouts of these behaviors are shown in Figure 19.1.

Bwana spent the largest percentage of time looking in the mirror and engaging in contingency-exploring actions. He repeatedly brought food to the mirror and watched himself eating it; he also pushed food through the mesh with his lips and made soft raspberry sounds while watching himself. After some time, he arranged his burlap bags into a nest while watching himself, then lay down and looked back over his head at the mirror. Bwana's calm, reflective attitude contrasted with Mkumbwa's mild agitation, expressed in repeatedly walking away from the mirror, banging the bars, and remaining away for longer periods. Zura, the subadult female spat, at her own image from two perspectives, which could have been a social response or at least in its repetition from a different position, an exploratory action.

Observations on chimpanzees

In October 1989 I observed the mirror-stimulated behavior of chimpanzees at the Oakland Zoo. At the time of my observations, the group was composed

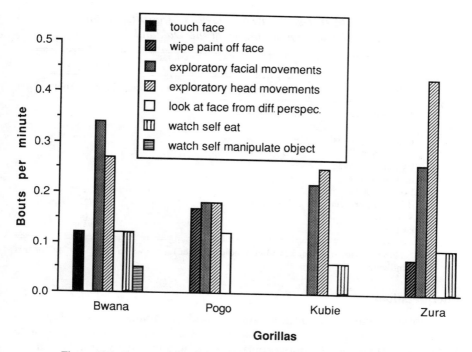

Figure 19.1. Frequencies of mirror stimulated behaviors in gorillas at the San Francisco Zoo based on individual observations (focal animal sampling).

of two adult males, Teddy and Larry; an adult female, Fran; her son Frolic; and a subadult female, Abby. As in the first observations of the gorillas, the animals all had free access to a 15×60–inch mirror that was propped up outside the outdoor cage for approximately half an hour. Consequently, as with the gorillas, I took ad-lib notes of all the behaviors directed at the mirror or mirror image.

On the first occasion the adult male, Larry, and the subadult female, Abby, displayed the greatest interest in the mirror. Abby watched herself, moved, wrist-waved, and clapped at her image, repeatedly approaching the mirror and backing up while watching. She also made head and mouth movements, and put her finger in her eye (a common tic of hers) while watching her image. Larry stared at his reflection and also watched himself while making head and mouth movements. He brought food to the mirror and watched himself eating, and also lay on his back and looked over the top of his head at the mirror. Both he and Abby repeatedly glanced in the mirror while they wrestled and played together. On the second occasion Abby also watched herself eating a carrot. Larry, after looking behind the mirror, watched himself moving his head. The female Fran showed less interest but did make some head movements, and on one occasion spat at the mirror.

Table 19.1. *Mirror-stimulated behaviors of gorillas at the San Francisco Zoo and chimpanzees at the Oakland Zoo*

	Gorillas				Chimpanzees		
Behavior	Bwana	Pogo	Mkumbwa	Zura	Larry	Abby	Fran
Touching face	+	−	−	−	−	+[a]	−
Wiping paint off face	−	+	+	+	NA	NA	NA
Exploratory facial movements	+	+	+	+	+	+	+
Exploratory head movements	+	+	+	+	+	+	+
Looking at face from different perspectives	+	+	+	+	+	−	−
Watching self eating food	+	−	+	+	+	−	+

[a] This behavior may not have been indicative because it was a common "tic" of this individual.

Interpretations

It is notable that both the zoo gorillas and chimpanzees exhibited self-directed and contingency-testing behaviors toward their mirror images (Table 19.1), and that these behaviors occurred immediately and apparently without prior experience with mirrors. Likewise, human infants often display virtually immediate signs of MSR (Lewis & Brooks-Gunn, 1979; Priel & deSchonen, 1986), as do some chimpanzees (see Boysen, Bryan, & Shreyer, *SAAH*13).

The most striking aspect of these brief observations is the similarity in the responses of the gorillas and chimpanzees. Many of these responses, notably the apparent playing with contingent facial responses of the mirror image, are characteristic of human infants who pass the mark test (Lewis & Brooks-Gunn, 1979). Likewise, similar actions have been reported in zoo orangutans in their initial contacts with mirrors (Lethmate & Dücker, 1973). Although a few individuals of both species showed some social responses, many of their behaviors seemed to be exploratory. Even the social behaviors they displayed were calm and tentative, as if they too were testing the contingency of the image.

These behaviors are quite different from the agitated and overtly aggressive social responses that monkeys often display on initial contact to their mirror images (e.g., Anderson, *SAAH*21; Benhar, Carleton, & Samuel, 1975; Gallup, 1977; Parker, personal observation). Rather, they appear to be exploratory behaviors aimed at discovering the contingent effects of actions on the image's behavior. These behaviors were repeatedly performed in a calm manner, suggesting that the animals did not believe that the image was a strange conspecific.[1]

Although suggestive of the capacity for MSR, these behaviors are not definitive. Indeed, the absence of the kind of unambiguous self-addressed behaviors that have been described in the gorilla Koko, the orangutan Chantek, and various chimpanzees is striking (Miles, *SAAH*16; Patterson & Cohn, *SAAH*17; Swartz & Evans, *SAAH*11).

The behaviors most suggestive of MSR were those of the gorilla silverback male Bwana and the female Pogo, which I saw on the first of two occasions in 1978. On this occasion Pogo was apparently unaware that she had marked herself until she saw her face in the mirror, because she wiped at the paint only after she saw her image. Although Bwana was aware that I touched his face, and indeed wiped off the paint immediately without looking, he later looked at his face in the mirror and not only wiped at the paint, but used a piece of paper to do so two separate times. The use of paper as an implement suggests more than a reflexive action in response to the memory of being touched on the face; indeed, it implies an understanding of causality that emerges in the fifth stage of the sensorimotor period (Piaget, 1952).

Conclusions

The gorilla Pogo apparently displayed MSR in the form of mirror-guided wiping of a previously undetected mark. The gorilla Bwana probably displayed MSR in the form of mirror-guided wiping of a previously detected mark with an implement. Although none of the other zoo-living chimpanzees or gorillas described in this report displayed unambiguous evidence of MSR, and none of the animals displayed self-directed actions of the kind described in Koko (Patterson & Cohn, *SAAH*17), they all displayed what could be described as incipient MSR, that is, contingency testing behaviors that could lead to the construction of MSR (Mitchell, 1993; Parker, 1991). These observations are consistent with those of Swartz and Evans (*SAAH*11), who also failed to see MSR in some zoo chimpanzees.

Three points are significant in relation to the contention that Koko's abilities are the result of special rearing and that gorillas generally lack the ability to recognize themselves in mirrors (e.g., Povinelli, *SAAH*18):

1. Naive zoo gorillas displayed the same behaviors in response to mirrors as do naive zoo chimpanzees.
2. Both groups displayed behaviors similar to those of human children who are capable of MSR.
3. At least one naive zoo gorilla displayed mark-directed behavior when she saw herself in the mirror after inadvertently and unknowingly marking her own face.

(Points 1 and 2 are significant because chimpanzees are known to be capable of MSR.)

The most parsimonious interpretation of these observations, and those of Law and Lock (*SAAH*20), and of Swartz and Evans (*SAAH*11) is that, like chimpanzees and orangutans, most gorillas have the capacity to construct MSR. The alternative hypothesis that gorillas have lost the capacity for MSR that all three great apes inherited from their common ancestor but are capable of regaining it under special rearing conditions (i.e., cross-fostering language-training) (Povinelli, *SAAH*18) is unable to account for the growing number of individual gorillas who so far have been reported to display MSR.

Acknowledgments

I want to thank Mr. Saul Kitchner, former director of the San Francisco Zoo, for permission to study the gorillas, beginning in 1975, and John Alcaraz, former keeper of the gorillas, for his help in the first mirror observations in 1978. I also want to thank the Hooper Foundation of the University of California, San Francisco, for funding a small grant in 1975 (No. 2-50-6905-36523-3) to study great apes. My observations of gorillas at the S.F. Zoo were originally undertaken in early 1975 in collaboration with Suzanne Chevalier-Skolnikoff, who was principal investigator under the Hooper Foundation grant. I want to thank Dr. Phillip Arnold, acting director of the San Francisco Zoo for permission to restudy the gorillas in 1989, and Mary Kerr, keeper of Gorilla World, for her help marking, and for videotaping the gorillas. I also thank Dr. Joel Parrott, director of the Oakland Zoo, for permission to study the chimpanzees, as well as Loren Jackson, former head keeper, and Ruth Leth, docent of the Oakland Zoo, for their cooperation. I thank Bob Mitchell for his comments, and for making Figure 19.1. Finally, I want to thank Sonoma State University for giving me a semester's paid leave for research.

Note

1. Please note that contingeny testing differs from understanding and using contingent relationships in service of gaining a reward. All animals understand contingent relationships between some of their actions and some environmental outcomes, and will use this contingency when they are rewarded by using it; indeed, operant conditioning is based on this understanding. Only a few animals such as humans and great apes, however, find controlling the contingency relationship itself intrinsically rewarding (e.g., Watson, 1972). The pleasure in controlling contingencies seems to play a key role in imitation and mirror self-recognition (Parker, 1991). Although contingent behaviors – such as manipulating a joystick to move and chase a computerized image – by rhesus monkeys can be taught and externally rewarded (e.g., Rumbaugh, Richardson, Washburn, Savage-Rumbaugh, & Hopkins, 1989) they have no obvious implications for MSR.

References

Benhar, E., Carlton, P., & Samuel, D. (1975). A search for mirror-image reinforcement and self-recognition in baboons. In S. Kondo, M. Kawai, & A. Ehara (Eds.), *Contemporary primatology* (pp. 202–208). Tokyo: Japan Science Press.

Gallup, G. G., Jr. (1970). Chimpanzees: Self-recognition. *Science, 167,* 86–87.

(1977). Self-recognition in primates. *American Psychologist, 32,* 329–338.

Ledbetter, D. H., & Basen, J. A. (1982). Failure to demonstrate self-recognition in gorillas. *American Journal of Primatology, 2,* 307–310.

Lethmate, J., & Dücker, G. (1973). Untersuchungen zum Selbsterkennen im Spiegel bei Orang-Utans und einigen anderen Affenarten. *Zeitschrift für Tierpsychologie, 33,* 248–269.

Lewis, M., & Brooks-Gunn, J. (1979). *Social cognition and the acquisition of self.* New York: Plenum.

Mitchell, R. (1993). Mental models of mirror self-recognition: Two theories. *New Ideas in Psychology*, *11*, 295–325.

Parker, S. T. (1990). Mirror-directed behaviors of gorillas at the San Francisco Zoo: Evidence for self-recognition. Videotape available through the Wisconsin Regional Primate Research Center Library, Madison.

    (1991). A developmental model for the origins of self-recognition in great apes. *Human Evolution*, *6*, 435–49.

Piaget, J. (1952). *The origins of intelligence in children*. New York: Norton.

Priel, B., & deSchonen, S. (1986). Self-recognition: A study of a population without mirrors. *Journal of Experimental Child Psychology*, *41*, 237–250.

Rumbaugh, D., Richardson, W. K., Washburn, D., Savage-Rumbaugh, E., & Hopkins, W. (1989). Rhesus monkeys (*Macaca mulatta*), video tasks, and implications for stimulus–response spatial contiguity. *Journal of Comparative Psychology*, *103*, 32–38.

Suarez, S. D., & Gallup, G. G. (1981). Self-recognition in chimpanzees and orangutans, but not gorillas. *Journal of Human Evolution*, *10*, 175–188.

Watson, J. S. (1972). Smiling, cooing and "The Game." *Merrill-Palmer Quarterly*, *18*, 323–39.

20   Do gorillas recognize themselves on television?

Lindsay E. Law and Andrew J. Lock

Following demonstrations that chimpanzees react to video images of themselves in the same way as to their mirror images (Menzel, Savage-Rumbaugh, & Lawson, 1985; Savage-Rumbaugh & Rubert, 1986), we monitored the reactions of four western lowland gorillas (*Gorilla gorilla gorilla*) to video images of themselves and others at the Bristol Zoo, United Kingdom. The male, Jason, and female, Delilah, were wild-born, estimated to be 26 years old, and housed together. Another male, Jeremiah, born at Bristol and aged 5 years, was housed with the wild-born female, Diana, whose estimated age was 17 years. As these gorillas were housed in glass-fronted enclosures with glass-covered pictures on the opposite wall, they all had some experience of reflections. Portable Canon or Panasonic video-recording equipment was used, along with a 52-cm Sony television monitor, the screen of which was positioned close to the glass fronts of the enclosures, enabling the gorillas to approach to within 50 cm of it. They were presented with three conditions: (1) videotapes of unfamiliar gorillas, (2) videotapes of themselves in delayed playback (recorded the previous day), and (3) live video transmissions of themselves.

Behaviors exhibited while watching unfamiliar gorillas

As well as head tilts, scratching, brow lifting, lip smacking (only Jeremiah), chin rubbing, shifting of position, walking around the enclosure, tapping the hands against the glass, masturbation (only Jeremiah), playing with or manipulating objects (including seeds, toy rings, barrels, and crates), approaching the screen, standing or sitting alert at the screen, and supplantation of cagemates to apparently obtain a better view, one behavior evidenced only when watching unfamiliar gorillas was Jeremiah's rolling of his arms round one another with his lips in an "oo" shape. Such hand-circling gestures have been interpreted as a request for "gentle" behavior (Patterson & Tanner, 1988). This apparently social response, then, could serve as an indication that Jeremiah perceived the screen stimulus as a gorilla other than himself or his own cagemate. However, there were no noticeable vocalizations, and no

308

charging toward the screen occurred, "other-directed" behaviors shown by gorillas in response to their own mirror image (Ledbetter & Basen, 1982; Suarez & Gallup, 1981).

Behaviors exhibited while watching noncontingent self in delayed playback

Together with avid attention, common behaviors included lip smacks, head tilts toward the screen, shifts in body position, yawning and examining or picking at the hands (only Jeremiah), sliding the hands against glass, coprophagy (only Jeremiah), sitting alert at the screen, approaching the screen, climbing steps or fixtures, and cowering under one arm while looking at the screen (only Diana). Interactive behaviors were seen only in cagemates Diana and Jeremiah, and these involved moving close to one another and swiping at each other. An interesting head-turning behavior occurred $2\frac{1}{2}$ min into Jason's second presentation. He abruptly turned around to look briefly behind him before immediately resuming his position close to the screen. When the tape was checked, it was found that, at about that time, the taped Jason had just peered through the wire division between the on-show and off-show quarters, and then had turned around and approached the glass front of the enclosure, so that his face was featured in close-up. Thus, it could be that Jason possibly recognized his surroundings but not himself, and looked behind to see if another gorilla was present in the enclosure. Alternatively, he may have recognized himself, and the look behind was to confirm that the enclosure on screen was the same as his in reality.

Behaviors exhibited while watching contingent self

Of particular interest are the behaviors exhibited by the gorillas while watching their own live video image, and here we present evidence of interest, image testing, and face exploration. Avid attention was accompanied by approaches, head tilts, tongue showing, and lip smacks, standing or sitting alert at the screen, shifting position, bending over (whether watching or not), examining or picking at the hands (only Jeremiah), folding and jerking of the arms (only Jeremiah), and watching oneself touch the face and bite the nails.

An interesting incident that may imply self-recognition occurred with Diana. She rose from her sitting position to approach the glass front. Shortly afterward, Jeremiah pushed a large barrel down toward her. It appears from the videotape that Diana had already started to move out of the way without having looked behind her to see the barrel coming in her direction. Her eyes were on the screen and, although she could have sensed its imminent arrival upon her by some other means (such as sound), it is possible that she knew she must move sideways because she saw the action and object on-screen. This, in turn, would suggest that she had recognized the surroundings as her enclosure and the gorilla on-screen as herself.

Jeremiah produced many interesting behaviors, most of them not seen in the other gorillas. He made other facial movements, such as brow lifting and lip pursing, and also yawned and scratched more than when watching unfamiliar gorillas, or himself in delayed playback. In one instance, seemingly guided by the video image, he reached behind for a rubber ring on the floor; but, as he could not get hold of it, he gave up, seeming reluctant to take his eyes off the screen to actually look behind him.

Such behaviors as purposeful limb movements and bobbing with the eyes fixed firmly on the screen appear to involve the animals testing the contingency between their actions and those of the image (after Lewis & Brooks-Gunn, 1979). These behaviors are similar to what Parker (1991) describes as "contingent head and facial movements," which seem to involve "kinesthetic-visual matching" of the image's movements to those of self (Mitchell, 1993). In the first condition (unfamiliar gorillas), Jeremiah did produce glass banging in front of the monitor, but it was not accompanied by the continued eye contact with the image, seen here, nor by head movements and bending down, as if to get a better look. Other limb movements were produced with a fixed gaze on the live image, including another instance in which Jeremiah lifted and moved his left arm in a straight line while watching the screen intently. The conservative interpretation that such actions constitute "other-directed" social behaviors cannot be ruled out, but they do not appear to have this quality. It seems that the gorillas are exploring an aspect of what they are detecting as "strange" in the behavior of the image they are viewing, and that "strangeness" is the contingency between the image's and their own behaviour.

The self-exploration of the chimpanzee Austin, who used a live video image of himself to observe his lip movements and look inside his mouth (Savage-Rumbaugh & Rubert, 1986, pp. 310–311), made us look for similar actions in our gorillas. We were rewarded when Jeremiah did something very similar the first time he saw his live image. After yawning three times while sitting in front of and watching his image, he begins to look inside his mouth. The sequence of his behaviors recorded on video suggests that he noticed the contingency between the yawning in the image and his own yawning. Following the three yawns, both his mouth shape and the way he moved his head and held his mouth open while looking at the screen were quite different from his typical yawns. Jeremiah bent his head forward and opened his mouth, gradually widening it and seeming to roll his tongue. He used his right hand to support and sometimes manipulate his lower jaw, as if to keep his mouth open to enable viewing. When he closed his mouth after looking inside it, Jeremiah's eyes remained on the screen and, after only a second or so, he began to look at his mouth again, this time using his left hand to hold his chin area.

Both of the female gorillas seemed to feel along certain areas of their faces while looking distinctly at the screen of themselves. Delilah sat in her favored position, cross-legged against the back of the enclosure, watching herself on screen intently and often tilting her head quizzically. She then seemed to cover her eyes and brow ridge with her left hand while looking at the screen, appearing to feel along the top of her brow ridge with her thumb, before taking her hand away and continuing to look very curiously. Diana, too, touched her face while clearly maintaining her gaze on the screen. She scratched her nose, stared for a while, and then scratched it again, before feeling along the bottom of her brow ridge with her left index finger, all the while looking at herself on screen. It appeared as if the gorillas were checking to see if the screen gorilla moved simultaneously. It is possible that they kept looking at the screen while scratching so as not to miss anything, but the movements looked more deliberate, as if to ascertain correspondence to the image on the screen.

Conclusions

Overall, the results from this study can be seen as surprising in the context of previous studies that have given gorillas extensive exposure to their own images, and used much longer observational periods, yet failed to find any evidence for gorilla self-recognition. The gorillas seemed to be intrigued by the screen, and tended to keep glancing intermittently at it if they were engaged in another activity. With most of the behaviors being exhibited while the gorillas were looking at the screen, it would appear that they might be a form of response to it. Interpretation of behaviors is highly debatable; for instance, are certain movements such as yawning, scratching, and lip-smacking nervous displacement activities (Dixson, 1981; Hinde & Tinbergen, 1958)? Can virtually any apparently purposeful movement, especially those directed to areas such as the face, be interpreted as "image testing" (Savage-Rumbaugh & Rubert, 1986) or self-investigation (Gallup, 1970)? Although the "self-exploratory" behaviors of investigating the inside of the mouth and feeling areas of the face such as the brow ridge may not be sufficient to conclude that these animals recognize themselves, these and other behaviors evidenced in this study suggest that there is something about their live image that the gorillas find more interesting than the other material they were shown, including images of their cagemates and of their noncontingent selves, and that the apparently anomalous position of the gorilla among the great apes with respect to self-recognition may need to be reevaluated.

References

Dixson, A. F. (1981). *The natural history of the gorilla*. London: Weidenfeld & Nicolson.

Gallup, G. G., Jr. (1970). Chimpanzees: Self-recognition. *Science, 167*, 86–87.

Hinde, R., & Tinbergen, N. (1958). The comparative study of species-specific behaviour. In A. Roe & G. Simpson (Eds.), *Behaviour and evolution* (pp. 251–268). New Haven, CT: Yale University Press.

Ledbetter, D., & Basen, J. (1982). Failure to demonstrate self-recognition in gorillas. *American Journal of Primatology, 2*, 307–310.

Lewis, M., & Brooks-Gunn, J. (1979). *Social cognition and the acquisition of self.* New York: Plenum.

Menzel, E., Savage-Rumbaugh, E. S., & Lawson, J. (1985). Chimpanzee spatial problem-solving with the use of mirrors and televised equivalents of mirrors. *Journal of Comparative Psychology, 99,* 211–217.

Mitchell, R. W. (1993). Mental models of mirror self-recognition: Two theories. *New Ideas in Psychology, 11,* 295–325.

Parker, S. T. (1991). A developmental approach to the origins of self-recognition in great apes. *Human Evolution, 6,* 435–449.

Patterson, F. G., & Tanner, J. (1988). Gorilla gestural communication. *Gorilla, 12*(1), 2–5.

Savage-Rumbaugh, E. S., & Rubert, E. (1986). Video representations of reality. In E. S. Savage-Rumbaugh (Ed.), *Ape language: From conditioned response to symbol* (pp. 299–323). New York: Columbia University Press.

Suarez, S. D., & Gallup, G. G., Jr. (1981). Self-recognition in chimpanzees and orangutans, but not gorillas. *Journal of Human Evolution, 10,* 175–183.

Part IV

Mirrors and monkeys, dolphins, and pigeons

21   The monkey in the mirror:
A strange conspecific

James R. Anderson

Introduction

Several studies of how chimpanzees (*Pan troglodytes*) and orangutans (*Pongo pygmaeus*) respond to their reflections in a mirror have established that these primates share with humans the capacity for self-recognition (Gallup, 1970; Lethmate & Dücker, 1973; Miles, *SAAH*16; Suarez & Gallup, 1981). Recent data on gorillas (*Gorilla gorilla*) (Patterson & Cohn, *SAAH*17; Swartz & Evans, *SAAH*11) strongly suggest that these great apes also are capable of self-recognition. Although the relationship between mirror self-recognition (MSR) and self-awareness in the sense used by social psychologists remains to be clarified (see Mitchell, *SAAH*6), there is general agreement that self-recognition implies the existence of some kind of cognitive (as opposed to a merely kinesthetic) self-awareness. This may in turn allow for the expression of a range of behaviors reflecting a "theory of mind" (Crook, 1988; Gallup, 1982; Humphrey, 1984; Whiten, 1991), although the relationship between self-recognition and theory of mind (ToM) remains to be clarified, as certain subjects (e.g., autistic children, Baron-Cohen, 1992) may show the former but not the latter (Gergely, *SAAH*5; Mitchell, 1993).

In contrast to the numerous demonstrations of self-recognition in great apes, studies of reactions to mirror-image stimulation (MIS) in monkeys have consistently failed to find evidence of self-recognition (reviews: Anderson, 1984a; Gallup, 1987). It has been suggested that the contrasting performances between great apes and monkeys on tests of self-recognition reflect the existence of a fundamental difference in cognition – concerning self-awareness – between the self-recognizing Pongidae and other nonhuman primates (Gallup, 1982, 1987).

In both great apes and human infants, two sets of behavioral events indicate the presence of self-recognition during mirror tests:

1. the spontaneous development of mirror-mediated self-directed acts, such as examining parts of the body that are not directly visible, and
2. the diminution of – culminating in the disappearance of – social responses directed toward the mirror images.

Self-recognition may be confirmed by appropriate use of the mirror during the mark test devised by Gallup (1970) (see Anderson, 1993, for a more

315

detailed discussion of what constitutes acceptable evidence of self-recognition). Self-recognizing apes and humans will usually satisfy these two criteria reliably (i.e., upon repeated presentation of a mirror; see e.g., Calhoun & Thompson, 1988). Two recent suggestions that individual monkeys might have shown one sign of self-recognition (passing the mark test: Boccia, SAAH23; Itakura, 1987a) refer to fleeting instances of a marked monkey touching the mark in the presence of the mirror; neither have been concerned with the reliability of the behavior observed, or with spontaneous self-exploration. In fact, spontaneous mirror-mediated, self-directed acts have never been reported in monkeys, while the tendency to respond to the reflection in a social manner persists. For example, although they gradually habituate to the reflection, macaques retain a readiness to respond to the mirror socially: A resurgence of social responding may occur when the animal's physical appearance (and therefore that of the animal in the mirror) is altered, as during the mark test of self-recognition (e.g., Anderson, 1983a), or when the mirror is either relocated or temporarily removed and replaced (Eglash & Snowdon, 1983; Gallup & Suarez, 1991; Suarez & Gallup, 1986).

Exploring species differences

Because of its relevance to the issue of the evolution of self-awareness, self-recognition – and the phylogenetic distribution of the capacity for self-recognition – has become an important and sometimes controversial topic in primatology. As with other areas in the domains of social and self-cognition (e.g., Byrne & Whiten, 1988; Cheney & Seyfarth, 1990), reliable, experimentally obtained behavioral data are essential to support claims about a given species's capabilities (Kummer, Dasser, & Hoyningen-Huene, 1990; see also Premack, 1988). There are several ways of evaluating the available evidence concerning monkeys' reactions to their reflection, and of improving our appreciation of possible reasons for the species differences obtained concerning self-recognition. One approach is to compare different species in terms of other capacities that are somehow linked to self-recognition, such as kinesthetic–visual matching or imitation (see Mitchell, 1993; Parker, 1991). Another approach is to intensify assessment of the effects of varying conditions of mirror exposure on monkeys' spontaneous behaviors along with subsequent reactions in tests of self-recognition (Anderson, 1984a; Gallup, 1987). This approach should include efforts to arrange conditions so as to make the emergence of self-directed responses more likely. A related approach consists of analyzing in more detail the extent to which monkeys can correctly comprehend reflected environmental information – the affordances of mirrors (Loveland, 1986) – and can use reflections of objects or of themselves to aid performance on search tasks (Anderson, 1986 [see Figure 21.1]; Itakura, 1987a,b; see also Eglash & Snowdon, 1983). Platt, Thompson, and Boatwright (1991) have recently made a number of useful suggestions in this context.

Figure 21.1. A juvenile *Macaca tonkeana* makes use of his reflection in a mirror to locate a raisin stuck on the outside of his cage. Two out of nine macaques tested by Anderson (1986) learned to use reflected information in this way. In the photo a nonlearner looks on.

Yet another approach – the one taken in the present chapter – is to weigh up the evidence that a monkey looking at itself in a mirror "thinks" its reflection is a conspecific. A thorough analysis of the social stimulus properties of mirrors is an important component in the broader search for self-recognition in primates. As described below, in both monkeys and many nonprimate species a variety of phenomena normally induced by a conspecific may occur when the conspecific is replaced by a mirror. It can be argued that, in the absence of convincing evidence for self-recognition in monkeys, such phenomena reinforce the view that monkeys, like most other animals, do not progress beyond a "social" stage when perceiving their own reflections.

Overt social responses to mirrors

When confronted with their mirror image, many animal species, including arthropods, reptiles, fish, birds, rodents, dogs, and cats, as well as monkeys, show social responses (for reviews, see Anderson, 1981; Gallup, 1968). Such social responses in monkeys may include facial expressions (e.g., lip smacking, open-mouth threats), vocalizations, and attempts to touch or reach behind the mirror as if to grab the "other animal" (for reviews, see Anderson, 1984a; Gallup, 1968, 1975). Self-recognizing primates (i.e., great apes and humans) also pass through a phase of social responding to their reflections before self-recognition emerges and social tendencies drop out (Amsterdam, 1972; Anderson, 1984b; Gallup, 1979; Lewis & Brooks-Gunn, 1979; Lin, Bard, & Anderson, 1992). Interestingly, elderly humans suffering from senile dementia show a reverse process of increasingly frequent misidentification of and social responding to their own reflection, before all responsiveness drops out (see Biringer & Anderson, 1992; Biringer, Anderson, & Strubel, 1988).

Influencing factors

Species, age, and sex. Social responses to a mirror are highly variable across primate species, ranging from short-lived, low-intensity posturing in some prosimians (*Lemur* spp.; Fornasieri, Roeder, & Anderson, 1991), to a full aggressive penile display in squirrel monkeys (*Saimiri sciureus*; MacLean, 1964). There have not yet been enough studies to allow firm conclusions regarding the nature of such species differences, but it seems possible that there may be an inverted U-shaped function between social responsiveness and cerebral cortical development.

Mirror-elicited social responses in monkeys also vary with age and gender both within and across species. Infant stump-tailed macaques (*Macaca arctoides*) predominantly huddled against and played with their reflections (Anderson & Chamove, 1986). Infant rhesus monkeys (*M. mulatta*) showed less social responding overall to the mirror than did their mothers (Gallup, Wallnau, & Suarez, 1980). Subadult capuchin monkeys (*Cebus* spp.) showed more mirror-directed social responses (especially play faces) than either adults or juveniles; social responses by adults consisted mostly of threats (Collinge, 1989). Juvenile patas monkeys (one male and one female), as well as an adult female, investigated a mirror, whereas an adult male repeatedly and aggressively yawned at it (Hall, 1962). In one study in which a group of Japanese macaques (*Macaca fuscata*) were given long-term exposure to a freestanding mirror, adults directed almost no social responses toward it (Platt & Thompson, 1985).

There are also striking differences in responses to MIS among different species of fish. For example, adult male Siamese fighting fish (*Betta splendens*) react with aggressive displays, as do male paradise fish (*Macropodus opercularis*) (Davis, Harris, & Shelby, 1974; Thompson, 1963). In contrast, goldfish (*Carassius auratus*) are nonaggressive in the same situation (Gallup & Hess, 1971).

Early rearing. Monkeys reared in isolation or with one other monkey do not react to mirrors in the same ways as more socially sophisticated monkeys. Isolation-reared rhesus monkeys preferred to look at their own reflection than at a live conspecific, whereas the opposite was true for wild-reared monkeys (Gallup & McClure, 1971). Isolates also learned to pull a chain to produce visual stimulation, and again showed a preference for the mirror. Mother-only–reared rhesus monkeys spent more time near a mirror at 6 and 12 months of age than did group-reared infants (Spencer-Booth & Hinde, 1969). Responsiveness to live stimulus animals, mirrors, and films of conspecifics also varied in stump-tailed macaques reared under different conditions (Anderson & Chamove, 1984). In particular, infants reared with only one peer as a social partner were strongly attracted to a mirror, whereas infants with extensive early experience of mirrors preferred to watch a film of an unfamiliar monkey.

Early rearing with nonconspecifics may attenuate monkeys' social responses to reflections. One Japanese macaques (*M. fuscata*) reared among pigtailed macaques (*M. nemestrina*) did not perform operant responses for MIS (Gallup, 1966); reduced responsiveness to a mirror was also noted by Collinge (1989) in hand-reared as opposed to group-reared capuchin monkeys (*Cebus* spp.). Diminished responsiveness to mirrors following cross-fostering is also reported in nonprimates: Dogs reared with cats in early life showed unusually subdued responsiveness to their reflections (Fox, 1969). Finally, paradise fish reared with nonconspecifics showed less frequent mirror-directed aggressive displays than control or isolated fish (Kassel & Davis, 1975).

Viewing conditions. The intensity of social reactions displayed in the presence of both a mirror and a live conspecific vary with the distance between the stimulus and the observing monkey (Anderson, 1983b; Anderson & Roeder, 1989; Hall, 1962; Masataka & Fujii, 1980). Distance from the mirror or conspecific can similarly affect responses in Siamese fighting fish (Bronstein, 1983) as well as in domestic chicks (Tolman, 1965). Capuchin monkeys viewing angled mirrors, with no possibility of face-on confrontation, showed considerably reduced social responding in this condition (Anderson & Roeder, 1989). Finally, Siamese fighting fish were more aggressive toward their own mirror image than toward a fish displaying at its own reflection on the other side of a one-way mirror (Meliska, Meliska, & Peeke, 1980).

The predictive value of responses to mirrors

If MIS constitutes a social stimulus, it can be asked whether responses to mirrors might predict behavior in other situations. The data currently available do not allow firm conclusions to be drawn. Confrontation with a mirror was included in a battery of tests designed to assess behavioral characteristics in rhesus monkeys (Spencer-Booth & Hinde, 1969). At both 6 months and 12 months of age, mother-only–reared infants spent more time near a mirror than did group-reared infants, but the relation between behavior in mirror tests and in other situations was not clear. What might be called the "external validity" of mirror tests has been investigated more thoroughly in nonprimate species. For example, a combination of quantitative and qualitative data indicated consistency across mirror tests in individual yellow-bellied marmots (*Marmota flaviventris*), and also indicated a positive relation between behavior when exposed to a mirror (e.g., avoidance) and that observed in natural surroundings (e.g., solitary ranging; Svendsen & Armitage, 1973). Dominant Siamese fighting fish perform more operant responses than subordinates when MIS is contingent upon such responses (Baenninger, 1984), but one study of whether relative dominance in pairs of Siamese fighting fish could be predicted on the basis of prior aggressivity shown toward a mirror gave at best inconclusive results (Bronstein, 1985). It would be interesting to examine the extent to which mirror image reactions can predict subsequent dominance positions in monkeys.

Overt social responses: Conclusions

There is no doubt that mirrors elicit a broad range of overt social responses in monkeys. However, there remains considerable scope for improved understanding of the responses elicited by MIS. In particular, there have been few direct comparisons of mirror-induced responses with their equivalents shown in the presence of a live conspecific or other social stimuli, although such comparisons exist for certain types of fish and birds (e.g., Figler, 1972; Vallortigara & Zanforlin, 1986). More careful comparisons with behavior shown toward live conspecifics and approximations to live conspecifics are required. For example, in some cases mirrors induce reactions that are more intense than those elicited by live conspecifics, which has led to MIS being described as a "supernormal" stimulus (Gallup, 1975). In other cases, however, MIS is clearly inferior to a live conspecific in eliciting or maintaining social responses (see later in this chapter; Anderson & Chamove, 1986; Bronstein, 1983). Indeed, the social abnormality of MIS may evenutally lead to avoidance of the reflection by socially sophisticated monkeys (Anderson & Bayart, 1985). Finally, there is also a need to clarify the influence of individual and contextual factors in responses to mirrors (e.g., Francis, 1990; Holder, Barlow, & Francis, 1991). As far as monkeys are concerned, such factors may well play a role not only in social responsiveness but also in the animal's understanding of the image as a representation of itself (see Platt et al., 1991).

Visual reinforcement

Social interaction with the mirror image may be reinforcing for monkeys. Contingent exposure to a mirror in squirrel monkeys supported operant responding, but only in monkeys that responded with aggression to the mirror (MacLean, 1964). Siamese fighting fish also show an association between aggressive display and operant responding for MIS (Thompson, 1963). However, it is clear that not all operant responding for MIS is related to aggressive motivation, in either monkeys or fish (Anderson, 1984a; Baenninger & Mattleman, 1973; Gallup & Hess, 1971).

Both exposure to MIS and to a live conspecific are reinforcing for macaques. When separated from others they prefer to view a live conspecific over other visual incentives (Butler, 1954) and will push open a door to view their mirror reflection (Gallup, 1966). Mirror-maintained operant responding is also present in fish (for review, see Hogan & Roper, 1978) and birds (e.g., Gallup, Montevecchi, & Swanson, 1972; Thompson, 1964).

The mirror image as a stand-in for social stimuli

In addition to overt social responses such as those described above and the visual reinforcing effects of MIS, a range of other, more indirect social phenomena can be obtained using mirrors rather than true social stimuli.

A mirror as a rearing companion

One mirror-rearing experiment with macaques indicates that MIS has sufficient social content to persistently elicit a range of species-specific social behaviors during infancy, and that the presence of a mirror can partly compensate for the negative consequences of being reared in the absence of conspecifics. However, exclusive experience with MIS is clearly inferior to experience with a live and fully accessible conspecific.

Stump-tailed macaques reared with their own reflections as the only source of social stimulation during most of the first year of life behaved differently from their peer-reared and isolation-reared conspecifics during a period of several months (Anderson, 1981; Anderson & Chamove, 1986). The reflection persistently elicited a range of social responses including affiliation, play, and aggression, but overall social responsiveness was considerably greater toward a live, fully accessible cagemate than toward the mirror. Figure 21.2 illustrates the difference between the two stimuli in the amount of play behaviors elicited. In subsequent tests, mirror-reared infants displayed fewer appropriate responses to different kinds of pictures of conspecifics than did socially reared controls (Anderson & Chamove, 1984).

Anderson and Chamove (1986) also reported that separating infant stump-tailed macaques from their live rearing partner, but not a mirror-image "rearing partner," resulted in increased vocalization rates. This suggests the formation of a stronger attachment to the real peer than to the reflection. One final interesting finding of this study was that isolation syndrome behaviors, including autoeroticism, self-clasping, stereotyped movements, and bizarre posturing, were slightly less frequent in mirror-reared than in isolation-reared infants. Therefore, mirrors might serve as an enrichment technique for individually housed monkeys.

Social facilitation

Social facilitation is "an increase in the frequency or intensity of responses or the initiation of particular responses already in an animal's repertoire, when shown in the presence of others engaged in the same behavior at the same time" (Clayton, 1978). Some authors use the term *coaction* (Zajonc, 1965) to describe the same behavioral phenomena. For animals that treat their reflection as another animal, MIS should constitute the perfect coaction situation, in that the reflection does only what the observing animal does and nothing else. It is interesting, therefore, that mirror-induced social facilitation of drinking bouts has been shown in adult male and female macaque monkeys (*M. arctoides*, *M. fascicularis*, *M. tonkeana*) tested in individual cages and presented with a bottle of orange juice (Anderson & Bayart, 1985; Bayart & Anderson, 1985; Straumann & Anderson, 1991) (see Figure 21.3). On the other hand, the presence of the mirror had no effect on adult female stump-tailed macaques' manipulation of novel objects (Straumann & Anderson, 1991); nor did the presence of a mirror influence the drinking performances of adult females of three species of lemur (*Lemur catta*, *L. fulvus*, *L. macaco*; Fornasieri, Roeder, & Anderson, 1991). Conceivably, MIS lacks some as-yet unidentified (possibly chemical) constituent necessary for the elicitation of such a social phenonemon in the more olfactorily dependent prosimians.

Once again, similar findings of mirror-induced social facilitation exist for nonprimate species, concerning food pecking (Montevecchi & Noel, 1978; Tolman, 1965) and the tonic immobility response (Gallup, 1972) in domestic chicks, as well as runway performance and maze learning in mice: Runway performance was better, learning was faster, and extinction was slower in the mirror condition than in a live-audience condition (Hamrick, Cogan, & Woolam, 1971).

Modulation of separation-induced protest

Further evidence of the powerful social stimulus properties of a reflection for monkeys comes from observations of the monkeys' reactions during separation from their partners. The presence of another infant is known to reduce agitation during separation from the partner in peer-reared macaques (Boccia, Reite, Kaemingk, Held, & Laudenslager, 1991). Pair-reared infant stump-tailed macaques had higher locomotion and vocalization scores during

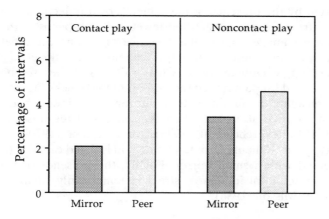

Figure 21.2. Social-partner-directed play activity in mirror- and peer-reared infant stump-tailed macaques. Although the mirror elicits play, the live partner is a more effective stimulus, especially for contact forms of play. (After Anderson & Chamove, 1986.)

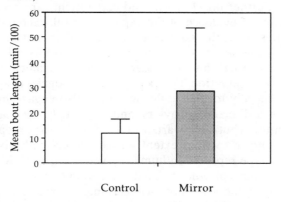

Figure 21.3. Mean duration of drinking episodes during control and mirror-image stimulation conditions, showing a social facilitation effect. Data are from seven adult female *Macaca arctoides*. (After Straumann & Anderson, 1991.)

10-min "total" separations (no contact with the partner) than during separations involving each partner on one side of a transparent partition, or in separations during which a large mirror was present (Anderson, 1983b). In that experiment the mirror, but not the familiar conspecific, elicited lip smacking, indicating that the reduction in separation-induced agitation was not merely due to the subjects interpreting the reflection as being the separated cagemate. A second experiment confirmed the calming effect of the mirror during separations in pair-reared, but not in group-reared infants.

In nonprimate species that form filial attachments to salient conspecifics, protest reactions to involuntary separation from the attachment figure can also be attenuated by the presence of the subject's own reflection. Domestic chicks separated from their peers emitted fewer isolation peeps in the presence of their mirror image than in the presence of a conspecific behind a transparent partition or when in an empty chamber (Gallup et al., 1972; Montevecchi, Gallup, & Dunlap, 1973). Clearer effects of the presence of a mirror were also found for pair-reared than group-reared chicks (Montevecchi & Noel, 1978), recalling the finding with infant macaques mentioned above. The presence of a mirror also reduced chicks' emotional responses to fear-evoking stimuli (Montevecchi et al., 1973), while Keiper (1970) reported reduced stereotyped movements ("route-tracing") in isolated canaries (*Serinus canarius*) presented with a small mirror. Finally, the presence of a mirror reduced some physiological indices of stress in temporarily isolated sheep (Parrott, Houpt, & Misson, 1988).

Conclusions: Can MIS reveal monkeys' inner selves?

The search for self-recognition in nonpongid primates, as well as in other animals, should continue. Further observations and experiments are necessary, not least because absence of evidence of the capacity for self-recognition or other cognitive capacities is not the same as evidence of absence (Desmond, 1979). In the context of the debate over self-recognition and its relation to self-awareness, several converging lines of evidence constitute a much stronger argument than any single observation. For example, self-recognizing apes not only cease responding socially to their reflection, they also engage in spontaneous mirror-aided, self-directed behaviors, and will satisfy the mark-test criterion of self-recognition (but see Swartz & Evans, 1991, for exceptions). Although it is not yet known to what extent mirrors induce reflective self-awareness effects of the type reported in humans (Anderson, 1993; Mitchell, 1993), convincing observations of other behaviors reflecting a theory of mind are rapidly accumulating in the great apes, concerning, for example, deception (Miles, 1986; Mitchell, 1986, 1991; Premack, 1988; de Waal, 1986), imitation (Miles, 1990; Mitchell, 1987), and attribution of states of knowledge (Povinelli, Boysen, & Nelson, 1990).

In contrast, the findings reviewed in the present chapter – on mirror-elicited overt social responses, visual reinforcement, social facilitation, and other social phenomena – constitute a body of convergent evidence suggesting that monkeys, unlike apes, persist in perceiving their reflection as a conspecific, albeit a strange one. Furthermore, results from experiments using totally different paradigms indicate that monkeys are markedly inferior to chimpanzees in the ability to attribute states of knowledge to others (Cheney & Seyfarth, 1990; Povinelli, Parks & Novak, 1991); their deceptive acts often appear to involve lower-level mechanisms (Whiten & Byrne, 1991), and there is little good evidence that monkeys can imitate (Visalberghi & Fragaszy, 1990). All of these data are in good agreement with the consistent failures to find self-recognition in monkeys.

In conclusion, although attempts to find self-recognition in monkeys have been going on for more than 20 years, there is no need to abandon the search. More individuals, more species and more innovative experimental procedures are necessary. At the same time, continuing study of mirror-induced social effects will contribute additional information on social and perceptual mechanisms. Along with other approaches, including research on imitative abilities (Mitchell & Anderson, 1993), these combined research efforts will clearly lead to more insights into the evolution of self- and social cognition in primates.

References

Amsterdam, B. (1972). Mirror image reactions before age two. *Developmental Psychobiology, 5*, 297–305.

Anderson, J. R. (1981). *Mirror image stimulation and behavioural development in stumptail macaques.* Unpublished doctoral dissertation, University of Stirling, Scotland.

(1983a). Responses to mirror image stimulation and assessment of self-recognition in mirror- and peer-reared stumptail macaques. *Quarterly Journal of Experimental Psychology, 35B*, 201–212.

(1983b). Mirror-image stimulation and short separations in stumptail monkeys. *Animal Learning and Behavior, 11*, 138–143.

(1984a). Monkeys with mirrors: Some questions for primate psychology. *International Journal of Primatology, 5*, 81–98.

(1984b). The development of self-recognition: A review. *Developmental Psychobiology, 17*, 35–49.

(1986). Mirror-mediated finding of hidden food by monkeys (*Macaca tonkeana* and *M. fascicularis*). *Journal of Comparative Psychology, 100*, 237–242.

(1993). To see ourselves as others see us. *New Ideas in Psychology, 11*, 339–343.

Anderson, J. R., & Bayart, F. (1985). Les effets du miroir sur le comportement de macaques; Habituation, évitement et facilitation social. *Biology of Behaviour, 10*, 159–167.

Anderson, J. R., & Chamove, A. S. (1984). Early social experience and responses to visual social stimuli in young monkeys. *Current Psychological Research and Review, 3*, 32–45.

(1986). Infant stumptailed macaques raised with mirrors or peers: Social responsiveness, attachment, and adjustment. *Primates, 27*, 63–82.

Anderson, J. R., & Roeder, J.-J. (1989). Responses of capuchin monkeys (*Cebus apella*) to different conditions of mirror-image stimulation. *Primates, 30*, 581–587.

Baenninger, R. (1984). Consequences of aggressive threats by *Betta splendens. Aggressive Behavior, 10*, 1–9.

Baenninger, R., & Mattleman, R. A. (1973). Visual reinforcement: Operant acquisition in the presence of a free mirror. *Animal Learning and Behavior, 1*, 302–306.

Baron-Cohen, S. (1992). The theory of mind hypothesis of autism: History and prospects of the idea. *The Psychologist: Bulletin of the British Psychological Society, 5*, 9–12.

Bayart, F., & Anderson, J. R. (1985). Mirror-image reactions in a tool-using, adult male *Macaca tonkeana. Behavioural Processes, 10*, 219–227.

Biringer, F., & Anderson, J. R. (1992). Self-recognition in Alzheimer's disease: A mirror and video study. *Journal of Gerontology: Psychological Sciences, 47*, 385–388.

Biringer, F., Anderson, J. R., & Strubel, D. (1988). Self-recognition in senile dementia. *Experimental Aging Research, 14*, 177–180.

Boccia, M. L., Reite, M. L., Kaemingk, K., Held, P., & Laudenslager, M. L. (1991). Social context and reaction to separation in peer-reared pigtail macaques: Some preliminary observations. *Primates, 32*, 255–263.

Bronstein, P. M. (1983). Onset of combat in male *Betta splendens*. *Journal of Comparative Psychology*, *97*, 135–139.

(1985). Predictors of dominance in male *Betta splendens*. *Journal of Comparative Psychology*, *99*, 47–55.

Butler, R. A. (1954). Incentive conditions which influence visual exploration. *Journal of Experimental Psychology*, *48*, 19–23.

Byrne, R., & Whiten, A. (Eds.) (1988). *Machiavellian intelligence: Social expertise and the evolution of intellect in monkeys, apes, and humans*. Oxford University Press (Clarendon Press).

Calhoun, S., & Thompson, R. L. (1988). Long-term retention of self-recognition by chimpanzees. *American Journal of Primatology*, *15*, 361–365.

Cheney, D. L., & Seyfarth, R. M. (1990). *How monkeys see the world: Inside the mind of another species*. Chicago: University of Chicago Press.

Clayton, D. A. (1978). Socially facilitated behavior. *Quarterly Review of Biology*, *53*, 373–392.

Collinge, N. E. (1989). Mirror reactions in a zoo colony of *Cebus* monkeys. *Zoo Biology*, *8*, 89–98.

Crook, J. H. (1988). The experiential context of intellect. In R. Byrne & A. Whiten (Eds.), *Machiavellian intelligence* (pp. 347–362). Oxford University Press (Clarendon Press).

Davis, R. E., Harris, C., & Shelby, J. (1974). Sex differences in aggressivity and the effects of social isolation in the anabantoid fish, *Macropodus opercularis*. *Behavioral Biology*, *11*, 497–509.

Desmond, A. (1979). *The ape's reflexion*. London: Blond & Briggs.

Eglash, A. R., & Snowdon, C. T. (1983). Mirror-image responses in pygmy marmosets (*Cebuella pygmaea*). *American Journal of Primatology*, *5*, 211–219.

Figler, M. H. (1972). The relations between eliciting stimulus strength and habituation of the threat display in male Simaese fighting fish, *Betta splendens*. *Behaviour*, *42*, 63–96.

Fornasieri, I., Roeder, J.-J., & Anderson, J. R. (1991). Les réactions au miroir chez trois espèces de lémuriens (*Lemur fulvus, L. macaco, L. catta*). *Comptes Rendus de l'Académie des Sciences de Paris*, *312*(3), 349–354.

Fox, M. W. (1969). Behavioral effects of rearing dogs with cats during "the critical period of socialization." *Behaviour*, *35*, 273–280.

Francis, R. C. (1990). Temperament in a fish: A longitudinal study of the development of individual differences in aggression and social rank in the midas cichlid. *Ethology*, *86*, 311–325.

Gallup, G. G., Jr. (1966). Mirror-image reinforcement in monkeys. *Psychonomic Science*, *5*, 39–40.

(1968). Mirror-image stimulation. *Psychological Bulletin*, *70*, 782–793.

(1970). Chimpanzees: Self-recognition. *Science*, *167*, 86–87.

(1972). Mirror-image stimulation and tonic immobility in chickens. *Psychonomic Science*, *28*, 257–259.

(1975). Towards an operational definition of self-awareness. In R. H. Tuttle (Ed.), *Socioecology and psychology of primates* (pp. 309–341). The Hague: Mouton.

(1979). Self-recognition in chimpanzees and man: A developmental and comparative perspective. In M. Lewis & L. A. Rosenblum (Eds.), *The child and its family* (pp. 107–126). New York: Plenum.

(1982). Self-awareness and the emergence of mind in primates. *American Journal of Primatology*, *2*, 237–248.

(1987). Self-awareness. In G. Mitchell & J. Erwin (Eds.), *Comparative primate biology, Vol. 2B: Behavior, cognition, and motivation* (pp. 3–16). New York: Liss.

Gallup, G. G., Jr., & Hess, J. Y. (1971). Preference for mirror-image stimulation in goldfish (*Carassius auratus*). *Psychonomic Science*, *23*, 63–64.

Gallup, G. G., Jr., & McClure, M. K. (1971). Preference for mirror-image stimulation in differentially reared rhesus monkeys. *Journal of Comparative and Physiological Psychology*, *75*, 403–407.

Gallup, G. G., Jr., Montevecchi, W. A., & Swanson, E. T. (1972). Motivational properties of mirror-image stimulation in the domestic chicken. *Psychological Record*, *22*, 193–199.

Gallup, G. G., Jr., & Suarez, S. D. (1991). Social responding to mirrors in rhesus monkeys (*Macaca mulatta*): Effects of temporary mirror removal. *Journal of Comparative Psychology, 105,* 376–379.

Gallup, G. G., Jr., Wallnau, L. B., & Suarez, S. D. (1980). Failure to find self-recognition in mother–infant and infant–infant rhesus monkeys pairs. *Folia Primatologica, 33,* 210–219.

Hall, K. R. L. (1962). Behaviour of monkeys towards mirror images. *Nature, 196,* 1258–1261.

Hamrick, C., Cogan, D., & Woolam, D. (1971). Social facilitation effects on runway and maze behavior in mice. *Psychonomic Science, 25,* 171–173.

Hogan, J. A., & Roper, T. J. (1978). A comparison of the properties of different reinforcers. In J. S. Rosenblatt, R. A. Hinde, C. Beer, & M.-C. Busnell (Eds.), *Advances in the study of behavior* (Vol. 8, pp. 155–255). New York: Academic Press.

Holder, J. L., Barlow, G. W., & Francis, R. C. (1991). Differences in aggressiveness in the midas cichlid fish (*Cichlasoma citrinellum*) in relation to sex, reproductive state and the individual. *Ethology, 88,* 297–306.

Humphrey, N. (1984). *Consciousness regained: Chapters in the development of mind.* Oxford University Press.

Itakura, S. (1987a). Use of a mirror to direct their responses in Japanese monkeys (*Macaca fuscata fuscata*). *Primates, 28,* 343–352.

(1987b). Mirror guided behavior in Japanese monkeys (*Macaca fuscata fuscata*). *Primates, 28,* 149–161.

Kassel, J., & Davis, R. E. (1975). Early behavioral experience and adult social behavior in the paradise fish, *Macropodus opercularis. Behavioral Biology, 15,* 176–184.

Keiper, R. R. (1970). Studies of stereotypy function in the canary (*Serinus canarius*). *Animal Behaviour, 18,* 353–357.

Kummer, H., Dasser, V., & Hoyningen-Huene, P. (1990). Exploring primate social cognition: Some critical remarks. *Behaviour, 112,* 84–98.

Lethmate, J., & Dücker, G. (1973). Untersuchungen zum Selbsterkennen im Spiegel bei Orangutans und einigen anderen Affenarten. *Zeitschrift für Tierpsychologie, 33,* 248–269.

Lewis, M., & Brooks-Gunn, J. (1979). *Social cognition and the acquisition of self.* New York: Plenum.

Lin, A. C., Bard, K. A., & Anderson, J. R. (1992). Development of self-recognition in chimpanzees (*Pan troglodytes*). *Journal of Comparative Psychology, 106,* 120–127.

Loveland, K. A. (1986). Discovering the affordances of a reflecting surface. *Developmental Review, 6,* 1–24.

MacLean, P. D. (1964). Mirror display in the squirrel monkey, *Saimiri sciureus. Science, 146,* 950–952.

Masataka, N., & Fujii, H. (1980). An experimental study on facial expressions and interindividual distance in Japanese macaques. *Primates, 21,* 340–349.

Meliska, C. J., Meliska, J. A., & Peeke, H. V. S. (1980). Threat displays and combat aggression in *Betta splendens* following visual exposure to conspecifics and one-way mirrors. *Behavioral and Neural Biology, 28,* 473–486.

Miles, H. L. (1986). How can I tell a lie? Apes, language, and the problem of deception. In R. W. Mitchell & N. S. Thompson (Eds.), *Deception: Perspectives on human and nonhuman deceit* (pp. 245–256). Albany, NY: SUNY Press.

(1990). The cognitive foundations for reference in a signing orangutan. In S. T. Parker & K. R. Gibson (Eds.), *"Language" and intelligence in monkeys and apes: Comparative developmental perspectives* (pp. 511–539). Cambridge University Press.

Mitchell, R. W. (1986). A framework for discussing deception. In R. W. Mitchell & N. S. Thompson (Eds.), *Deception: Perspectives on human and nonhuman deceit* (pp. 3–40). Albany, NY: SUNY Press.

(1987). A comparative-development approach to imitation. In P. P. G. Bateson & P. H. Klopfer (Eds.), *Perpectives in ethology* (Vol. 7, pp. 183–215). New York: Plenum.

(1991). Deception and hiding in captive lowland gorillas (*Gorilla gorilla gorilla*). *Primates, 32,* 523–527.

(1993). Mental models of self-recognition: Two theories. *New Ideas in Psychology, 11*, 295–325.

Mitchell, R. W., & Anderson, J. R. (1993). Discriminative learning of scratching, but failure to obtain imitation and self-recognition in a long-tailed macaque. *Primates, 34*, 301–309.

Montevecchi, W. A., Gallup, G. G., Jr., & Dunlap, W. P. (1973). The peep vocalization in group-reared chicks (*Gallus domesticus*): Its relation to fear. *Animal Behaviour, 21*, 116–123.

Montevecchi, W. A., & Noel, P. E. (1978). Temporal effects of mirror-image stimulation on pecking and peeping in isolate, pair- and group-reared domestic chicks. *Behavioral Biology, 23*, 531–535.

Parker, S. T. (1991). A developmental approach to the origins of self-recognition in great apes. *Human Evolution, 6*, 435–449.

Parrott, R. F., Houpt, K. A., & Misson, B. H. (1988). Modification of the responses of sheep to isolation stress by the use of mirror panels. *Applied Animal Behaviour Science, 19*, 331–338.

Patterson, F. (1991). Self-awareness in the gorilla Koko. *Gorilla, 14*(2), 2–5.

Platt, M. M., & Thompson, R. L. (1985). Mirror responses in a Japanese macaque troop (Arashiyama West). *Primates, 26*, 300–314.

Platt, M. M., Thompson, R. L., & Boatwright, S. L. (1991). Monkeys and mirrors: Questions of methodology. In L. M. Fedigan & P. J. Asquith (Eds.), *The monkeys of Arashiyama* (pp. 274–290). Albany, NY: SUNY Press.

Povinelli, D. J., Boysen, S. T., & Nelson, K. E. (1990). Inferences about guessing and knowing by chimpanzees (*Pan troglodytes*). *Journal of Comparative Psychology, 104*, 203–210.

Povinelli, D. J., Parks, K. A., & Novak, M. A. (1991). Do rhesus monkeys (*Macaca mulatta*) attribute states of knowledge and ignorance to others? *Journal of Comparative Psychology, 105*, 318–325.

Premack, D. (1988). "Does the chimpanzee have a theory of mind?" revisited. In R. Byrne & A. Whiten (Eds.), *Machiavellian intelligence* (pp. 160–179). Oxford University Press (Clarendon Press).

Spencer-Booth, Y., & Hinde, R. A. (1969). Tests of behavioural characteristics for rhesus monkeys. *Behaviour, 33*, 179–211.

Straumann, C., & Anderson, J. R. (1991). Mirror-induced social facilitation in stumptailed macaques (*Macaca arctoides*). *American Journal of Primatology, 25*, 125–132.

Suarez, S. D., & Gallup, G. G., Jr. (1981). Self-recognition in chimpanzees and orangutans, but not gorillas. *Journal of Human Evolution, 10*, 157–188.

(1986). Social responding to mirrors in rhesus macaques (*Macaca mulatta*): Effects of changing mirror location. *American Journal of Primatology, 11*, 239–244.

Svendsen, G. E., & Armitage, K. B. (1973). Mirror-image stimulation applied to field behavioral studies. *Ecology, 54*, 623–627.

Swartz, K. B., & Evans, S. (1991). Not all chimpanzees (*Pan troglodytes*) show self-recognition. *Primates, 32*, 483–496.

Thompson, T. I. (1963). Visual reinforcement in Siamese fighting fish. *Science, 141*, 55–57.

(1964). Visual reinforcement in fighting cocks. *Journal of the Experimental Analysis of Behavior, 7*, 45–49.

Tolman, C. W. (1965). Emotional behaviour and social facilitation of feeding in domestic chicks. *Animal Behaviour, 13*, 493–496.

Vallortigara, G., & Zanforlin, M. (1986). A newborn chick's companion. *Monitore Zoologica Italia, 20*, 63–73.

Visalberghi, E., & Fragaszy, D. M. (1990). Do monkeys ape? In S. Parker & K. Gibson (Eds.), *"Language" and intelligence in monkeys and apes: Comparative developmental perspectives* (pp. 247–273). Cambridge University Press.

de Waal, F. B. M. (1986). Deception in the natural communication of chimpanzees. In R. W. Mitchell & N. S. Thompson (Eds.), *Deception: Perspectives on human and nonhuman deceit* (pp. 221–244). Albany, NY: SUNY Press.

Whiten, A. (Ed.) (1991). *Natural theories of mind: Evolution, development, and simulation of everyday mindreading*. Oxford: Blackwell Publisher.

Whiten A., & Byrne, R. W. (1991). The emergence of metarepresentation in human ontogeny and primate phylogeny. In A. Whiten (Ed.), *Natural theories of mind: Evolution, development and simulation of everyday mindreading* (pp. 267–281). Oxford: Blackwell Publisher.

Zajonc, R. B. (1965). Social facilitation. *Science, 149*, 269–274.

22   The question of mirror-mediated self-recognition in apes and monkeys: Some new results and reservations

Robert L. Thompson and Susan L. Boatright-Horowitz

It is the peculiar power of mirrors to show you
what is not otherwise there.
           – E. L. Doctorow, *Billy Bathgate*

This paper reports further observations, experiments, and methodological considerations concerning mirror-mediated self-recognition and mirror-correlated behavior in pigtailed monkeys (*Macaca nemestrina*) and chimpanzees (*Pan troglodytes*). We also describe from videotape an instance of apparent self-recognition by a monkey. Last, we discuss behavioral processes we think applicable to the "passing" of the Gallup mark test and to the larger issues of self-awareness and a self-concept.

The empirical facts of mirror-mediated/mark-directed (MM/MD) responding in some chimpanzees and orangutans, but not in gorillas[1] nor in the dozen or more monkey species tested by the well-known Gallup (1970) procedure (hereafter, the "mark test"), have been widely discussed (e.g., Anderson, 1984; Gallup, 1987). A positive mark test is said to imply – even to operationally define – self-recognition, self-awareness, a self-concept, and, ipso facto, consciousness. From there, questions of proximate causes, function, development, and evolution are pursued; species discontinuities are postulated; comparisons with human cognition are made, and searches are prompted for indications of empathy, deception, attributions of mental states, and more. If sentience or self-awareness is attributed to any nonhuman animals, ethical and legal questions of human obligations toward animals are made even more urgent. In short, the importance of this range of issues should find us disquieted that experimental investigations of self-recognition in primates have relied so strongly on tests with mirrors (see discussion by Cheney & Seyfarth, 1990; commentary by Povinelli & deBlois, 1992). For some of our suggestions for evaluating mirror skills and identifying criteria for self-recognition or self-awareness see Platt, Thompson, and Boatright (1991). Opposing views on the question of self-conception can be sampled in Gallup (1987, 1991) and in Epstein and Koerner (1986).

Our long-range aim has been to try to discover what is necessary and

330

sufficient for chimpanzees to "pass" the mark test and be credited with self-recognition. Next, given that monkeys seem never to pass the mark test under the conditions that some chimpanzees do, we want to determine what monkeys actually do and can achieve in the realm of mirror-mediated or -correlated behavior. Finally, we want to find what it takes to bring monkeys up to passing the mark test. To these ends, we and our students have looked critically at the methodology and interpretation of mirror studies, and have experimented formally and informally with many properties and parameters of experience with mirrors and video images. Our perspective is from behavior analysis, conditioning theory, primate ethology, and general experimental psychology. For the paper at hand, we will present our views of selected problem areas under headings dealing with

1. the replicability of mirror self-recognition tests within and between chimpanzee subjects;
2. whether or not monkeys have demonstrated competent use of mirrors;
3. whether or not monkeys are unable to demonstrate mirror self-recognition; and
4. concluding thoughts on analysis and interpretation of the self-recognition problem.

Reliability and generality of mark tests with chimpanzees

Is it true that chimpanzees reliably demonstrate self-recognition as defined by MM/MD responses in the mark test? To date, our group has mark tested and videotaped 12 chimpanzees at the Laboratory for Experimental Medicine and Surgery in Primates (LEMSIP) in Tuxedo, New York (Boatright-Horowitz, 1992; Calhoun, 1983; Calhoun & Thompson, 1988; Thompson & Weisbard, unpublished data). For 10 subjects nothing went awry in the video recordings and mark tests. We have seen the classic Gallup phenomenon only once, with a red dye, rhodamine-B, odorless and without tactile qualities, applied under anesthesia; the subject is identified as chimpanzee James, age about 3 years. We also accept as positive four other cases (Barash, Mona, Amy, and Reggie) but with variants on the Gallup (1970) procedure, principally the use of white typewriter correction fluid diluted with 70% ethyl alcohol as the target mark. The typewriter correction fluid itself was odorless and contrasted better with the skin color of most of our chimps. When applied under anesthesia we presume that its tactile properties would habituate well before test time. When we applied it to our own foreheads, we soon became unaware of it, a condition that remained for hours afterward. When touched by the awake animal it provides some tactile feedback (as might a scab from a wound) and results in more persistent and possibly "interested" activity. In several cases we have had the impression that chimps were uninterested in the red rhodamine marks.

Chimpanzee Barash was not positive until a third test applied without anesthesia[2] and after some six weeks of enriched social experience. Chimpanzee

Mona showed only one mark-directed response, and that was some 18 sec after turning away from the mirror. (Mona was judged negative by Calhoun [1983]. Whether mirror orientation, appearing to look at the mark[s], and mark-directed responses must always occur concurrently or in a tight sequence to meet criterion is generally not specified in the literature.) Amy and Reggie emitted mark-directed responses in the premirror control period, but not in excess of later MM/MD responses. Retesting a positive responder within days of the first test seems to yield diminishing returns; the animals rapidly lose interest in the mark. On the other hand, after about a year had elapsed since mirror exposure and the last mark test, James and Barash both demonstrated clear MM/MD responses (Calhoun & Thompson, 1988).

Mark test failures

Of those who failed the mark test, Maho, age 13 months, may be pardoned as immature; Spike, age about 4 years, was regarded by his caretakers as somewhat "dull"; Rocky, age about 5 years, occasionally exhibited repetitive activity such as body flips and spinning in place. (While such repetitive activity is sometimes taken as a symptom of a cognitively handicapped ape, it may have normal determinants in exercise or play, and might also be a product of unintended reinforcement from the attention of entertained caretakers.) Chatter and Melissa, each about 4 years old, appeared normal. Although aspects of the LEMSIP cage environment at the time this work was done could be considered somewhat stultifying to nonhuman primate cognition, all of the animals were in good health, alert, and active. Some were wild born, others laboratory born; some raised to weaning by the natural mother, some raised by human caretakers. All were individually caged; none were raised in isolation, but the amount of critical social experience undoubtedly varied and could not be usefully characterized by us. All showed at least some self-directed activity while oriented to the mirror.

Swartz and Evans (1991, *SAAH*11) reported no MM/MD responses in 10 of 11 chimpanzees they tested in modified versions of the mark test. Gallup accounted for his negative cases as a result of inadequate social experience with conspecifics and thus an inadequate concept of self (Gallup, McClure, Hill, & Bundy, 1971). But even among Gallup's (1970) positive responders individual differences were suggested, although group data were emphasized. (The rules for counting the number of discrete MM/MD responses are vague. If, following eye contact with the mirror, discrete wipes or touches to the mark are made in quick succession without lowering the arm, are these to be counted as one response, several responses, or a bout?) We must agree with Swartz and Evans that MM/MD responding is not as robust a finding as has been suggested, and strongly recommend that individual rather than group data be reported hereafter.

Gallup wrote, "I do not think their [chimpanzees'] sense of identity or self-concept in any way emerges out of experience with a mirror. A mirror simply

represents a means of mapping what the chimpanzee already knows..."
(1977, p. 335). We do not doubt that some chimpanzees have the special stuff
– the requisite genes and experience – to pass the mark test unaided. Nevertheless, do chimps who don't pass, like all the monkeys that don't pass, "lack
a cognitive category for processing mirrored information about themselves"
(Gallup, 1979, p. 420) and therefore have no self-sense?

Of course there is much more to say here. The mirror situation has some
features of an observational learning task with the subject–reflection interactions (contingencies) as the model; but observational learning itself requires
further analysis, and there are, as well, other approaches from a learning
perspective (see later in this chapter). If we ask the basic questions of learning tasks: What is learned, what is necessary and sufficient for the learning to
occur, and how the learning is made evident in performance, we have very
little to say. If the mirror "simply represents a means of mapping what the
chimpanzee already knows," it is not easy to describe in behavior-analytic
terms how to establish the act of mapping (but see later in this chapter, and
see Mitchell, 1993, for his approach). For now, we would urge investigators
to look carefully for the features or experiences that enable rapid learning of
mirror contingencies through exposure to mirrors alone, without additional
interventions. One such may have to do with the quality and quantity of
attention to the content of the mirror.

Mirror eye contact and the mark test

The frequency and duration of eye contact with her reflection were tabulated
for the first 10 min of several hour-long videotaping sessions for one chimpanzee (Amy) with a positive mark test (Boatright-Horowitz, 1992). Eye
contact was defined by a video camera behind a one-way mirror (0.74×0.74
m) mounted in a blind 1 m from the cage front. Figure 22.1 displays the data
for Sessions 1, 5, 12, 13 (first mark test) and 22 (second mark test). The total
number of eye contacts during each 10-min period on those days was 55, 10,
32, 25, 58, and 9, respectively. Median duration of eye contacts in seconds is
seen in the upper panel, and total duration of eye contacts in minutes is seen
in the lower. After the first (novel) session, in which eye contacts with a
median duration of 5 sec occupied 60% of the recording time, the total duration of eye contacts fell to less than 10% of the 10-min data recording
period, and the median duration was 1 sec, that is, a glance rather than a
stare. During the first positive mark test (Day 13), eye contact duration cu-
mulated to 40% of the session with a median duration of 3 sec. At the second
mark test (Day 22 of continuous exposure to the mirror), which was also
positive but with fewer MM/MD responses than the first, the total duration
of eye contacts was 0.35 min (3.5%) with a median duration of 2 sec. A third
mark test (not shown) on Mirror Day 55 was negative, but eye contacts
occupied 17% of the recording period.

The right-hand side of Figure 22.1 shows frequency distributions for eye-

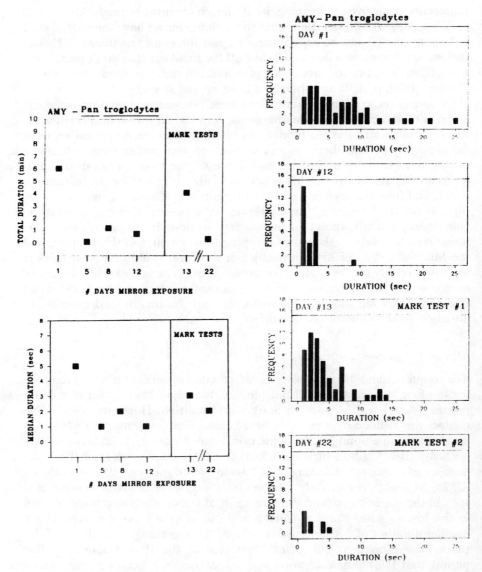

Figure 22.1. Temporal characteristics of mirror viewing (camera-defined eye-to-eye contacts) in chimpanzee Amy during the first 10 min of each of selected days of the Gallup mark test procedure. The first mark test occurred on Day 13, the second on Day 22. The upper left panel shows total duration (in minutes) of eye contacts; the lower panel shows median duration in seconds. The right panels show histograms of eye contact duration on selected days. Sessions were videotaped from a camera centered behind a one-way mirror mounted in a darkened blind.

contact durations for Mirror Days 1, 12, 13, and 22. It is clear that the number and duration of eye contacts with the mirror decreased over days, increased on the first mark test, then decreased again, remaining low on subsequent mark tests. In our experience, interest in marks on the brow and ears, even those having some tactile quality, diminished rapidly.

We are still some distance from uncovering the steps to passing the mark test. We have not yet analyzed eye-contact data from our other chimpanzees. We do not know where Amy's data stand in relation to the others, but for comparison with the monkey data later in this paper, note that the longest duration of eye contact recorded here was 25 sec on Day 1; after Day 1, most eye contacts lasted less than 5 sec, with a mode of 1 to 2 sec.

Monkeys and mirror competence

Is it true that monkeys, who as a taxonomic group fail mark tests, have otherwise been able to acquire competence in mirror use, that is, to find hidden food or to look at other monkeys? Results from various researchers have been mixed. We do not intend to provide a review here, but see, for example, Anderson (1984), Boatright-Horowitz (1992), or Gallup (1991). The paper by Brown, McDowell, and Robinson (1965) is often cited in support of mirror competence; and Anderson (1986) succeeded in obtaining mirror-mediated finding of hidden food in two of the nine macaques studied. Boatright-Horowitz (1992) described a pigtailed macaque trained to locate hidden food from mirrored cues. When the mirror was moved a few centimeters back from its training position, the animal's performance deteriorated. We have argued elsewhere (Platt et al., 1991) that the results above may be minimally relevant to issues of competence with mirrors because the tasks consist of simple visual discriminations with cues displaced from the response locus. The mirror could as well be replaced by a cue light. Anderson (1986) entertained a similar conclusion but apparently rejected it as insufficient because one animal was judged to exhibit mirror-guided reaching (see next paragraph).

Of what does mirror competence consist? The literature contains many examples of *mirror-* as a reflexive prefix, as in mirror-controlled, -using, -mediated, -directed, -correlated, -oriented, -guided, and -indicated. We have used *mirror-mediated* or *mirror-correlated* as general terms; *mirror-indicated* for the conditional discrimination task in which the mirror provides a cue to the location of another stimulus but does not participate in providing moment-to-moment feedback in reaching for, approaching, or otherwise responding at the site of the indicated stimulus; and *mirror-guided* when the subject's movements as viewed more or less continuously in the mirror determine the accuracy, effectiveness, or success of the response. Mirror-guided responding can also be called *tracking*. One of Anderson's (1986) subjects appeared to demonstrate mirror-guided reaching, although less complex mirror-indicated reaching was not ruled out. Itakura's experiments (1987a,b) provide one of the best examples of mirror-indicated reaching. Whether a case can be made for mirror-guided reaching in Itakura's subjects can be contested. Menzel, Savage-Rumbaugh, and Lawson (1985) described tasks that provided for both mirror-indicated and mirror-guided reaching in which rhesus monkeys were unsuccessful. Some competence in mirror-guided reaching may be essential to passing the mark test.

Platt et al. (1991) reasoned that valid competence in mirror use requires

1. a generalized concept of mirrors as bounded reflecting surfaces providing contingencies of appearance and movement;
2. mirror-mediated locating and reaching; and
3. instrumental behavior maintained by bringing the mirror or the self into appropriate positions for (2).

Mirror competence means recognizing that whatever its shape, size, and frame, or wherever it is positioned, it is nevertheless a mirror; that it is a tool to be used for locating objects and guiding movements as well as a source of visual stimulation under the control of the viewer, with consequences that may be reinforcing, neutral, or aversive. The disappearance of social (other-directed) responses through extinction or habituation is a minimal index of competence and a prerequisite for generalized use of the mirror as a tool. (On the other hand, some social responding to one's reflection need not indicate failure to recognize the source of the reflection. Do we not smile or frown at our reflections now and then?)

Unless something approaching general mirror competence can be demonstrated, interpretation of failure on the mark test is far from clear. "If the mirror cannot be validated as a tool for investigating self-recognition in a particular animal, it does not mean that self-recognition with a different operational definition cannot be demonstrated via the visual or some other modality" (Platt et al., 1991, p. 281). For a Gibsonian perspective on acquiring mirror competence see Loveland (1986).

Social responses or expressive behavior?

We found, as did Suarez and Gallup (1986), that moving a mirror to which monkeys had habituated to a new location within the animal's "personal space" initiated a transient increase in social responding, but not in every instance. Further, arranging for a nonsocial object, a red balloon, to appear first in a mirror, evoked jaw thrusting with lip protrusion and scalp retraction to both the reflection and the real object. The subject, a mature pigtailed macaque, oriented alternately to the mirror and to the real balloon (Boatright-Horowitz, 1992). Although jaw thrusts are often directed at conspecifics and, in many cases, may be called "social," they are not confined to social contexts. The balloon demonstration was replicated with a banana. Novel objects, and even novel contexts with familiar objects, may trigger facial displays.

Do solitary monkeys play? Little data on the baseline rates of specific facial displays in the solitary animal are known to us, nor are the criteria for identification of expressions of lesser amplitude like scalp and ear retractions. We view many primate facial displays as expressive as much as communicative, although these are not mutually exclusive. Studies of mirror relocation as an independent variable would benefit from a control condition with a nonsocial source of continuously available stimulation. For example, a video monitor to which a monkey is thoroughly habituated, playing continuous programs or just video noise ("snow," e.g., Calhoun, 1983), if moved to a new location within the animal's space may evoke jaw thrusts, lip smacking, brow retractions, ear retractions, and open-mouth "threats" when first perceived, or if the animal is witness to the commotion of making such a change. (Young children, though old enough to pass the mark test, may cry or display anxiety or anger when major pieces of furniture are rearranged in their room.) However, Suarez and Gallup (1991) reported no such effects when their mirror was turned away from the cage or when furnishings were moved about in the room. We are again left ignorant of the variables responsible for these differences. Some may be in the monkeys' competence with mirrors; some may be in the subjects' species or social group structure. The rhesus monkeys studied by Suarez and Gallup (1986, 1991) were a male–female pair, housed together. In general, the male was more reactive and emitted principally threat responses; the female emitted mainly appeasement gestures. Our pigtailed macaques were all females, housed individually.

It seems to us that too much emphasis is put on the conjecture that monkeys continue to see their reflection as that of another monkey. This suggestion has intuitive support in that the mirror reflection shares visual attributes with other real monkeys. Nevertheless, monkeys may well detect the several attributes of a mirror reflection that differ from a real conspecific or from a conspecific viewed through a window. In our laboratory a pigtailed monkey thoroughly adapted to laboratory procedures, and the presence of mirrors was systematically observed during alternating sessions in the presence of a mirror or another pigtailed monkey behind a window. Distance of image as well as luminance factors were matched as best we could between the two conditions. During the 6 days of this study, the test animal spent more time oriented to the stimulus monkey than to her own mirror image (Figure 22.2), suggesting discrimination between the two conditions. Similar results were obtained by Gallup and McClure (1971) in a differently designed experiment. Of course, the discrimination was not based on any concept of virtual image versus real live monkey. The stimulus monkey was relatively novel or more interesting to watch and to interact with.

Platt and Thompson (1985) reported low levels of social responding to mirrors by Japanese macaques housed in a group in an outdoor compound. Conspecifics, as well as the self, were usually visible in the mirror. The authors' impression was that the monkeys responded to their reflection neither as their own nor as if it were another monkey. They suggested that it may not be useful to impose a self-versus-other dichotomy on all descriptions of the animal's behavior toward its reflection.

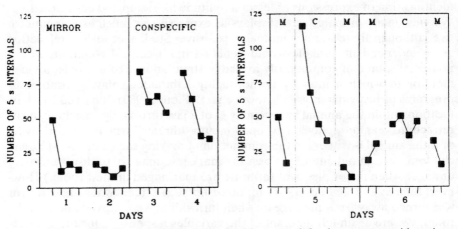

Figure 22.2. Number of 5-sec intervals with camera-defined eye contacts with a mirror (M) or conspecific (C). The subject was Roberta, a pigtailed macaque. The *x*-axis shows 15-min periods within daily sessions of 1 hour (Days 1–4) or 2 hours (Days 5–6).

Monkeys, mirrors, and mark tests

Can it be held that "monkeys lack a cognitive category for processing mirrored information about themselves" (Gallup, 1979)? Gallup's statement is but one of several speculative accounts of why monkeys have regularly failed the mark test. Among other accounts of what it is that interferes with learning the mirror contingencies we find gaze aversion (Premack, 1983), persistent social responding, or other activity interpreted as unconditioned responses to the monkey in the mirror (Epstein & Koerner, 1986), and a problem stemming from "Species or individual differences ... in the organization of visual spatial perception and ability to adapt to spatially displaced images" (Menzel, Savage-Rumbaugh, & Lawson, 1985). (Recent evidence indicates that monkeys are capable of using spatially displaced visual feedback to guide their actions. Rumbaugh, Richardson, Washburn, Savage-Rumbaugh, & Hopkins [1989] trained two rhesus monkeys to manipulate a joystick to respond in computer-generated tasks involving contact, chase, and pursuit of a target. The monkeys succeeded even though the joystick was located 9–18 cm from the video screen. See Gallup [1991] for a different argument in rebuttal to Menzel et al.)

An associative model

Our inclination has been to look for a problem in the domain of learning or performance, although this in no way excludes constraints on learning from perceptual, cognitive, or representational capacities. The relevant properties of a mirror must be learned. Interfering responses must be eliminated and sufficient attention paid to the somasthetic stimulation arising from the postures and movements of the head, face, and body, concurrently with the visual stimulation from the mirror, to enable an association between the somasthetic feedback from different parts of the body and the corresponding visual feedback (cf. Mitchell, 1993). The feel and the sight (except for the head parts) of what your own body is doing are paired with the sight of what you are doing in the mirror. This paired stimulation is a critical aspect of the contingencies of appearance and movement, and possibly the missing perceptual integration considered by Menzel et al. (1985). The associative requirements resemble a multiple intermodal sensory-preconditioning procedure, but one that has no precedent in laboratory studies of sensory preconditioning. It may also be likened to a conditional-discrimination procedure: If you feel your leg doing something, you will see a leg doing something if you look at a (your) leg in the mirror; but if you feel your leg doing something, you will not see a leg doing something if you look at your arm in the mirror. Stimuli arising from the head (and face) may be the most difficult to attend to (cf. Parker, 1991). Vision probably overshadows somasthesis in associating other body parts with their reflections, and cross-modal conditional-discrimination training is known to be difficult for monkeys. If associating somasthesis from the head, and visual stimulation from the mirror, is as difficult as we suppose, some intervention would help to give salience to the stimuli.

Given the very large number of stimulus–stimulus associations required here, some overriding process should come into play wherein each source of stimulation is categorized or abstracted (e.g., experiencing "own-body" *arm* movements and seeing *arm* movements in mirror), and a reflexive (not reflex) relation is established: an "if this, then that" relation in which the order of somatic and visual stimuli is interchangeable. Tactile and movement sensations from the body must be discriminated and classified as own-body sensations that yield visual sensations at corresponding points on the reflection. Similarly, visual indication of appearance or movement in the mirror should correlate with sensations from one's own body (see also Mitchell, 1993; Watson, *SAAH*8).

Once the intermodal associations between somasthesis from the head, and visual stimulation from the mirror, are strengthened (with or without intervention), some indicator response, some means to reveal or report the associative learning has also to be considered.[3] This is the role of the MM/MD response; but without some motivation to touch, groom, or otherwise contact an unfamiliar mark or a mark in an unfamiliar place on one's body, why should the response occur? Why must we leave this to chance, uncontrolled? However, care must be taken that the general criterion-response topography – touching the brow and ears – is not made so probable that it will mask any later mark-directed touching. We need be reminded that the self-recognition criterion is more accurately stated as a probability of MM/MD responses significantly greater than the probability of mark-directed responses in a control period in which the mirror is absent.

In either of the two processes – that having to do with establishing the intermodal associations, and that with identifying and motivating an appropriate reporting response – individuals may differ widely for a number of reasons. Not the least of these is that the training contingencies are essentially uncontrolled and very difficult to isolate and specify. Further, species differences, individual genetic differences, and behavioral histories alone or in combination, acting to determine degrees of associative "preparedness," can be expected to be a source of substantial variability. With the foregoing as the barest outline of an associative model of what it takes to pass the mark test, how can we proceed to shed some light on what monkeys do and can be made to do in preparation for the test?

Indirect induction of attention to the mirror

Without intervention, and under stable baseline conditions, how does the pattern of a monkey's eye contacts with a mirror compare with its eye contacts with a window of like size in which there is nothing to see but a video camera and the plywood walls of a blind? And how does the monkey's pattern of eye contacts with a mirror compare with that of a chimpanzee? If learning the mirror's contingencies of appearance and movement is essential to self-recognition, can attention to the mirror be increased by indirect or direct behavioral interventions? We tried indirect methods first, asking, If a monkey is reinforced for other behavior in the presence of a mirror, will it be motivated to come to terms with the monkey in the mirror?

When biologically important events (reinforcers or Pavlovian unconditioned stimuli) are scheduled at regular intervals of the order of seconds or minutes, certain of the available response classes tend to fall into orderly positions in the interreinforcement intervals, and newly induced or conditioned forms may appear as well, sometimes in considerable strength (see Staddon, 1977, for a review of schedule-induced behavior). Cohen and Looney (1973) described schedule-induced mirror responding in pigeons on fixed ratio schedules of food reinforcement. Hudson and Singer (1979) described the schedule induction of polydipsia in two of three crab-eating macaques given the opportunity to view a companion monkey for 10 sec once every minute (a fixed time [FT] schedule, designated *FT-1 min*). There is no response requirement in FT schedules. Hudson and Singer reported that visual/social deprivation was essential for the 10-sec views to be effective as reinforcers, and that the identity of the stimulus monkey was also important.

We asked if something like the reverse conditions would work: Would visually restricted, water-deprived (for 22 hours) monkeys receiving small drinks (1.2 ml of water) each minute show increased viewing of a mirror in the interreinforcement intervals? A window, through which was seen the video camera and the bare plywood walls of the camera blind, alternated with the mirror in more or less counterbalanced order within and between sessions. Between sessions, the window or mirror was continuously present, depending on the order of conditions in the next session. Our experiment had

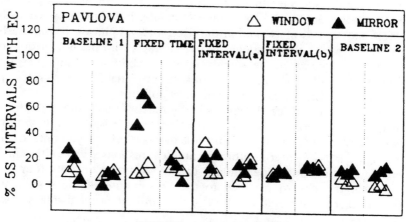

FIRST & LAST SESSIONS

Figure 22.3. Percent of 5-sec intervals with camera-defined eye contacts (EC) with the mirror reflection or with a camera behind a clear window. Data are for the first 10 min of each condition in each session. Data points represent means of blocks of three sessions, the first three blocks and last three blocks in each of the five phases of the experiment. Only during the first three blocks of the fixed time–1-min phase was there a clear difference between the window and mirror conditions. The subject was Pavlova, a pigtailed monkey. See text for details of the experiment.

five phases, in which the first and last were baseline conditions. The second was an FT-1 min of water availability. No work was required of the subjects. The third and fourth phases were fixed-interval 1-min schedules differing in a minor detail: Both required lever-pressing to provide the opportunity to drink. Each session was 100 min long. Within sessions, a 10-min adaptation period followed each change of conditions. Each phase was run until a stability criterion was met. Sessions were videotaped from the mirror/window blind, and one–zero sampling at 5-sec intervals was used to later record camera-defined eye contacts, facial displays, and self-directed touches to the monkey's body or head. Interobserver reliability measures were, with one exception, greater than 90. For complete details of the experiment see Boatright-Horowitz (1992).

Schedule-induced eye contact

The subjects were three pigtailed macaques: Pavlova, Bibi, and Roberta, ages 9, 18, and 9 years, respectively. All had more than 156 hours of exposure to mirrors and windows in various laboratory procedures. Over the five phases, Pavlova participated in a total of 133 sessions, Bibi in 143, and Roberta in 104.

Figure 22.3 shows the frequency of eye contacts (ECs) for one monkey, Pavlova, who was later positive on a mark test. Pavlova's results were intermediate between the other two subjects. Eye contacts with the mirror increased early under the fixed time schedule, but then dropped to baseline levels later in this phase and in the fixed interval phases. Except early in the fixed time phase, the frequency of eye contacts with the mirror did not differ from the frequency in the window (control) condition. Moreover, there was no indication of social facilitation of lever pressing. Response rates appeared the same in both window and mirror conditions.

The frequency of facial displays in the mirror condition ranged from 1–4% of the 5-sec sampling intervals and did not differ significantly within or between phases. Facial displays did not increase when eye contacts increased during periods of water reinforcement. Most responses were either jaw thrusts or brow retractions. Open-mouth threats occurred on only four occasions, of which three were confined to the first minute of the session in which they occurred. Subjects rarely, if ever, emitted facial displays toward the window.

Head and body touches, including autogrooming, were even less frequent, occurring in 0.5–1.5% of the sampling intervals. For each monkey we looked at the number of intervals in which eye contact and self-touching occurred together, expressed as a joint probability, and compared this with a statistically independent model based on the product of the absolute probabilities of eye contacts and touches. Although the frequency of each of the events of interest was small, only Pavlova showed a difference between the mirror and window conditions; self-directed touching accompanied eye contact with the mirror but not with the window (i.e., not with eye contact with the camera behind the window).

More associative learning

Did Pavlova learn something new about the monkey in the mirror (call her MIM)? Schedules of water reinforcement produced transient increases in the frequency of mirror looks, perhaps of the same sort as the brief and frequent looks toward a conspecific, referred to as "checking" by Kaufman and Rosenblum (1966). Under the arousal attending water deprivation and the increased "value" of sips of water, checking on MIM might well result in a revised evaluation, even some identification with MIM. Put more conservatively, the stimulus properties of MIM may be changed in the context of reinforcement. MIM drinks when you drink, so MIM may become a conditioned reinforcer or perhaps an "occasion setter" (Catania, 1992). If you feared that MIM might compete for your opportunities to drink, you nevertheless drink unchallenged, and so you find MIM harmless, not aversive; your fear may diminish or extinguish as you see MIM drinking only as you drink, thereby providing another contingency for conditioned reinforcement or at least neutralization of the stimulus MIM.

Why then doesn't mirror looking increase throughout the reinforcement conditions? Perhaps because MIM is absent in the window condition and you drink anyway. MIM is no longer correlated with reinforcement; MIM has been made irrelevant in this context and need not be attended to. The stimulus functions of MIM have changed and thus may become susceptible to change in still other ways. (Technically, it is not the stimulus that has changed, it is the organism, Pavlova.) It is also conceivable that such contingencies might potentiate the value, or otherwise modulate the stimulus functions, of MIM in a latent manner, to be activated in some other context. At least, we may suppose that some salience has been added to the stimuli arising within Pavlova and emanating from the mirror.

Differential reinforcement of gaze duration

The next stage of our work was direct intervention. Two monkeys, Pavlova and Roberta, participated. Duration of eye contact with the mirror (or camera in the window condition) was differentially reinforced (shaped) to sustain gazes of the order of 20 sec. This is a difficult task even for humans. We found it necessary to increase water deprivation to 45–46 hours, use up to 2.0 ml water per reinforcement, and run sessions every other day. Duration requirements had to be increased very gradually, and often lowered to maintain responding. Satiation within a session was often a problem. Following a session, the animal had ad-lib access to water and fruit for at least one hour. The subjects were monitored closely for any signs of dehydration, excessive weight loss, or other harmful effects.

The results for Pavlova are seen in Figure 22.4. Sessions 50 to 54 were criterion sessions in which only responses having a duration of at least 20 sec were reinforced. A histogram of responses in the mirror, and window conditions during the first 10 min of Session 54, appears in the lower panel. Sessions 55–57 were extinction sessions in which reinforcement was withheld for all responses (eye contacts). Long durations were reconditioned in Sessions 58 and 59. Clearly, gazes about as long as any recorded for chimpanzee Amy were realized. Further, it is clear that gaze duration can be controlled in operant fashion, by its consequences. The results for Roberta were much the same. Again, a full account of these experiments is to be found in Boatright-Horowitz (1992).

Apparently in order to sustain long gazes, both monkeys displayed swooping head movements. Based on our own subjective experience, these movements help contend with the aversiveness of sustained gaze. At the same time, they afford opportunities for the monkey to experience the movement contingency of the mirror.

A positive mark test

Mark tests were the next step. Under anesthesia, rhodamine-B in 70% ethyl alcohol was applied to each animal's left brow and right ear. The alcohol alone was applied to the otherwise unmarked brow and ear. Recovery from

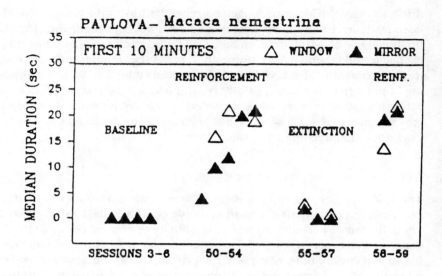

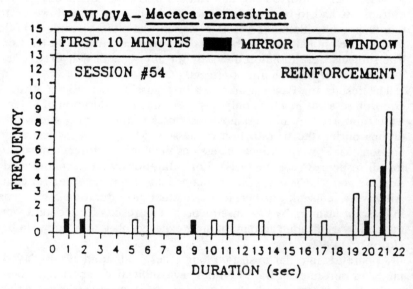

Figure 22.4. Differential reinforcement of eye contact with a mirror or window. After the baseline data, the last five sessions (50–54) shown had a criterion duration of at least 20 sec required for reinforcement. By Sessions 53 and 54 the median duration of gaze had reached 20 sec. Duration dropped rapidly in extinction sessions (55–57) and reconditioned rapidly to 20 sec in Sessions 58–59. The lower panel is a histogram of gaze durations in Session 54. Only the first 10 min in each condition within each session is shown.

anesthesia took place in the course of several hours in the window condition. The animal was monitored continuously. Videotaping began at the first sign of coordinated autogrooming and continued for at least 10 min before the mirror was introduced. Neither animal made any mark-directed responses during the window (control) condition. Videotaping in the mirror condition continued for one hour.[4]

Pavlova made one MM/MD response about 16 min after the mirror was made available. Except for one glance toward the mirror at about 10 min, she ignored it. She then sat with her back to the mirror for 6 min, looked up and stared at her reflection, jaw-thrusted toward the mirror, touched her left knuckle(s) to the small red mark on her brow, then in a continuous movement, rubbed the back of her left hand against the marked brow, inspected the back of her left hand, jaw-thrusted again, and turned rapidly away from the mirror. The sequence from stare to turning away lasted 3–4 sec. No other mark-directed responses occurred.

Shortly after the mirror was introduced, Roberta glanced briefly toward the mirror and, as her head was turning away, made a wiping response toward the marked brow. Roberta's posture made it virtually impossible to clearly classify this response as mark-directed.

In work that followed, that two monkeys were reinforced for touching red dots on the arms and head of the experimenter (SB-H) and other personnel, and red dots on the monkey's own arms, legs, and torso, but never on the head or face. A second and third mark test followed for each animal. Results from Pavlova were ambiguous at best; those from Roberta were clearly negative.

Have we demonstrated self-recognition in a monkey? The response topography was there. The single sequence of an MM/MD response followed by inspection of the hand was too well coordinated to be a chance event. The jaw thrusts may be taken as social responses directed at MIM, but we have argued that jaw thrusts may also occur in response to novel stimulation. Replication was disappointing, but so it is in chimpanzees too. Perhaps the mark test is just not made for monkeys. If it is a question of preparedness of the relevant response and stimulus categories, this should guide further attempts to train monkeys for the mark test. We hope that soon someone will train the touching of marks or sites on the head in concert with training mirror skills, just to see what will happen.

Conclusions

So, what more can we say? We are not comfortable with polarized expression of the issues such as "apes can do–monkeys can't do." Negative results should not be the basis of a strong inference, particularly if other, potentially converging operations, have not come near being exhausted. And if the object of the can do–can't do contest is itself a hypothetical construct, the garden path may lead through the looking glass. Anderson (1984), Gallup (e.g., 1987, 1991), and others, ourselves included, have suggested many variants on the mirror mark test that have yet to be tried, as well as other methods that should be pursued vigorously. Whatever problems we present monkeys and apes to solve will share a cognitive interpretation with what has been called (good naturedly), a "demoting operant explanation."

There is quite a bit of methodological and conceptual untidiness in some primate work because the subject are far more precious than rats and pigeons. Nevertheless, this should not cloud our judgment so that we uncritically welcome anecdotal accounts of cognitive accomplishment, nor bias us because our subject is identified as James rather than CH-175.

We have noted here several problem areas. Among those that vex us most is the frequent insistence that the mirror and behavioral interactions in preparation for the mark test take place "spontaneously," that is, without experimenter intervention. This contrasts human learning about mirrors with all the instruction, prompts, molding, modeling, and reinforcement attending it. Why not attempt to teach monkeys a self-concept, or at least train MM/MD responses? We could then experiment to see what they do with them. (This will oblige some of us to try to discover what it is that we want to instill in our subjects.) There is ample precedent in the apes, dolphins, seals, and parrots that have received language training.

We have difficulty finding the right demoting operant analysis of verbs like mapping and integrating, as well as nouns such as consciousness, awareness, and self-concept. We are encouraged by recent developments in the experimental analysis of behavior. The concepts of equivalence classes and emergent behavior provide a parsimonious basis for a better understanding of many kinds of associative relations that apply to "self" (Sidman, 1990; Sidman et al. 1982). In many areas, behavior analysis and cognitive psychology invigorate one another. Rule-governed behavior, contrasting with contingency-shaped behavior, provides a new dimension for comparative studies among all primates (Catania, 1992).

Our experience indicates that some chimps meet the mark-test criteria, and many do not. We see no reason why those chimpanzees, monkeys, and gorillas who do not, could not learn with a little help such as we humans get. It should not matter to the chimp's self-concept whether this is remedial or prosthetic. Further, it matters little if we speak in the language of behavior analysis or use perceptual terminology or cognitive constructs; but it helps to be consistent. The great apes and humans may have more of the right stuff "hard wired," but we don't think it is theirs exclusively.

We have labored for some years on the question of whether or not MM/MD responses necessarily or sufficiently imply self-recognition, and what, in turn, self-recognition implies. Our lingering skepticism, along with difficulty in finding an altogether convincing alternative model, is evident in our staying with the operational designation "MM/MD response" instead of the commonly used terms, "self-recognition" or "mirror self-recognition." The definition and implications of "self" have been widely discussed in the languages of various disciplines. It is recognized as a perfectly bad word in that the level at which it is to be understood is often ambiguous and controversial; but it is one for which we have no adequate substitute. That it is not alone as a bad word in the psychological lexicon is of little comfort.

What is it that is recognized in a self-recognition task? Minimally, a subject discriminates by naming, sorting into categories, or pointing to the source of certain classes of stimuli, and/or a certain repertory of behavior. As examples, consider "I feel X"; "he did X himself, I didn't"; "I did X"; "that's me." This doesn't come close to capturing all of the instances when "self" appears, however. Further, a recognition paradigm seldom deals with *all* of the properties of the stimulus. A red balloon that is recognized as red need not also be recognized as a balloon (and vice versa).

Let us close with a thought experiment devised by one of us (SB-H): Imagine a primate housed with an upright lamp with a lampshade. The items are rigid, unbreakable, and unmovable. They are up against the wall of the living space such that the part of the lampshade against the wall has never been seen directly by the primate. It is, however, visible in a mirror permanently mounted on one wall of the space. Assume further that this remote part of the lampshade has never been inspected by touch. Now, while the primate sleeps, a red mark is added to the lampshade in a place that is noticeable only by way of the mirror. After awakening, the primate sees the red mark in the mirror. Because the mark is new, and because primates are curious, the primate clearly demonstrates mirror-guided responses to touch and feel about the mark. Perhaps our subject even inspects its hand afterward. Shall we say that our primate has a well-integrated sense of a lampshade?

Acknowledgments

This paper is based on a presentation supplemented with videotapes at the Conference on Self-Recognition and Self-Awareness in Monkeys, Apes, and Humans held at Sonoma State University in August 1991. Portions of the research were supported by NIH grant RRO 8176, George H. Shuster Faculty Fellowship Awards to RLT, a City University Fellowship and APA Dissertation Award to SB-H, and the biopsychology program at Hunter College. We thank Drs. Elizabeth Muchmore and James Mahoney, and the staff at the Laboratory for Experimental Medicine and Surgery (LEMSIP); Barbara Wolin, Sonia Acevedo, and Ray Ferranti of the Hunter College Animal Facility; and Hunter College Audio Visual Services for their help. More students, colleagues, and staff than can be named provided help in lab routines, data analysis, and discussions of what it all means. Drs. Suzanne Calhoun and Charles Weisbard collaborated in the work at LEMSIP, as did Guy Harris. Gloria Brown was particularly helpful at Hunter College. Valeda Slade provided essential help with the manuscript. Gordon G. Gallup, Jr., brought this topic to the fore and provided the challenges that have made it exciting. In this chapter we use terms from learning theory and behavior analysis that may not be familiar to all readers. Catania (1992) provides a useful text and glossary.

Notes

1. Patterson and Cohn (*SAAH*17) present a compelling case for self-recognition in a sign-trained gorilla.
2. Experimenters have been inconsistent in the use of anesthesia prior to marking the animal. The dangers of anesthesia must be weighed against the possible cuing inherent in marking the awake animal, however surreptitiously. With or without anesthesia, our procedure has always sham-marked, with solvent alone, the unmarked brow and ear of all of our chimpanzee and monkey subjects.
3. In our analysis, discrimination of and reporting (to oneself or to others) stimuli arising from one's own body is the basis of self-awareness. Nevertheless, a problem remains in establishing the range and complexity of the stimuli, and the number and order of descriptive response classes asked of the subject (e.g., "I report that I have reported"). Not all that qualifies as self-awareness implies all that self-awareness implies.
4. A copy of videotaped segments of various phases of this work is being prepared for the Wisconsin Primate Library.

References

Anderson, J. R. (1984). Monkeys with mirrors: Some questions for primate psychololgy. *International Journal of Primatology*, 5, 81–98.

(1986). Mirror-mediated finding of hidden food by monkeys (*Macaca tonkeana* and *M. fascicularis*). *Journal of Comparative Psychology*, 100, 237–242.

Boatright-Horowitz, S. L. (1992). Mirror behavior and "self-hood" among primates. *Disseration Abstracts International*, 53, 2047B. (University Microfilms No. DA9224796).

Brown, W. L., McDowell, A. A., & Robinson, E. M. (1965). Discrimination learning of mirror cues by rhesus monkeys. *Journal of Genetic Psychology*, 106, 123–128.

Calhoun, S. (1983). The question of contingent image self recognition in apes and monkeys. *Dissertation Abstracts International*, 44, 1621B. (University Microfilms No. DA8319749).

Calhoun, S., & Thompson, R. L. (1988). Long-term retention of self-recognition by chimpanzees. *American Journal of Primatology*, 15, 361–365.

Catania, A. C. (1992). *Learning* (3rd ed.). Englewood Cliffs, NJ: Prentice-Hall.

Cheney, D. L., & Seyfarth, R. M. (1990). *How monkeys see the world*. Chicago: University of Chicago Press.

Cohen, P. S., & Looney, T. A. (1973). Schedule-induced mirror responding in the pigeon. *Journal of the Experimental Analysis of Behavior*, 19, 395–408.

Epstein, R., & Koerner, J. (1986). The self-concept and other daemons. In J. Suls & A. Greenwald (Eds.), *Psychological perspectives on the self* (Vol. 3, pp. 28–53). Hillsdale, NJ: Erlbaum.

Gallup, G. G., Jr. (1970). Chimpanzees. Self-recognition. *Science*, 167, 86–87.

(1977). Self-recognition in primates: A comparative approach to the bidirectional properties of consciousness. *American Psychologist*, 32, 329–338.

(1979). Self-awareness in primates. *American Scientist*, 67, 417–421.

(1987). Self-awareness. In G. Mitchell & J. Erwin (Eds.), *Comparative primate biology, Vol. 2, Part B: Behavior, cognition, and motivation* (pp. 3–16). New York: Liss.

(1991). Toward a comparative psychology of self-awareness: Species limitations and cognitive consequences. In J. Strauss & G. R. Goethals (Eds.), *The self: An interdisciplinary approach* (pp. 122–135). New York: Springer Verlag.

Gallup, G. G., Jr., & McClure, M. K. (1971). Preference for mirror-image stimulation in differentially reared rhesus monkeys. *Journal of Comparative and Physiological Psychology*, 75, 403–407.

Gallup, G. G., Jr., McClure, M. K., Hill, S. D., & Bundy, R. A. (1971). Capacity for self-recognition in differentially reared chimpanzees. *Psychological Record*, 21, 69–74.

Hudson, R., & Singer, G. (1979). Polydipsia in the monkey generated by visual display schedules. *Physiology and Behavior*, 22, 379–381.

Itakura, S. (1987a). Mirror-guided behavior in Japanese monkeys (*Macaca fuscata fuscata*). *Primates, 28,* 149–161.

(1987b). Use of a mirror to direct their responses in Japanese monkeys (*Macaca fuscata fuscata*). *Primates, 28,* 343–352.

Kaufman, I. C., & Rosenblum, L. (1966). A behavioral taxonomy for *Macaca nemestrina* and *Macaca radiata. Primates, 7,* 206–258.

Loveland, K. A. (1986). Discovering the affordances of a reflecting surface. *Developmental Review, 6,* 1–24.

Menzel, E. W., Jr., Savage-Rumbaugh, E. S., & Lawson, J. (1985). Chimpanzee (*Pan troglodytes*) spatial problem solving with the use of mirrors and televised equivalents of mirrors. *Journal of Comparative Psychology, 99,* 211–217.

Mitchell, R. W. (1993). Mental models of mirror self-recognition: Two theories, *New Ideas in Psychology, 11,* 295–325.

Parker, S. T. (1991). A developmental approach to the origins of self-recognition in great apes. *Human Evolution, 6,* 435–449.

Platt, M. M., & Thompson, R. L. (1985). Mirror responses in a Japanese macaque troop (Arashiyama West). *Primates, 26,* 300–314.

Platt, M. M., Thompson, R. L., & Boatright, S. L. (1991). Monkeys and mirrors: Questions of methodology. In L. M. Fedigan & P. J. Asquith (Eds.), *The monkeys of Arashiyama: Thirty-five years of research in Japan and the West* (pp. 274–290). Albany, NY: SUNY Press.

Povinelli, D. J., & deBlois, S. (1992). On (not) attributing mental states to monkeys: First, know thyself. *Behavioral and Brain Sciences, 15,* 164–166.

Premack, D. (1983). Animal cognition. *Annual Review of Psychology, 34,* 351–362.

Rumbaugh, D. M., Richardson, W. K., Washburn, D. A., Savage-Rumbaugh, E. S., & Hopkins, W. D. (1989). Rhesus monkeys (*Macaca mulatta*), video tasks, and implications for stimulus–response spatial contiguity. *Journal of Comparative Psychology, 103,* 32–38.

Sidman, M. (1990). Equivalence relations: Where do they come from? In D. E. Blackman & H. Lejeune (Eds.), *Behaviour analysis in theory and practice: Contributions and controversies* (pp. 93–114). Hillsdale, NJ: Erlbaum.

Sidman, M., Rauzin, R., Lazar, R., Cunningham, S., Tailby, W., & Carrigan, P. (1982). A search for symmetry in the conditional discriminations of rhesus monkeys, baboons, and children. *Journal of the Experimental Analysis of Behavior, 37,* 23–44.

Staddon, J. E. R. (1977). Schedule-induced behavior. In W. K. Honig & J. E. R. Staddon (Eds.), *Handbook of operant behavior* (pp. 125–152). Englewood Cliffs, NJ: Prentice-Hall.

Suarez, S. D., & Gallup, G. G., Jr. (1986). Social responding to mirrors in rhesus macaques (*Macaca mulatta*): Effects of changing mirror location. *American Journal of Primatology, 11,* 239–244.

(1991). Social responding to mirrors in rhesus monkeys (*Macaca mulatta*): Effects of temporary mirror removal. *Journal of Comparative Psychology, 105,* 376–379.

Swartz, K. B., & Evans, S. (1991). Not all chimpanzees (*Pan troglodytes*) show self-recognition. *Primates, 32,* 483–496.

23   Mirror behavior in macaques

Maria L. Boccia

Introduction

Although various forms of self-knowledge and self-awareness are thought to be present in human infants (Neisser, 1988; Stern, 1985), the question of self-awareness in nonhuman primates has been overshadowed by discussion of their capacity for mirror self-recognition (Gallup, 1977a, 1985; Suarez & Gallup, 1981). Evidence for mirror self-recognition (MSR) via passing the mark test and/or examining body parts not visible without a mirror have been obtained for each great ape species (Gallup, 1970; Hyatt & Hopkins, *SAAH*15; Miles, *SAAH*16; Patterson & Cohn, *SAAH*17; Suarez & Gallup, 1981). In contrast, evidence for MSR in monkeys is much more questionable (Anderson, *SAAH*21; Gallup, *SAAH*3): Self-examination using the mirror is typically absent, and mark-directed behaviors during the mark test are infrequent and different in form than that seem in the apes (e.g., Itakura, 1987; Thompson & Boatright-Horowitz, *SAAH*22).

The rationale for using the mark test in the assessment of self-recognition is that the individual must have a mental representation of the self that he or she understands to be reflected in the mirror (Gallup, 1988). The presence of the mark is a violation of that internal representation, and therefore is a novelty to which the individual's behavior is directed.

Macaques may have failed to provide evidence for MSR for reasons other than a failure to have an internal representation of the self. First, few subjects have been used in the tests of self-recognition, ranging from one (Gallup, 1977a) to four (Gallup, 1970) of any particular species. (Although Anderson [1983] tested ten stump-tailed macaques, these infants had abnormal rearing conditions, including isolation.) If the mark test is at the upper limit of monkeys' cognitive capacities (see later in this chapter), it would be necessary to test more than a few to observe the mark-directed response in any monkeys. Second, the amount of mirror exposure (ranging from 60 to 2,400 hours) may have been too little to be effective (Benhar, Carlton, & Samuel, 1975; Gallup, 1970; Gallup, Wallnau, & Suarez, 1980). Third, researchers have emphasized that the development of self-recognition requires social experience (Gallup, 1977a; Hobson, 1990; Mead, 1934.) Two studies that used lifetime exposure to mirrors also involved abnormal social experience (isolation housing

350

[Anderson, 1983] and peer rearing [Suarez & Gallup, 1986]). Hill, Bundy, Gallup, & McClure (1971) have demonstrated that even chimpanzees reared in isolation fail to develop self-recognition as defined by the mark test. Fourth, the opportunity to observe others behaving in conjunction with their mirror images may be a crucial experience in the development of mirror self-recognition in monkeys. That is, the opportunity to observe other monkeys' behavior concurrently with their mirror image may provide important environmental scaffolding for the development of the understanding the correspondence between their own behavior and the behavior of their mirror image. In studies with macaques, however, individual monkeys have been exposed to the mirror in isolation from others. Although Gallup et al. (1980) tested rhesus monkeys in a social setting, four of the five animals involved were age 6 months or younger, and infants of this age would not be expected to be cognitively mature enough to express this capacity.

There are reasons for believing that MSR, although possible in macaques, is likely to be uncommon. Passing of the mark test by human infants is correlated with a Piagetian Stage 6 sensorimotor intelligence, as indicated by understanding of object permanence (Bertenthal & Fischer, 1978). Whether monkeys are capable of this level of cognitive functioning is debatable. Although some research indicates that they may be capable of an incomplete Stage 6 on Piaget's sensorimotor object-permanence series (Gibson, 1977; Parker, 1977), this has been disputed (Natale & Antinucci, 1989). Thus, Stage 6 understanding of object permanence, and its correlated ability for MSR is, at best, at the upper limit of macaques' cognitive capacity. In such a case, the best performance will require greater environmental scaffolding than with a task not at the upper limit (Fischer, 1980). In addition, a task at the upper limit of a cognitive capacity will require the testing of a substantial number of monkeys to find a few that exhibit the skill, due to individual differences in that capacity. This suggests that, given the right circumstances, some macaques at least may be able to pass the mark test of self-recognition in the mirror tasks.

To date, only a few studies have taken into account all of these considerations in the design of a mirror test for macaques (e.g., Anderson, 1986). The present study takes these factors into consideration and utilizes several mirror-behavior tasks that require a range of levels of cognitive demand. Bertenthal & Fischer (1978) developed a scale of mirror behavior to study the development of self-recognition in human infants. There are five tasks included in this scale (see Table 23.1). When they presented these tasks to human infants, they found a Guttman scale, with the tasks emerging in sequence, and cooccurring with the sensorimotor object-permanence stages indicated. Four of these tasks (excluding the naming task) were adapted for use with monkeys for the present study, but following a procedure different from that of Itakura (1987).

The scale of mirror behavior tasks developed by Bertenthal & Fischer (1978) was chosen because these tasks represent a range of difficulty in their

Table 23.1. *Self-recognition tasks used in this study, and corresponding object-permanence stage*

Task name	Corresponding object-permanence stage	Behavioral requirements
1. Tactual exploration of mirror image	3	Coordination of own reaching with image seen in mirror
2. Hat task: find object attached to self	4	Coordinate own reaching with image seen in mirror of movement of own body
3. Toy task: find object independent of self	5	Coordinate body movements with mirror image and object which is independent of self
4. Mark test: compare image of self to schema of self	6	Coordinate body movements with mirror image and schema of own face

Source: After Bertenthal & Fischer (1978).

cognitive demands. That is, the mark test in this scale represents the second-highest level of cognitive difficulty, the highest level being the naming tasks (in which the child produces his or her own name when asked who is in the mirror). We therefore considered that more could be learned about macaques' understanding of mirrors if a range of tasks were given that represented different levels of cognitive challenge. The corresponding object-permanence task performance is not intended to convey any implications about a causal relationship between object permanence and mirror self-recognition.

In the procedures of the study described here, all of the above-cited limitations were minimized:

1. Mirror exposure was lifetime, ranging from 3 to 15 years (i.e., 17,000–58,000 hours of exposure).
2. The subjects were all socially reared, so there are no abnormal rearing histories involved.
3. The mirror exposure occurred in a social group, thus allowing the individuals to observe the behavior of others and their reflections during the mirror exposure, providing additional opportunities to observe the properties of mirrors.

These circumstances provided more environmental scaffolding than previous studies, and thus, the monkeys may have been able to exhibit their best possible performance.

Methods

The subjects were 15 female pigtail macaques (*Macaca nemestrina*), 8 adults (mean age 10.2 years) and 7 subadults (mean age 4.0 years), housed at the University of Colorado Health Sciences Center Primate Laboratory. They were laboratory-born and had lived their entire lives in front of mirrors in social groups composed of an adult male, adult females, infants, and juveniles.

One-way mirrors were present in the front wall of each pen to provide opportunity for unobtrusive observation of the groups. These windows measured 0.95×0.70 m. Several large pens had two such windows. The doors also contained one-way mirrors measuring 0.50×0.75 m. These windows were placed halfway up the wall, such that a monkey sitting on the shelving (their preferred location, where they spent a great majority of their time) could look directly into the mirror and observe themselves or other monkeys sitting on the shelves. The monkeys were habituated to the test chamber by 1 week of daily hour-long exposures prior to any mirror tests. The mirror used in this study measured 0.76×0.92 m, and was large enough to nearly cover an entire wall of the test chamber thus allowing the monkey to see her whole body. All test sessions were videotaped for later scoring. Testing for each task in the mirror behavior scale was conducted as follows:

> *Tactual exploration*: Each monkey was placed in the test chamber with the mirror in place, and behavior was videotaped for 30 min.
>
> *Hat task*: Each monkey was fitted with a leather vest to which a rod constructed of aluminum flashing was attached. The monkey was habituated to the vest and rod for four days prior to testing. On the day of testing she was anesthetized with hexamethazone, and a bright red or blue light weight Nerf ball was attached to the rod. She was allowed to recover from anesthesia in the test chamber in the dark. When she was awake and alert, the light was turned on and responses videotaped. Although test duration was planned to be limited to 30 min, no subject required longer than 3 min to respond after the lights were turned on, and the trial was terminated after the response.
>
> *Toy task*: A specially designed box with opaque door and sides was affixed to the top of the test chamber. A bright red-and-white toy was contained within the box. The monkey was placed in the test chamber, and when she was looking in the mirror the box's door was opened, and the toy was dropped behind her. In Toy task control, the same toy was presented in the same manner as the toy task, except that it was presented outside the test chamber on the other side of the Plexiglas panel, without the mirror present. The monkey was placed in the test chamber; when she was looking out the Plexiglas, the toy was presented.
>
> *Mark test*: The monkey was anesthetized, and an odorless red dye was placed on the left brow, right ear, and left wrist. She was allowed to recover from the anesthesia in the dark test chamber. When she was awake and alert, the lights were turned on and 30 min of data were collected without a mirror present. The mirror was then placed in the test chamber and 30 min of data were collected with the mirror.

For each task, a pass or fail was scored for each monkey, using Bertenthal & Fischer's (1978) criteria:

> *Tactual exploration*: Touching the mirror image.
>
> *Hat task*: Looking in the mirror at the Nerf ball, followed by looking up to the actual ball and attempting to reach it.

Toy task: Looking at the toy in the mirror, followed by turning around and searching for the toy behind her.

Toy task control: This was used to determine whether the monkey understood the difference between the mirror (reflecting space behind her) and clear Plexiglas panel (revealing space in front of her). She was scored as understanding this distinction if, when the toy was presented, she did not turn around and search behind her.

Mark test: Touching only the wrist in the no-mirror condition, and directing touches to the face while looking in the mirror during the mirror condition.

In addition to the preceding behaviors, several other behaviors were scored during tactual exploration and the mark test in order to facilitate interpretation of the results: duration and frequency of

1. looking in the mirror;
2. contingency testing (looking intently at the mirror image while slowly moving the head or a limb, which suggests a limited understanding of the relationship between the mirror image's and the body's movements, and the beginnings of self-recognition; Lewis & Brooks-Gunn, 1979);
3. mirror-directed social behaviors (including affiliative behaviors such as lip smack, jaw thrust, and social present; and aggressive behaviors such as threat and dominance display); and
4. other behaviors (including locomotion, cage exploration, stereotypy).

Analyses of variance were conducted on these behaviors with one factor: age (adolescent vs. adult).

Results

The adolescent and adult monkeys performed similarly on the mirror tasks: All but two passed all the tasks up to and including the toy task. One evidenced mark-directed and contingency-testing behavior. Three other monkeys exhibited contingency-testing behavior without any mark-directed behaviors in the mark test.[1]

Table 23.2 presents the results for all 15 monkeys' performance on the mirror-behavior tasks. All but two of the monkeys passed up to the toy task. None of the monkeys in the toy task control turned to search behind her for the toy when it was dropped beyond the Plexiglas barrier.

One monkey appears to have passed the mark test; that is, she looked in the mirror, swiped her hand across her face in the vicinity of the mark, and then looked at her hand. She performed this behavior four times in the test period with the mirror, examining her hand after each swipe, as outlined in Table 23.3. She never exhibited this behavior in the no-mirror mark-test control condition. On one additional occasion, she looked intently in the mirror, then grabbed the top edge of her marked ear precisely where the mark was located (there was a 0.1-sec delay between her look in mirror and the grabbing of her ear). She also exhibited contingency testing during both the tactile exploration and mark tests.

Table 23.2. *Pigtailed macaque responses to mirror-behavior tasks*

Age category	Animal ID	Age (yr)	Mirror	Hat	Toy	Mark
Adults	P49.7	8.8	+	+	+	−
	P53.7	9.4	+	+	+	+
	P36.6	9.6	+	+	+	−
	P16.7	9.7	+	+	+	−
	P23.3	10.6	+	+	−	−
	P8.5	11.0	+	+	+	−
	P83.1	11.2	+	+	+	−
	P42.3	11.4	+	+	+	−
Adolescents	P83.1.3	3.3	+	+	−	−
	P34.1.5	3.5	+	+	+	−
	P49.11	3.8	+	+	+	−
	P49.2.4	3.9	+	+	+	−
	P8.5.3	4.2	+	+	+	−
	P42.3.2	4.9	+	o	+	−

Note: +, pass the test; −, fail the test; o, missing data.

Table 23.3. *Description of each mark-directed touch by P53.7*

Look no.	Delay between look in mirror and face touch	Behavior
1	1.9 sec	Touched brow, slid hand down to top of nose, licked hand
2	2.3 sec	Touched bridge of nose, licked & groomed hand
3	0.0 sec	Touched side of face, including one eye orbit, looked at hand
4	0.0 sec	Touched bridge of nose, licked hand

Six of eight adults and six of seven adolescents inspected the dye mark on their wrists during the test period. Only 4 of the 15 monkeys – 2 adolescents and 2 adults – exhibited any contingency testing. One of these adults was also the female who touched her face in the mark test.

Other behaviors examined during the tactile exploration and mark tests revealed few difference between individuals. Not surprisingly, as these monkeys had grown up in front of mirrors, none of them exhibited frequent social behaviors toward the mirror. The social behaviors that were seen were primarily affiliative (typically jaw thrusting) rather than aggressive. The only age difference found was that adolescents looked at the mirror more than the adults did.

Discussion

The study of self-recognition using the mark test across the primate order has led some researchers to the conclusion that MSR is limited to humans and great apes (Gallup, *SAHH*3); but as noted above, many of the studies of macaques have been flawed. Thus, it has not been possible to definitively state whether nonhominoid primates could perform the self-recognition task.

This study tested a species of Old World monkeys under conditions designed to maximize the likelihood of eliciting mirror self-recognition in animals that are capable of it: The subjects studied here had a lifetime of mirror exposure (up to 58,000 hours) in a social context; a relatively large number of them were tested (15), so that if this task was at the limit of their cognitive capacity, some evidence might be found in a few of them; and they all had normal rearing histories and lived in mixed-age and -sex social groups. Evidence of MSR was limited to one subject who showed both mark-directed and contingency behaviors. Three other subjects exhibited contingency-testing behavior, although they did not exhibit the critical mark-directed behavior. If, as Bertenthal and Fischer (1978) suggest, ability to perform on the mark test correlates with Stage 6 sensorimotor intelligence as assessed by object permanence, and this form of intelligence is at the upper limit of their capacity, then this finding is to be expected: Only a few animals will exhibit the capacity, and one must test a relatively large number of them to observe it. In this case 1 of 15 monkeys displayed the capacity. Itakura (1987) found that one of the four Japanese macaques he tested exhibited mark-directed behavior. He discounted the results, however, because the other three did not.

Gallup (1970; 1988) has argued that the mark test is merely confirmatory of the spontaneous behavioral evidence of self-recognition, which is mirror-guided, self-directed behavior. Four of the 15 subjects exhibited what might be classified as an incipient form of such behavior (contingency-testing behavior) although they did not meet the criteria outlined by Gallup. Thus, even under these optimal conditions, the self-recognition capacities of macaques, as measured by mirror behavior, appear to be limited. Lewis and Brooks-Gunn (1979) argue that contingency-testing behavior represents some understanding of the nature of reflections and is related to an emerging self-recognition. Thus, the three monkeys who tested but did not touch the mark may have been on the verge of understanding.

Their performance on the cognitively simpler tasks indicated that nearly all of the monkeys were capable of passing the foregoing mirror tasks associated with Stage 5 object permanence (the toy task). This, and the failure to search during the toy task control condition, suggests that they understand that the mirror reflects space behind them, and that they can coordinate that knowledge with knowledge of their own position in space. These results indicate that monkeys can locate their bodies in space and relate that location to the location of other objects, suggesting the presence of internal representations of both the self and the environment. Furthermore, they must understand the nature of the reflective surface of the mirror sufficiently to perform these tasks.

It has been suggested that macaques lack such an internal representation because direct eye contact is a threat gesture (Fischer, personal communication). Consequently, monkeys fail to look at their faces in the mirror and hence fail to develop a schema for their own face. The rates of looking at themselves in the mirror measured in this study, however, indicate that they do not have any qualms about looking at their faces.

Anecdotal observations suggest another factor that may contribute to the failure of pigtail, Japanese, and rhesus macaques to pass the mark test. In studying grooming behavior in macaques (Boccia, 1983, 1986, 1989; Boccia, Reite, & Laudenslager, 1989; Boccia, Rockwood, & Novak, 1982), I have been struck by an interesting phenomenon. Whether grooming themselves or one another, these macaques often pass over objects (e.g., a piece of wood-chip bedding) in the fur. This inattention suggests that the macaques lack a critical cognitive piece necessary for mirror self-recognition: Interest in their body surface and its deviations from an internal representation of self. Humans and chimpanzees exhibit a great deal of interest in the conditions of their bodies, placing on and removing objects from them (as do orangutans, the other great ape whose MSR capacity is undisputed). Macaques simply do not show this interest; that is, macaques routinely examine familiar parts of their bodies and ignore visual anomalies produced by foreign objects in their fur. Because they have been observed doing this daily with parts of their bodies that are visible to them, it seems unlikely that they would attend to the anomaly produced by the mark on otherwise invisible parts of their bodies reflected in the mirror. Since they are red, the marks more resemble wounds, which monkeys do show an interest in grooming. Wounds, however, have other, nonvisual properties (most notably pain or discomfort) that may provide a basis for their interest that would not be shared by marks. Thus, it is possible that a macaque might possess an internal representation of self and "self-awareness" (such as by those who exhibited testing behavior in this study), but not be interested in a foreign object on the body (the mark).

Mirror-mediated social interactions

Anecdotal observations of pigtail macaques in the University of Colorado Health Sciences Center Primate Laboratory are consistent with the foregoing experimental evidence that these monkeys have the capacity to understand and utilize mirror reflections (see also Anderson, 1984, 1986; Gallup, 1982). In our laboratory, all animal enclosures are fronted with one-way mirrors. Animals are thus exposed to mirrors on a continuous basis for as long as they live in the laboratory. In the context of observations for a number of other studies, I have observed monkeys utilizing the mirrors to mediate social interactions. For example, in one social group, the breeding male and one of the females utilized the mirrors during courtship. The two animals would

stand on ledges at opposite ends of the pen, each directing jaw thrusts, bobbing, and lip smacking toward the mirror image of the other. When the
female would present toward the mirror, the male would approach and mount
her. This interaction occurred repeatedly for the entire time I observed them.
In a different social group of pigtail macaques, I observed an interaction
involving the adult male and an infant less than a year old. The infant was
sitting on the ledge between the male and the mirror. Both were oriented
toward the mirror. The infant made a play face directed toward the mirror
image of the male, then turned around, jumped on the male, and began
wrestling with him; the male responded with play. In a third pigtail group, the
alpha female watched other members of her group in the mirror, and would
threaten the others via the mirror. She appeared to be utilizing the mirror to
surreptitiously observe other group members before directing a threat to
them. Often she would threaten another animal via her reflection and then
turn and threaten it directly. These observations suggested that monkeys
were capable of understanding and utilizing the reflective properties of mirrors
and were capable of recognizing individuals in the mirror image.

Conclusion

It is clear that the evidence for MSR provided in this study is quantitatively,
and probably qualitatively, different from that reported for great apes. In the
reports presented in this book, the frequency of mark-directed touching in
the chimpanzees who performed this behavior was typically greater than 50
in a short test period. This high frequency contrasts with the four reported
in one monkey herein. Furthermore, it is clear from the great-ape data that
their contacts are qualitatively different from those reported here and in
Itakura (1987) and Thompson & Boatright-Horowitz (*SAAH*22): The apes
typically gaze intently into the mirror while tactilely exploring the mark. The
macaques merely make a passing swipe at the mark, and then look at their
hand. One may, in fact, refer to the macaque behavior as incipient recognition or protorecognition rather than as full self-recognition.

All of this suggests that mirror self-recognition is not an all-or-none
phenomenon. Rather, understanding of mirror contingency and correspondence, interest in body surface, internal representation, and the recognition
of anomalies may be additive phenomena that produce a gradual or developmental emergence if this capacity. Further comparative and developmental studies may elucidate these processes.

Several theories of the development of self in human infancy have been
proposed suggesting that there may be different kinds of self that must be
considered in studies of nonhuman primates (Mitchell, *SAAH*6; Neisser, 1988;
Stern, 1985). Research to date in nonhuman primates has focused almost
exclusively on MSR. Recognition of one's body as self is, however, only one
part of self-awareness (Damon, 1983; Mitchell, 1993, *SAAH*6). Although this
is a necessary area of investigation, it by no means represents a complete

description of self-awareness in primates. In fact, it remains to be determined if MSR is indicative of self-awareness. Self-awareness is a global concept that may include several different types of self, and that has been given different definitions by different researchers. For example, to some it is a metacognition; to others it is not. What is needed are clearer definitions of the type of cognition being spoken of, and other tests of self-recognition independent of mirror behavior that can serve as converging operations on the phenomenon of self-awareness (Gallup, *SAAH*3).

Acknowledgments

This research was supported by Developmental Psychobiology Research Grant No. 157 from the Grant Foundation, and by PHS grant MH44131. I would like to thank Amy Schmitt for her assistance in data collection and scoring. I would like to thank Kurt Fischer for his helpful discussions and comments on the development of this study. This research was conducted at the University of Colorado Health Sciences Center Primate Laboratory, Martin Reite, director.

Note

1. A videotape that includes a sample adult and sample adolescent response to each task, plus all four occurrences of mark-directed behavior by P53.7, has been archived at the Wisconsin Primate Center Library. Interested individuals may view this tape by contacting Larry Jacobsen (University of Wisconsin Primate Center, 1220 Capitol Court, Madison WI 53715; phone: 608-263-3512, e-mail: Jacobsen@primate.wisc.edu).

References

Anderson, J. R. (1983). Responses to mirror-image stimulation and assessment of self-recognition in mirror- and peer-reared stumptail macaques. *Quarterly Journal of Experimental Psychology*, 25(B), 201–212.

 (1984). Monkeys with mirrors: Some questions for primate psychology. *International Journal of Primatology*, 5, 81–98.

 (1986). Mirror-mediated finding of hidden food by monkeys (*Macaca tonkeana & M. fascicularis*). *Journal of Comparative Psychology*, 100, 237–42.

Benhar, E. E., Carlton, P. L., & Samuel, D. (1975). A search for mirror-image reinforcement and self-recognition in the baboon. In S. Kondo, M. Kawai, & A. Ehara (Eds.), *Contemporary primatology* (pp. 202–208). Tokyo: Japan Science Press.

Bertenthal, B. L., & Fischer, K. W. (1978). Development of self-recognition in the infant. *Developmental Psychology*, 14, 44–50.

Boccia, M. L. (1983). A functional analysis of social grooming patterns through direct comparison with self-grooming in rhesus monkeys. *International Journal of Primatology*, 4(4), 399–418.

 (1986). Grooming site preferences as a form of tactile communication and their role in the social relations of rhesus monkeys. In D. M. Taub and F. A. King (Eds.), *Current perspectives in primate social dynamics* (pp. 505–518). New York: Van Nostrand Reinhold.

 (1989). A comparison of the physical aspects of social grooming in two species of macaques (*Macaca nemestrina* and *M. radiata*). *Journal of Comparative Psychology*, 103, 177–183.

Boccia, M. L., Reite, M., & Laudenslager, M. (1989). On the physiology of grooming in a pigtail macaque. *Physiology and Behavior*, 45, 667–670.

Boccia, M. L., Rockwood, B., & Novak, M. A. (1982). The influence of behavioral context and social characteristics on the physical aspects of social grooming in rhesus monkeys. *International Journal of Primatology, 3*(1), 91–108.

Damon, W. (1983). *Social and personality development.* New York: Norton.

Damon, W., & Hart, D. (1982). The development of self-understanding from infancy through adolescence. *Child Development, 53,* 841–864.

Fischer, K. W. (1980). A theory of cognitive development: The control and construction of hierarchies of skills. *Psychological Review, 87,* 477–531.

Gallup, G. G., Jr. (1970). Chimpanzees: Self-recognition. *Science, 167,* 86–87.

(1977a). Absence of self-recognition in a monkey (*Macaca fascicularis*) following prolonged exposure to a mirror. *Developmental Psychobiology, 10,* 281–284.

(1977b) Self-recognition in primates: A comparative approach to the bidirectional properties of consciousness. *American Psychologist,* 329–338.

(1982). Self-awareness and the emergence of mind in primates. *American Journal of Primatology, 2,* 237–248.

(1985). Do minds exist in species other than our own? *Neuroscience and Biobehavioral Reviews, 9,* 631–641.

(1988). Self-awareness. In G. Mitchell and J. Erwin (Eds.), *Comparative primate biology, Volume 2, part B: Behavior, cognition, and motivation* (pp. 3–16). New York: Liss.

Gallup, G. G., Jr., Wallnau, L. B., & Suarez, S. D. (1980). Failure to find self-recognition in mother–infant and infant–infant rhesus monkey pairs. *Folia Primatologica, 33,* 210–219.

Gibson, K. R. (1977). Brain structure and intelligence in macaques and human infants from a Piagetian perspective. In S. Chevalier-Skolnikoff & F. E. Poirier (Eds.), *Primate biosocial development: Biological, social, and ecological determinants* (pp. 113–158). New York: Garland.

Hill, S. D., Bundy, R. A., Gallup, G. G., & McClure, M. K. (1970). Responsiveness of young nursery-reared chimpanzees to mirrors. *Proceedings of the Louisiana Academy of Sciences, 33,* 77–82.

Hobson, R. P. (1990). On the origins of self and the case of autism. *Development and Psychopathology, 2,* 163–181.

Itakura, S. (1987). Use of a mirror to direct their responses in Japanese monkeys (*Macaca fuscata fuscata*). *Primates, 28,* 343–352.

Lewis, M., & Brooks-Gunn, J. (1979). *Social cognition and the acquisition of self.* New York: Plenum.

Mead, G. H. (1934). *Mind, self, and society from the standpoint of a social behaviorist.* Chicago: University of Chicago Press.

Mitchell, R. W. (1993). Mental models of mirror self-recognition: Two theories. *New Ideas in Psychology, 11,* 295–325.

Natale, F., & Antinucci, F. (1989). Stage 6 object-concept and representation. In F. Antinucci (Ed.), *Cognitive structure and development in nonhuman primates* (pp. 97–112). Hillsdale, NJ: Erlbaum.

Neisser, U. (1988). Five kinds of self-knowledge. *Philosophical Psychology, 1,* 35–59.

Parker, S. T. (1977). Piaget's sensorimotor series in an infant macaque: A model for comparing unstereotyped behavior and intelligence in human and nonhuman primates. In S. Chevalier-Skolnikoff & F. E. Poirier (Eds.), *Primate biosocial development: Biological, social, and ecological determinants* (pp. 43–112). New York: Garland.

Patterson, P. (1984). Self-recognition by *Gorilla gorilla gorilla. Gorilla, 7*(2), 2–3.

Stern, D. (1985). *The interpersonal world of the infant,* New York: Basic.

Suarez, S. D., & Gallup, G. G. (1981). Self-recognition in chimpanzees and orangutans, but not gorillas. *Journal of Human Evolution, 10,* 175–188.

(1986). Social responding to mirrors in rhesus macaques (*Macaca mulatta*): Effects of changing mirror location. *American Journal of Primatology, 11,* 239–244.

24   Evidence of self-awareness in the bottlenose dolphin (*Tursiops truncatus*)

Kenneth Marten and Suchi Psarakos

Dolphins are big-brained, socially sophisticated mammals, socially and cognitively comparable to monkeys and apes in memory capacity, language comprehension, and other cognitive abilities (Herman, 1980). Their remarkable capacities suggest that dolphins, like apes, may come to recognize themselves in mirrors (e.g., Anderson, 1984; Gallup, 1970, 1982; Lethmate & Dücker, 1973; Suarez and Gallup, 1981). This chapter reports on the results of a series of studies utilizing a mirror to assess whether dolphins recognize contingent representations of themselves or use the mirror to examine an area of the body not otherwise visible that has been marked with a highly salient substance. We adapted the mirror mark test (Gallup, 1970) for use with the bottlenose dolphin. We employed several control conditions, including mirror without mark, no mirror and no mark, and first encounter between unfamiliar dolphins through a barrier. We also devised and conducted several new tests for self-recognition, tailored for dolphins rather than primates. These tests, which utilize self-view television and video playback, are summarized here. This chapter focuses on interpreting mirror-directed behavior (both marked and unmarked) by comparing it to the control data. We address the central question of whether the dolphins' mirror-directed behavior is social or self-examination. We also discuss the role of environmental, social, and individual influences on the test results.

Methods

Subjects and setting

Five dolphins, 6–14 years old, living at Sea Life Park, served as subjects of this study (Table 24.1). The groups of dolphins changed throughout the course of the research. The dolphins dwell in a two-tank complex (Figure 24.1). The laboratory has five underwater windows looking into the large tank. One window is a 1.2-m-diameter circle, which can be made into a large one-way mirror. Another window is 0.6 m × 0.6 m and houses a 20-in. Sony Trinitron color television that the dolphins can watch. Data were collected by video-taping the animals through the window or a one-way mirror with a Minolta

361

Table 24.1. *Mark test subjects and methods*

Subject	Age (yr)	Sex	Origin[a] and subspecies	Mirror exposure before first mark test	Mirror exposure before first mark test (hr)	Mark test mirror size (m)[b]	Body part marked	Marking substance[c]
Keola	14	male	CB, *truncatus*	1	<1	1.2	side	ZO
				2	225	1.2	fluke	ZO
				3	260	1.2	rostrum/ stomach	Ich/GV
Hot Rod	6	male	CB, *gilli*	1	<1	1.2	side	ZO
				2	225	1.2	pec fin	ZO
				3	260	1.2	melon/ stomach	ZO/GV
Okoa	13	female	WC, *gilli*	1	270	1.2	side	ZO
Laukani	14	female	WC, *gilli*	1	80	0.8×0.7	side	ZO
				2	84	1.2	melon	ZO
Itsi Bitsi	13	female	CB, *truncatus*	1	75	0.8×0.7	behind pec fin	GV
				2	78	1.2	side	ZO
				3	86	1.2	side	ZO

[a] CB, captive born; WC, wild caught.
[b] Single dimension is diameter.
[c] ZO, zinc oxide; GV, gentian violet; Ich, ichthammol.

S-VHS Series V-200 video camera. Notes on behavior were also recorded on an ongoing basis during testing. No food rewards were provided.

Apparatus

The mirror was a reflective aluminum-coated polyester Mylar film made by DuPont, which we applied to the window with soap and water. Viewed from the tank, it makes a perfect mirror, yet because of its one-way properties we were able to videotape the dolphins' mark tests through the same mirror from inside the lab. In some tests we covered the entire 1.2-m-diameter circular window with the reflective Mylar, while in others, we used a smaller 0.8 m × 0.7 m rectangle of Mylar in the same window (Table 24.1).

Procedures

Mirror mark tests. All dolphins had some mirror exposure prior to their first mark test. The dolphins were not isolated during their mark tests, and sometimes more than one animal was marked at a time. All but one of the dolphins had multiple mark tests. Table 24.1 presents information regarding the

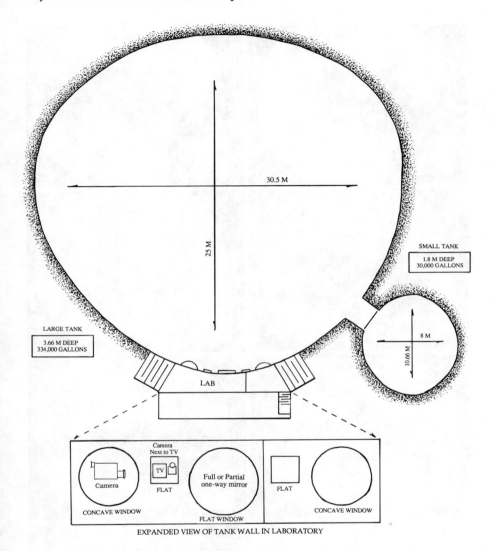

Figure 24.1. Two-tank complex (top view) and underwater viewing laboratory, Sea Life Park, Hawaii. Videotaping of the body mark tests was done through a one-way mirror on the large, flat window. A camera next to a 20-in. television in the small square window provided the frontal self-view on the television (mirror mode) for the television tests summarized in the text.

hours of mirror exposure prior to each mark test, the body part marked, and other relevant information.

Unless noted otherwise, all animals were marked on their sides with approximately $\frac{3}{4}$ oz of zinc oxide (Figure 24.2). Zinc oxide is a tactile as well as visual stimulus. (We could not find an appropriate stain without tactile stimulus.) Other stains used were gentian violet, a purple topical antiseptic,

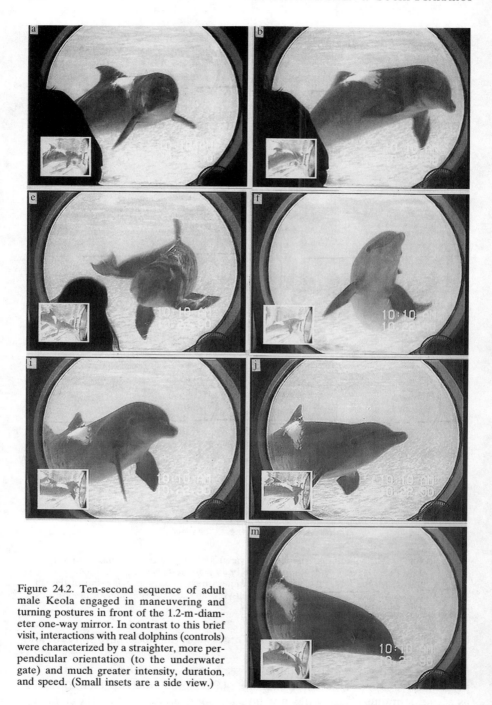

Figure 24.2. Ten-second sequence of adult
male Keola engaged in maneuvering and
turning postures in front of the 1.2-m-diam-
eter one-way mirror. In contrast to this brief
visit, interactions with real dolphins (controls)
were characterized by a straighter, more perpendicular orientation (to the underwater
gate) and much greater intensity, duration,
and speed. (Small insets are a side view.)

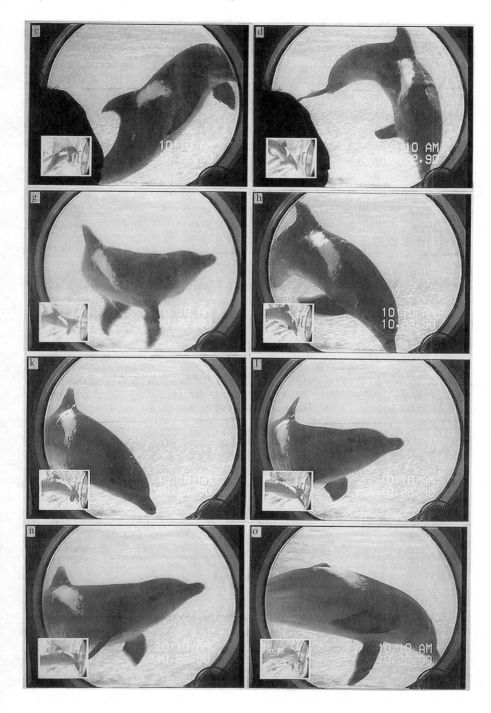

and ichthammol, a black antiseptic ointment. One subject, Itsi Bitsi, was sham marked (with the mirror present) with Vaseline six times prior to her first zinc oxide mark test. Mark locations were selected for their likelihood of being visible only with the aid of a mirror. Bottlenose dolphin eyes point laterally and cant forward and down. This creates a binocular field in front of and below the animal, with a blind area behind and above, where the marks were located.

Each dolphin was observed and filmed after being marked, and recorded for latency to approach the mirror, time in front of the mirror, frequency of approaches to the mirror, and behavior while in front of the mirror. Behaviors at other locations in the tank were also noted. In some cases if a marked dolphin did not approach the mirror for approximately half an hour, we tapped on the window to attract it to the mirror area. Although enticing is undesirable, the important data was not whether the dolphin came to the mirror, but what it did in front of it once it saw its mark.

Was the dolphin examining its mark? Guidelines for "passing" the mirror mark test. The purpose of the tests presented here was to evaluate objectively if the dolphin treated its mirror image as itself or as another dolphin. These guidelines were used to ascertain whether a given dolphin used the mirror to examine its mark:

1. The dolphin engaged in maneuvering, turning, or twisting postures that brought the mark into view. (The dolphins also occasionally performed similar postures when not marked. These were most likely instances of the animal simply using the mirror to examine the unmarked lateral area.)
2. The dolphin visited the mirror and made its mark visible to itself within 10 min of being marked.
3. The dolphin made at least one visit of 10 sec or more to the mirror with the mark visible to itself.
4. The dolphin's physical distance to the mirror during a visit (including the one or more that lasted at least 10 sec) was no more than a foot.
5. The dolphin's eye was distinctly oriented toward the mirror reflection of its mark (optional: this was not noted in all cases, could only be seen in person, and was too subtle to show up on videotape data).
6. Sea Life Park staff familiar with these dolphins and their behavior, as well as colleagues who have seen videotape of the tests, had the subjective impression that the dolphin was using the mirror to examine its mark.

Control conditions. Three control conditions were used with some subjects for comparison with the above-described mark tests:

1. The mirror was available, but the animals were not marked.
2. The animals were not marked and there was no mirror available.
3. The animals were exposed to unfamiliar dolphin(s) through a metal grate, under varying conditions.

Results were videotaped, and samples of the data were analyzed. For control Conditions 1 and 2, we videotaped the unmarked dolphins' behavior both with and without the one-way mirror in place. We analyzed 24- and 35.5-hour samples from two different groups of dolphins (Table 24.2). The samples were picked to be close in time to the mark test. We looked especially for the maneuvering postures (Figure 24.2) we later saw in mark tests. We counted these postures and generally observed and characterized the dolphins' behavior. For control Condition 3 we videotaped behavior from the surface whenever new individuals met each other at Sea Life Park.

Results

Mirror mark test results

The dolphins varied considerably in their interest in the mirror during the mark test and at other times. There was a brief period of heightened tank activity shortly after the marks were applied to Keola and Hot Rod, when all four dolphins in the tank engaged in fast, circular swimming for 2 min. All of the animals tested engaged in behaviors suggesting that the dolphins were examining the zinc oxide mark during the mirror mark test. In addition, all of the subjects played with objects in front of the mirror, suggestive of self-examination. Finally, each animal's behavior was unique, but gave evidence of utilizing the mirror in various ways. Results described here are not exhaustive, and are meant only to convey test highlights.

Keola, 14-year-old adult male Atlantic bottlenose. In Keola's first mark test, he swam to the mirror while we were still in the process of putting it up. As he approached, we were producing the noises the mirror-application process makes. As he arrived we stopped the application process and made no further sounds. He positioned himself so that his mark was probably visible to him and engaged in some extreme "maneuvering" postures and contortions for 13 secs (Figure 24.2). In his second mark test, Keola started removing his mark right away by rubbing against the tank. Keola passed by the mirror many times during this and his third mark tests, but did not give the mark or the mirror as much special attention as he did in his first test. The gentian violet that was applied during his third mark test washed away in the tank water almost immediately.

In a later mark test done with the 0.8 m × 0.7 m rectangular mirror, on at least one pass by the mirror Keola swam so that his mark moved in front of the mirror, then stopped just as the mark reached the edge of the mirror. He remained relatively motionless for 15 sec, with the mark still visible near the edge of the mirror, and then calmly swam on.

Okoa, 13-year-old adult female Pacific bottlenose. Within 5 min of being marked Okoa swam by the mirror in such a way that the mark was visible to her, but she did not stop and engage in any "maneuvering" postures. Thirteen minutes later she returned and stayed for 11 sec, less than a foot from the mirror and roughly parallel to it, in a position where the mark was most

Table 24.2. *Mark test results*

Subject	Mark test	No. of deliberate visits to mirror	Total time in front of mirror with mark visible (sec)	Maxi. single-visit time with mark visible (sec)	Latency to approach mirror after mark (min)	Mirror–no mark control: Length of sample analyzed (hr)	Mirror–no mark control sample: Maneuvering posture rate during sample (times/hr)	Did dolphin examine its mark in first mark test?
Keola	1	5	13	13	5	24	0.08	yes
	2	0	0	0	NA			
	3	17	35.5	15	11			
Hot Rod	1	16	18.5	4	4	24	0.12	probably
	2	3	7.5	3	70			
	3	23	54.5	9	7			
Okoa	1	4	25	11	5	24 & 34.5	0.25/hr & 0	yes
Laukani	1	12	14	12	10	34.5	0.06	yes
	2	3	11.5	8	48			
Itsi Bitsi	1	15	57	12	20	34.5	0.03	possibly
	2	0	0	0	NA			
	3	4	14	12	46			

likely visible to her. Her behavior included an unusual lifting of the midsection, a behavior that we scored as a maneuvering posture. Nineteen minutes later, upon being enticed to the mirror area from another part of the tank by our tapping on the mirror, she engaged in another 11-sec visit close to the mirror with the mark visible to her. This time she exhibited very deliberate maneuvering and turning postures, positions similar to those of Keola (Figure 24.2), and similar as well to six occurrences of maneuvering postures we counted during the 24-hour sample of unmarked control data for her.

Laukani, 14-year-old adult female Pacific bottlenose. In her first mark test, Laukani did not spontaneously go to the mirror like Keola and Okoa, but needed to be enticed to the mirror by tapping on the underwater window. Three times she was positioned within a foot of the mirror in such a way that her mark could have been visible to her. The first instance, 10 min after the mark was applied, lasted only 2 sec, but included a brief maneuvering posture. The second instance lasted 20 sec: She stayed horizontal and parallel to the mirror, only a few inches in front of it, with her eyes straining backward (toward her mark). The third was similar to the second and lasted 3 sec, until movement inside the laboratory inadvertently caused her to leave.

Itsi Bitsi, 13-year-old adult female Atlantic bottlenose. Although she visited the mirror several times during her gentian violet mark test, and was positioned so that the mark was probably visible to her, she did not engage in any obvious behavior or postures that seemed to favor viewing the marked area.

Itsi Bitsi did not go to the mirror at all during her first zinc oxide mark test, but instead aggressively chased the other dolphins in the tank. In her second zinc oxide mark test she visited the mirror four times, but we had to entice her there by tapping on it. Once, for 3 sec, she was positioned within a foot of the mirror so that her mark was probably visible to her. At this time she engaged in a brief, subtle maneuvering posture.

Hot Rod, six-year-old juvenile male Pacific bottlenose. During his first mark test, while the mirror was being applied to the window, Hot Rod swam by the mirror with his mark facing it, but did not stop. Four min later he engaged in maneuvering postures virtually identical to those of Keola in Figure 24.2, but with his *unmarked* side facing the mirror, then turned around immediately so that the mark was visible to him for 2 sec. Five min after that, he stopped in front of the mirror with the mark visible for 3 sec. In his third mark test, 7 min after receiving the mark he stopped in front of the mirror, where he stayed relatively motionless, with the marked side facing the mirror, for 10 sec. Two min later he did this for another 4 sec.

Control condition results

Mirror, no mark. During the mirror–no mark control condition, the dolphins approached the mirror regularly. In one control sample, for the first 5 days the dolphins moved slowly and acted passive in front of the mirror, with long, stationary watches. After 5 days (50 hours of mirror exposure) their behavior shifted to more active body movements, mouth movements, bubble blowing, and open-mouth behaviors. During mirror visits, the dolphins engaged in the same type of posturing seen during mark tests: In a 24-hour sample, twice for Keola, three times for Hot Rod, six times for Okoa; in a different 34.5-hour sample, twice for Laukani and once for Itsi Bitsi (Table 24.2). (These counts are minimums.)

Interactions while facing the mirror were characterized by mouth movements, bubbles, vocalizations, and melon butts against the window; they appeared more social, aggressive, and interactive than sideways interactions. During sideways interactions, in which the dolphin's body is parallel to the mirror, the dolphins were more passive and relatively motionless, engaging in occasional body twisting, posturing, and slight humping of the midsection. In these interactions the dolphins seemed self-absorbed rather than interactive, and appeared to be maneuvering either to bring certain parts of their bodies into view or to stay in place in front of the mirror. Once, after approximately 200 hours of mirror exposure, Okoa remained motionless in front of the mirror and released a continuous, slow, fine stream of bubbles for approximately $1\frac{1}{2}$ min. She repeated this behavior a short time later for a similar period of time, although the bubble release was not continuous. Her eye movement in both episodes indicated she was watching the bubbles rise to the surface.

During mirror exposure, the dolphins occasionally approached the mirror with an object and manipulated it. This play was sometimes characterized by the dolphin relocating the object in front of the mirror if the object moved out of view. After being fed, the dolphins often spit fish out while in front of the mirror.

Adult male Kamalii, who was never marked, was exposed to a mirror on one occasion for 2 hours. He approached the mirror several times during this test and engaged in a variety of behaviors. Once, he blew a bubble, and moved quickly through a series of maneuvering postures while whistling. Two min later he came close to the mirror, and while oriented perpendicular to it, engaged in rapid, rhythmic, repetitive shallow pumping of his mid and lower section, punctuated by an abrupt, audible sideswipe of his flukes at the end. The movement was accompanied by a continuous click train. The movement sequence lasted approximately 10 sec and contained two bouts of the pump and sideswipe. Later, he returned and repeated the behavior, accompanied by the same rapid click train.

No mirror, no mark. When no mirror was present, posturing in front of the window occurred only occasionally, and was subtle and brief. In addition, the dolphins did not approach the window with the same frequency or regularity as when the mirror was up. Playing with toys occurred with no mirror, but infrequently. Spitting out fish also occurred with no mirror.

Encountering an unknown dolphin. If dolphins do not recognize their mirror reflection as representing themselves, then one might expect them to respond to the mirror in the same way they respond to encountering an unknown dolphin. (We found no literature on dolphin encounters with unfamiliar dolphins, although Ostman, 1985, and Overstrom, 1983, present some ethological and acoustic information.) In the third control condition we tested how dolphins respond to encountering other dolphins for the first time. The adult females (Okoa, Laukani, and Itsi Bitsi) were introduced to an unfamiliar adult male, Kamalii, separated by an underwater barred gate with bars 15 cm apart. These females spent time inside the channel at the underwater gate during most of the 40 min of the experiment. There were often two or three females at the gate together. The dolphins appeared excited, but were not physically hyperactive (i.e., did not create a lot of turbulent water). All three females, but most frequently Okoa, blew bubbles. All four of the dolphins were usually oriented perpendicular to the gate. Laukani and Kamalii engaged in three bouts of open-mouth interaction with each other. During two of these, they touched rostra. During one, Kamalii engaged in a brief vertical undulating motion, much like simultaneously swimming and pushing his rostrum through the gate. In the third open-mouth episode, both Laukani and Kamalii looked as if they were taking small, rapid bites at one another. At one point, Kamalii swam slowly and closely by the gate and turned sideways, presenting his ventral aspect to Okoa, who released a large bubble. He immediately circled back and faced her.

Interactions with strangers of a closely related species through an underwater barrier. Four bottlenose dolphins, including juvenile male Hot Rod, encountered two false killer whales through an underwater barred gate. Although some of the dolphins had previously lived with a false killer whale (and one Sea Life Park dolphin has successfully mated with one, producing a fertile hybrid), they had never met these two. The dolphins gave the strangers their undivided attention for the entire 1-hour observation period. There was a brief period of prolonged, rapid, and loud airborne vocalizing during the initial confrontation across the gate. Unlike in mirror mark tests, the dolphins oriented almost exclusively perpendicular to the gate (there is much parallel orientation to the mirror in mark tests.) There appeared to be little or no body posturing. Their occasional open-mouth behavior was not accompanied by rhythmic head movements as it often is in front of a mirror.

No underwater barrier. Watching stranger dolphins interact without the restraint of an underwater barrier between them shows how the dolphins would behave if they were swimming freely and not separated by glass (as a subject and its reflection are). Keola, the adult male with the most convincing mark test, was put in a tank with Maka, an adult male Keola had never seen or heard before. The result was very unlike Keola's mirror behavior: high-speed synchronous swimming around the edge of the tank and quick, intense movements. Keola repeatedly put his rostrum near Maka's genitals. Keola's intensity and undivided attention with respect to Maka lasted for most of a day (i.e., much longer than a mark test).

Summary of new tests: Self-view television[1]

Using television to distinguish between social behavior and self-examination

When a dolphin looks in a mirror it often opens its mouth and moves its head around in some rhythmic fashion. In a mark test, as it does not have a hand to touch its mark, the dolphin maneuvers its body in various postures to see it. In our opinion, the mark tests and controls presented in this chapter suggest that these behaviors are contingency-checking and self-examination, respectively. However, another possible interpretation is that the dolphin is using postures to interact "socially" with what he perceives to be another dolphin in the mirror. We therefore conducted additional tests to distinguish self-examination from social behavior in the context of dolphin–mirror interaction. Space limitations do not allow them to be included in full in this chapter, but they are summarized here.

Television tests: Real-time self-view versus playback of same

In previous experiments we determined that dolphins respond to a televised image as if real (Marten & Psarakos, 1992). To distinguish self-examination from social behavior, we put a video camera aimed into the tank next to the dolphins' 20-in. television monitor. This setup is referred to as "mirror mode," or real-time, self-view television. A real-time image is displayed on the dolphin's television, allowing the dolphin to look at itself as in a mirror; this material is videotaped. When we later play back the videotape of a mirror mode session to the dolphins on their television – "playback mode" – the movements of the taped dolphin on the television no longer mirror those of the dolphin watching. The dolphins were exposed to alternating sessions of 10 min of mirror mode and 10 min of playback mode. One would expect the self-aware dolphin to behave differently in the two television modes: perhaps to utilize the television image for self-examination in mirror mode, and merely to observe in "playback mode." If the dolphin perceives its mirror-mode television image as another dolphin, then whether it is watching mirror mode or playback mode, the dolphin should act the same way – possibly interacting socially with the "television dolphin," or perhaps just observing it, but behaving the same in both modes. The results for the adult male Keola were clear-cut: In mirror mode he spent quite a bit of time opening his mouth wide and moving his head in various rhythmic ways, whereas he never did this during playback mode. The results for juvenile Hot Rod and babies Maui and Tinkerbell, however, did not show such clear-cut differences between the viewing modes, although the babies seemed to spend more time in front of the television during mirror mode.

Television mark tests

Some of the dolphins who participated in the mirror mark tests were also exposed to the mirror mode–playback test described above. The video camera was located next to the television, to the dolphins' left, and thus favored showing the left side when they looked at the television. We marked them on their right sides to see if they would counter this bias by maneuvering the marked side into view. Indeed, presentation of the marked (right) side during mirror mode was three times what it was in playback mode, and 2.5 times what it was in the unmarked control.

Mouth mark tests

We wondered if the wide-open-mouth behavior that the adult male Keola exhibited in mirror mode, but not during playback, was Keola examining the inside of his mouth. To test this, we marked his teeth with colored denture adhesive and compared his open-mouth behavior in front of the television before and after he was marked. After his teeth were marked, he came to the self-view television twice and repeatedly opened his mouth extremely wide. However, in 20 min of control data immediately preceding his mark, he visited the television five times and engaged in open-mouth behavior during three of them. Nevertheless, he never opened his mouth as wide or as long as when he had the dental adhesive on his teeth.

Turn tests

We showed the dolphins a real-time frontal self-view on their television, then suddenly changed it to a side view to see if this would cause them to turn. If so, it would suggest that the dolphin might know that the image it is watching is itself. Two-year-old Maui's turn rate during the "turn trials" was five times what it was in control tests with no view-switching, and 3-year-old Tinkerbell, who never turned in the control situation (no view-switching), turned five times in the 39 turn trials of her last two sessions.

Discussion

Mirror mark tests

We have concluded from the behavior of the dolphins during their mark tests that three of the adults, Keola, Okoa, and Laukani, used the mirror to visually inspect their bodies where the mark was located. The locations the dolphins chose to take in front of the mirror, and the body orientation they assumed during these inspections, suggested they were looking toward the mark. Their behavior while in front of the mirror also suggested this. For example, during one close approach to the mirror, Laukani moved her eyes far back, seeming to strain to see the reflection of the mark behind her head.

For some subjects, the data were harder to interpret in terms of self-directed behaviors. Hot Rod spent as much time as the adults in front of the mirror with his mark visible to him, but his movements were not as suggestive of examination as those of Keola, Okoa, and Laukani. The presence of the mirror and the mark did not appear to alter Itsi Bitsi's behavior. Interestingly, she was the animal who had received the vaseline sham marks prior to her zinc oxide mark tests. She had the shortest visit to the mirror of all the dolphins who were mark tested, and she was the only adult who did not spend more than 10 sec directly in front of the mirror with the zinc oxide mark visible to her. Perhaps the novel tactile stimulus of the zinc oxide without sham marking is needed to elicit longer visual investigation in the mirror.

Interpretation of combined results from mirror mark and television tests

One of the most important results from this research comes from a combination of the mirror mark tests and the television tests. Keola, the only adult to have both tests, behaved significantly differently when the television was in mirror mode as compared to playback mode. His behavior strongly suggests that he did not perceive the television image as another dolphin in either mode and was not interacting with it "socially." When he was marked with zinc oxide, he almost immediately positioned himself within a foot of the mirror so that the mark could have been visible to him and then engaged in rapid, extreme, postures. It is very likely that he was examining his mark. The only reasonable alternative is that he might have been posturing socially to what he believed to be another dolphin in the mirror. Since this "social" explanation is rendered not very likely by the television tests, the data suggest that this dolphin examined his mark in the mirror during his mark test, and took his mirror image to be himself.

Controls

The comparison of the three control conditions with the mark test suggest that the dolphins' behavior during the mark tests was unique. Although behaviors suggesting postural adjustments to enable viewing otherwise invisible parts of the body occurred during the mirror–no mark condition, the dolphins, particularly Keola and Okoa, engaged in this behavior at a much higher rate during their mark tests. The virtual absence of these postural adjustments during the no-mirror condition further supports a self-examination interpretation of these postures.

Although there were similarities, Kamalii and the females' behavior at the underwater gate differed substantially from their behavior at the mirror. The biggest difference was their attention span: the females crowded at the gate, and one or more was almost always at the gate throughout the 40-min experiment. In contrast, when the mirror was present in their tank, single individuals made only occasional visits of a few seconds. Although open-mouth behavior occurred in front of the mirror, we never saw the type of rapid, open-mouth biting that Laukani and Kamalii engaged in across the gate. Bubble blowing occurred both at the gate and at the mirror.

As for play behavior, which occurred both with and without the mirror, it is our impression that mirror play behaviors were qualitatively different from play behaviors elsewhere in the tank, and that the dolphins appeared to utilize the reflective properties of the mirror while playing.

Conclusion from encounters with unknown dolphins

Overall, when we compare dolphins' mirror behavior to their behavior with unfamiliar dolphins, there are more differences than similarities. Mirror-directed behavior is different from social behavior in these experiments. The high rate of vocalization when the four dolphins encountered the false killer whale through the underwater gate was never duplicated by a dolphin in front of the mirror. Dolphins pay close, rapt attention to a real stranger for virtually 100% of a test period, even when it lasts hours. They pay close attention to their mirror image for less than 1% of test periods. This critical and important difference between the two types of conditions supports the interpretation that dolphins' mirror behavior is not social, and that they recognize their mirror images as themselves.

Rate of "maneuvering postures" in mark test compared to no-mark controls

An examination of the column labeled "Mirror–no mark control sample: maneuvering posture rate" in Table 24.2 shows rates of 0–0.25 per hour (average = 0.09/hour). Thus, the control data would predict these postures to be an unlikely event in a 1–3-hour mark test. However, during their mark tests, Keola and Okoa spent the majority of their time in front of the mirror engaging in such maneuvering postures; the postures occurred as well in Hot Rod's, Laukani's, and Itsi Bitsi's mark tests, although at a lower rate and more subtly. If this twisting behavior is considered to be social behavior, we see no reason to expect it to increase during a mark test; if it were self-examination, such an increase would be expected. The virtual absence of the maneuvering postures in the no-mirror condition further supports a self-examination interpretation of these postures.

Comparison with primates: Amount of time spent in front of mirror

Dolphins spend less time in front of mirrors than do primates. Children attend to mirrors and contingent video images on the order of 50% of their test time (Lewis & Brooks-Gunn, 1979); but a dolphin swimming in a tank is very different from a baby sitting in front of a mirror. The baby is more or less physically bound to remain sitting in front of the television, and is thus more likely to watch it. Dolphins, being primarily mobile, are less likely to remain in one spot for an extended period of time. This difference between dolphins and primates forced us to resort to enticing marked dolphins to the mirror; however, we view this as equivalent to putting a mirror in front of a marked primate. (As mentioned, dolphin attention changes virtually to 100% when faced with a real stranger, not a mirror, as in the unfamiliar dolphin controls.)

Furthermore, most studies of mirror behavior are conducted with the subject isolated from conspecifics. Our subjects were tested in social groups, and this may account for differences in behavior. Animals that were tested in larger social groupings tended to spend less time with the mirror. In this context, the dolphins were more interested in social interactions than in mirror interactions.

Dolphins appear to respond to the mirror quite rapidly, in comparison to data reported for primates. The explanation may lie in the presence of windows in their tanks. The quality of the reflections vary, due to time of day, but appear to be adequate to provide experience to the dolphins with reflective surfaces. The lack of novelty of reflections may also explain the relatively brief attention span of the dolphins during mirror exposure: Responses to mirror mark tests might diminish as the novelty of the new "good" mirror wears off.

Although dolphins have good vision, the relative importance of visual stimuli and visual images to them is open to question. In his first mark test Keola rushed over to the mirror as soon as it was being put up and appeared to look at his mark, but then did not return to the mirror for the rest of the 2-hour test. He did, however, eventually rub the zinc oxide off, on the side of the tank. This suggests he may have been more concerned with the tactile stimulus than the visual stimulus. We have the impression from our research that visual images do not have as high a priority for dolphins as they do for primates.

Contingency checking compared with primates

Primate researchers deal with their subjects as if the animals can recognize their own static images. We must guard against this kind of approach for dolphins, as the role of movement could outweigh visual form. It appears that in primates, both self-recognition and contingency checking occur in the great apes, but are absent in monkeys (Gordon Gallup, personal communication, 1991; contra Boccia, *SAAH*23). In humans, contingency play and self-recognition (as measured by a rouge mark test) appear together developmentally, whether the contingency play is in a mirror or on self-view television. Lewis and Brooks-Gunn (1979, p. 109) feel that this supports their belief "that contingent play is not a precursor but rather an indicator of self-knowledge." We observed contingency play in the mirror in all of the dolphins tested including the youngest (1.5 years). Anyone who watches tapes of dolphins from our laboratory looking at themselves in mirrors cannot help but be impressed by the rhythmic head and mouth movements and sounds involved. In dolphins, contingency checking may be necessary for self-recognition.

Environmental, social, and individual influences on test results

Environmental. Our tests were conducted in a situation in which the dolphins are accustomed to seeing people on the other side of the window used for the one-way mirror. They know it is people habitat, not dolphin habitat. Also, we were present during tests, so although the dolphins could not see us, they could hear us if we made any sounds. On the other hand, the test in Marino, Reiss, and Gallup (*SAAH*25) had the one-way mirror located in a channel leading to another tank. The dolphins were used to encountering strange dolphins and interacting with them there, and during these tests they could hear other dolphins behind the mirror. These differences in the test situation would make it much more likely for dolphins in that study to be deceived into thinking they were seeing other dolphins in the mirror than those tested here. Furthermore, when we resorted to tapping on the window to attract the dolphins to the mirror area, we reminded the dolphins that there were people on the other side of the mirror, not dolphins. Most of the mirror-directed behavior we saw did not appear to be the kind of social behavior that generally accompanies encounters with other dolphins, such as hyperactivity and pronounced social cohesion.

Social. Dolphins are intensely social creatures, and the influence of social factors on these tests should not be overlooked. When there were not many dolphins in the tank (two or three), they interacted with the mirror and television quite a bit. When more dolphins were present (four or five), the dolphins were so busy with each other that they paid less attention to their mirror or television image. Social interactions could have influenced the animals' ability to focus attention on the test. A given animal's position within the social hierarchy may have contributed as well. For example, Itsi Bitsi ("top" dolphin at the time) failed to go to the mirror on her first mark test, and instead chased other dolphins, as if immediate hierarchical imperatives prevailed. In addition, our inability to isolate the animals may have affected the unknown-dolphin control test. The ratio of "unknown" to known animals is 1 : 1 in a mirror setup, whereas Kamalii was faced with three new females. This imbalance could have altered Kamalii's behavior. The gender of the individuals involved in a first-time encounter could possibly influence their behavior. The females' being of opposite sex to Kamalii (as well as of different race, in Laukani and Okoa's case) was a confounding variable that could have contributed to differences in behavior through the gate, compared to the mirror.

Individual variation. It is worth mentioning that positive and negative responses to tests like these may reflect individual personality differences rather than intellectual characteristics of the species. There may be a correlation between a natural proclivity for looking in the mirror and the use of it for self-examination. While unmarked, Keola was very interested in the mirror, and he gave intense results when he was marked. Throughout their mirror exposure when unmarked, Itsi Bitsi and Laukani showed no special interest or preference for the mirror, and had to be enticed over to the mirror when marked. The seemingly negative result of an adult animal being mark tested for self-awareness may indicate more about that animal's likes and dislikes than it does about self-awareness. Dolphins who show no natural interest in mirrors may not be good candidates for mirror mark tests.

Conclusion

No single test presented here proves self-recognition in bottlenose dolphins. The tests were developed mainly from primate research paradigms, and their limitations for interpretations of dolphin behavior are apparent. Nevertheless, the data taken together make a compelling case for self-recognition in this species. Four of five dolphins apparently examined their marks in a mirror; most brought objects to the mirror and played with them in front of it, even moving the object back when it drifted out of view; and most of the mirror-mode television tests designed to distinguish self-examination from social behavior suggested self-examination. Not only did dolphins attend to their mirror (or television) images less than 1% of the time as compared to 100% for real dolphins, but they engaged in different behavior with mirrors than they did with other dolphins. The results obtained in the experiments presented here are consistent with the hypothesis that these animals are using the mirror to examine themselves. More definitive results, however, will have to come from methodologies developed specifically for dolphins.

Acknowledgments

We express special appreciation to Don White and Dexter Cate for their dream of an underwater laboratory that would probe the dolphin mind by allowing dolphins to interact with modern human technology, and for their founding work on the lab. In addition, we recognize Don White for creating the project, conceiving the dolphin–video interaction system, and working to secure funding for the project. His thoughtful and valuable ideas, discussions, and suggestions have made this work possible. Sea Life Park, Hawaii, also made this research possible. The project is a cooperative effort between the conservation organization Earthtrust and Sea Life Park. Earthtrust supplies the research staff; Sea Life Park supplies the dolphins, their care, feeding, and maintenance, as well as the laboratory space. We greatly appreciate Sea Life Park's participation in general, as well as help with the mark tests from Sea Life Park curator Marlee Breese and trainers Stephanie Vlachos, Susan Rodgers, Roberta Horne, Keana Pugh, and Rich Nunes. Trainer Carol Chang provided valuable information on the dolphin Kamalii. We would also like to express appreciation and admiration for the tremendous help received from Lori Marino. Shortly after we began exposing dolphins to self-views and delayed playback, Lori and Diana Reiss independently exposed two juvenile males to a one-way mirror and conducted a zinc oxide control-mark and mark-test experiment at Marine World in California. Later, once we knew about each other's work through Dr. Reiss, our research, especially the methodology for systematic tests, benefited greatly from Lori's input. Gordon Gallup gave valuable advice for the mirror mark tests. Maria Boccia, Marlee Breese, Gordon Gallup, Emily Gardner, Don Griffin, Chris Johnson, Lori Marino, Bob Mitchell, Kati Nagy, Sue Parker, Diana Reiss, Don White, and Jana Wolff reviewed the manuscript. Dave Hack provided the schematic rendering of the tank. Soltech in Honolulu, Hawaii, and Security Glass in Pleasanton, California, donated the one-way mirror polyester film. Oceanic Institute provided facility support. The research was supported by Mollie Malone, Wendy Grace, the Boudjakdji Family, the Sarah Stewart Foundation, the Barbara Gauntlett Foundation, Outrigger Hotels, and the members and supporters of Earthtrust.

Note

1. The television tests are described in detail in a paper entitled, "Using self-view television to distinguish between self-examination and social behavior in the bottlenose dolphin (*Tursiops truncatus*)," submitted to the journal *Consciousness and Cognition*.

References

Anderson, J. (1984). Monkeys with mirrors: Some questions for primate psychology. *International Journal of Primatology*, 5(1), 81–98.

Gallup, G. G., Jr. (1970). Chimpanzees: Self-recognition. *Science*, *167*, 86–87.

    (1982). Self-awareness and the emergence of mind in primates. *American Journal of Primatology*, 2, 237–248.

Herman, L. M. (1980). Cognitive characteristics of dolphins. In L. M. Herman (Ed.), *Cetacean behavior: Mechanisms and functions* (pp. 363– 429). New York: Wiley.

Lethmate, J., & Dücker, G. (1973). Untersuchungen zum Selbsterkennen in Spiegel bei Orang-Utans und einigen anderen Affenarten. *Zeitschrift für Tierpsychologie*, *33*, 248–269.

Lewis, M., & Brooks-Gunn, J. (1979). *Social cognition and the acquisition of self*. New York: Plenum.

Marten, K., & Psarakos, S. (1992). Do dolphins perceive television as reality? *Earthtrust Chronicles* (November), 8.

Ostman, J. (1985). *An ethogram for dolphin social behavior, and observations on changes in aggressive and homosexual behavior among two subadult male bottlenose dolphins (Tursiops truncatus) in a captive colony*. Unpublished master's thesis, San Francisco State University.

Overstrom, N. (1983). Association between burst-pulse sounds and aggressive behavior in captive Atlantic bottlenose dolphins (*Tursiops truncatus*). *Zoo Biology*, 2, 93–103.

Suarez, S. D., & Gallup, G. G., Jr. (1981). Self-recognition in chimpanzees and orangutans, but not gorillas. *Journal of Human Evolution*, *10*, 173–188.

25 Mirror self-recognition in bottlenose dolphins: Implications for comparative investigations of highly dissimilar species

Lori Marino, Diana Reiss, and Gordon G. Gallup, Jr.

To date, the only species to show compelling evidence of self-recognition are humans, chimpanzees, and orangutans. (Only one gorilla, Koko, has displayed self-recognition; see Patterson & Cohn, *SAAH*17). An important question is whether self-awareness is a uniquely primate (great ape–human) ability or a potential outcome of any sufficiently intelligent system. In order to address this question we can compare tests of self-awareness in primates with another taxonomic group that has diverged extensively from them yet displays a comparable level of neurobehavioral complexity, namely, the Cetacea. Comparisons of primates and cetaceans, therefore, form the basis for a test of convergent cognitive evolution and could set the stage for subsequent tests of the generality of Gallup's (1982) model of self-awareness.

Dolphins are provocative candidates for self-awareness for a number of reasons. They share those neurological, cognitive, and social characteristics with great apes and humans that are generally regarded as having been important for the development of self-awareness in primates. Humans, great apes, and dolphins show similarities across several measures of encephalization, including a high brain-weight / body-weight ratio (Glezer, Jacobs, & Morgane, 1988; Jerison, 1982), extensive cortical surface area (Jerison, 1982), and a high neocortical-volume / total-cortical-volume ratio (Glezer et al., 1988). The bottlenose dolphin is capable of high levels of performance on a variety of auditory learning, artificial language comprehension, and memory tasks comparable to those mastered by chimpanzees (for a review of this literature see Herman, 1986). Moreover, dolphins are extensive imitators of others' actions, a behavioral capacity that some view as necessary for mirror self-recognition (e.g., Mitchell, 1992). Finally, many dolphin species are similar to chimpanzees in social structure and in group capture of prey (Bradbury, 1986; Connor & Norris, 1982; Johnson & Norris, 1986; Norris & Dohl, 1988; Tyack, 1986). Nonkin-related altruism and interspecific helping has also been informally observed in a number of dolphin species (Connor & Norris, 1982; Pryor, 1973; Tyack, 1986). Dolphins' apparent social intelligence may, like that of great apes and humans, have been a selective factor in their evolution (Humphrey, 1983) that led to self-awareness (Gallup, 1982).

380

In this chapter we describe the responses of two captive bottlenose dolphins with whom we attempted to replicate the mirror self-recognition paradigm originally applied to chimpanzees (Gallup, 1970).

Method

Subjects and apparatus

Because this was the first formal attempt to assess self-recognition in a cetacean species and much was already known about the operational aspects of this procedure (Gallup, *SAAH*3), we attempted to technically replicate the primate mirror paradigm. Subjects were two 7-year-old captive-born male bottlenose dolphins, Pan(ama) and Delphi. The dolphins were half-brothers that had resided together since birth. They were housed in a circular pool, 60 ft in diameter and 18 ft deep, at Marine World Africa USA in Vallejo, California. This pool is connected to an identical pool by a channel 20 ft long and 12 ft deep. No mirrors or underwater windows were available to the dolphins prior to the study.

The dolphins were exposed simultaneously to a 3×5-ft underwater one-way Plexiglas mirror placed at the opening of the channel in the dolphins' pool. The mirror was either uncovered or covered with a 3×5-ft sheet of opaque, nonreflective Plexiglas that slid, on a track in front of the mirror. The dolphins' behavior and sound production were recorded by an underwater Sony Video 8 Handycam CCD-M8u on the transparent side of the mirror, a Navy Sonabuoy hydrophone, and an above-water Sony Trinitron video camera.

Procedure

Baseline behavior was recorded during covered-mirror conditions for 1.5 hours on the day before initial mirror exposure, and for 0.5 hour of mirror exposure on Day 1, 1 hour on Day 7, 0.25 hour on Day 8, 0.25 hour on Day 9, 1 hour on Day 10, and 1 hour on Day 11. The dolphins were exposed to the uncovered mirror for 1 hr on Days 1–4, and for 4.5 hrs on Day 5, 6 hrs on Day 6, 5.5 hrs on Day 7, 5 hrs on Day 8, 3 hrs on Day 9, 4 hrs on Day 10, and 4 hrs on Day 11 (the day of the mark test): a total of 36 hours of mirror exposure. Sham marking was used instead of anesthetization. Both dolphins previously had been trained to go to and remain at their individual stations by the side of the pool when called, and to leave when released with a hand signal. During sham marking, each dolphin, positioned on his side at his station, was rubbed from dorsal fin to the base of the pectoral fin with either a clean towel or a bare hand. This usually occurred during a feeding session. Delphi was sham-marked on Days 2, 8, 10, and 11; Pan was sham-marked on Days 2, 7, 8, and 11. In addition, both dolphins were marked with Neo-Blue (an iodinelike medicinal liquid) on Day 10. This amounted to a sham marking because the dolphins were already habituated to Neo-Blue application; moreover, the substance does not visually contrast well with their skin, and so does not represent a true mark. On Day 11 the dolphins were called to their stations and zinc oxide (a white tactile substance that contrasted with their skin color) was applied by hand to the left side of each dolphin from the base of the pectoral fin back toward the base of the dorsal fin. The dolphins were then released. Pan was marked in the morning session and Delphi was marked in the afternoon session.

Behavioral analysis

The dolphins' videotaped behavior was coded using the behavioral ethogram in Table 25.1. Scoring involved a frequency count of behaviors in Table 25.1 for each animal for the entirety of the videotapes. Behaviors were more generally categorized as "suggestive" (self- or mark-directed), social/aggressive, sexual, or other (Table 25.2). *Suggestive* behaviors were defined as those movements and postures observed only during exposure to the mirror and/ or when the mark was exposed to the mirror. These excluded typical social behaviors and were more narrowly defined as those behaviors in which an animal would position itself in front of the mirror and exhibit repetitive movements of the mouth, head, or body, as though checking the contingency of these movements, or perform apparent self-examination of the inside of the mouth or of body parts, or exhibit related behaviors. Mean viewing time was also obtained for each dolphin for the entirety of the videotaped sessions. Reliability scores for behavioral scoring were not quantified. Scoring was done primarily by one investigator (LM), a subset of whose scored behaviors were confirmed by one of the other authors (DR) in order to reach agreement about the interpretation of behaviors.

Results

Viewing time

Viewing time included only the times when each subject was positioned in front of the mirror with its head oriented to the mirror, and/or swam by with its head turned toward the mirror. The viewing-time patterns of both dolphins, measured as the number of seconds spent viewing for each 5-min period in the entirety of the tapes, were strongly positively correlated, $r = .85$ ($p < .001$) during the ten days prior to the mark test, as well as during the mark test, $r = .94$ ($p < .001$). (See Figures 25.1 and 25.2.)

Specific responses to the mirror

The dolphins exhibited highly synchronized behaviors at the mirror, such as repetitive head movements and posturing, as well as other repetitious

Table 25.1. *Behavioral ethogram*

Category	Code[a]	Behavioral description
Approach	BUA	Backups and advances
	CH	Charge (fast swim straight ahead)
	SBC	Swim-by – clockwise
	SBX	Swim-by – counterclockwise
	ST	Stationing (stationary position)
	BR	Barrel roll (corkscrew forward)
	CL	Circling
	AB	Absent
Body position	HO	Head oriented at mirror (rostrum perpendicular to mirror)
	LRO	Lateral right orientation (right side to mirror)
	LLO	Lateral left orientation
	VO	Ventral orientation (ventral side to mirror)
	MS	Marked side to mirror (during mark test)
	UMS	Unmarked side to mirror
	IVS	Inverted swim (upside-down position)
Sound production	E	Echolocation (click trains)
	W	Whistle (modulated pure tones)
	SQ	Squawk (low-freq. pulsed sounds)
	UV	Underwater sounds (other)
	PID	Positive ID of animal who made sound
Air production	BB	Bubble burst or ring
	BS	Bubble stream
Contact	CMF	Contact w/ mirror frame (of any part of body)
	CMS	Contact w/ mirror surface
	RPD	Rostrum pull down (vertical rubbing of mirror or frame w/ rostrum)
Sexual	ER	Erection
	DPIA	Intromission attempt – Delphi to Pan
	PDIA	Intromission attempt – Pan to Delphi
	BG	Beak-to-genital contact
	DMFV	Dead man's float – vertical (head up or down suspended float signaling sexual receptivity)
	DMFH	Dead man's float – horizontal (body parallel to surface)
Movements	HJU	Head jerk upward
	HJD	Head jerk downward
	HSV	Head shake – vertical (up–down)
	HSH	Head shake – horizontal (sideways)
	RHMV	Repetitive head movement – vertical (slower than head shake)
	RHMH	Repetitive head movement – horizontal
	TS	Tail slap (on surface of water)
	TKP	Tail "kerplunk" (deep slap creating vertical column of bubbles under water)
	HCO	Head cocking
	RHCL	Repetitive head circling
	BR	Body rocking
	PM	Pectoral fin movement
	HP	Head pass (in front of other animal)
	DIP	Dipping head below mirror – coming up suddenly
Open mouth	OM	Open mouth
	WOM	Wide open mouth
	JC	Jaw clap (open mouth followed by rapid closing of mouth – sound made)

[a] An "S" was added to codes to indicate that the given behavior was synchronously performed by both subjects.

Table 25.2. *Specific behavioral responses*[a]

Social/aggressive	Sexual	Suggestive
Squawk	Erection	Open mouth
Jaw clap	Intromission attempt	Bubble streams (while watching them
Tail slap	Beak-to-genital contact	in mirror)
Tail kerplunk	Dead man's float	Head dipping
Charge		Any repetitive head and/or body
		movements

[a] Three categories of behaviors considered important for evaluating the responses of our subjects to the mirror.

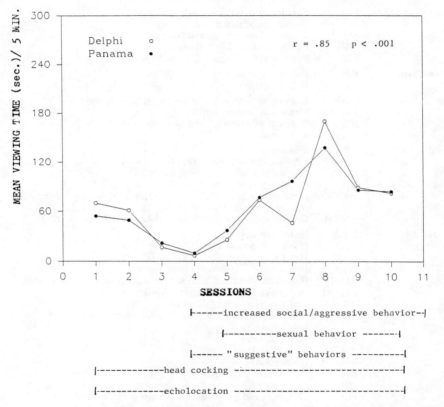

Figure 25.1. Mean viewing time per 5-min intervals (in sec) across 10 days of mirror exposure for both subjects. The presence of more specific behaviors throughout this period is also indicated below the graph.

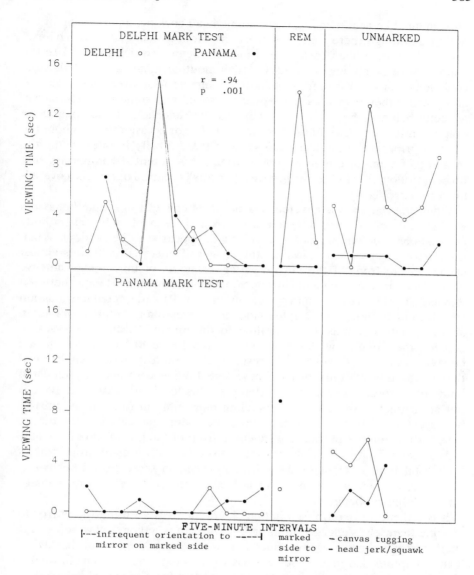

Figure 25.2. Mean viewing time per 5-min intervals (in sec) during the mark test, mark removal phase (REM), and unmarked conditions for both subjects independently.

movements like head circling, body tilting, head cocking, and opening of the mouth. Although there were numerous instances of "suggestive behaviors," such as repetitive head cocking and mouth opening (which looked like the animals were examining the insides of their mouths), many of these behaviors could not be distinguished from open-mouth threats. Moreover, these occurred along with other responses that appeared social and sometimes aggressive. Mimetic behavior between a dolphin and the reflection of the companion dolphin in the mirror was observed on a number of occasions throughout the 10 days prior to the mark test. In spite of the difficulty in interpreting the meaning of some of the above observations, it is potentially important that no behaviors similar to the "suggestive behaviors" occurred during the covered-mirror control condition.

When the mirror was covered, the dolphins engaged in sexual behaviors with each other in many different areas around the pool and in many body orientations. The pattern of these behaviors appeared to be random. When the mirror was exposed, however, they engaged in sexual behaviors almost exclusively in front of the mirror. After 8.5 hours of mirror exposure we observed an increase in tactile interaction and intromission attempts between Pan and Delphi. In one 30-min period on this day, Pan attempted intromission with Delphi 24 times and Delphi made 19 intromission attempts with Pan; in all cases both dolphins were oriented to the mirror. If, during the sexual activity, their bodies would move out of the frame of the mirror, sexual behavior ceased and would only resume when both dolphins repositioned themselves in front of the mirror. These sexual sequences were long, repeated bouts of reciprocal intromission attempts. This localized sexual behavior is not an artifact of the dolphins' spending more time in front of the mirror, because the frequency of these sexual behaviors increased dramatically, whereas the amount of total time spent in front of the mirror did not change substantially. That is, the dolphins spent a substantially greater proportion of their total time in front of the mirror engaged in these sexual behaviors. However, we cannot rule out a social facilitation effect of the mirror image on the dolphins' behavior.

It is difficult to determine whether the dolphins were using the mirror to observe themselves engaged in these behaviors or whether they were viewing their mirror images as conspecifics and were engaged in social displaying. These dolphins had much prior experience viewing other animals through a gate where the mirror was now located, and this form of sexual behavior or sexual displaying was not previously observed, based on anecdotal information.

Some behaviors were enacted exclusively in front of the mirror. After approximately 8 hours of mirror exposure, Pan and Delphi were positioned in front of the mirror; during a 1-minute sequence Pan oriented close to the mirror, emitting various types of bubbles and bubble streams while producing a variety of sounds. He oriented his blowhole toward the mirror while doing this, and appeared to be monitoring the bubbles. (This was atypical behavior for these dolphins.) Still orienting to the mirror surface, he appeared to touch the bubbles rising above his head by continuously monitoring the reflected image. Without a mirror, these dolphins have been observed to blow bubble rings and visually follow them to the surface (Reiss, 1990), but Pan's behavior during this sequence differed in his attention to the mirror image, apparently to locate the rising bubbles.

Responses to the mark

The behavior most compelling as evidence of self-recognition occurred during the mark test. Although both dolphins were in the pool during the test, only one dolphin was marked at a time. After one of the dolphins was marked and given a release signal, they both broke into an hour-long fast swim that ended when they were called to their feeding stations to have the zinc oxide removed with a towel. During the removal of the mark, the marked subject independently (and without being given the release signal) left his station and swam directly to the mirror, repeatedly exposing the marked side of his body to the mirror, and then returned to his station (again, without being given a signal to do so), where more of the mark was removed. Furthermore, when each dolphin swam to the mirror, the very first and last positions at the mirror were those that exposed the mark to the mirror. This suggests that the purpose of visiting the mirror was to inspect the mark. This behavior was independently exhibited by both dolphins three times, and was exhibited by only the marked dolphin in each session. This response was observed only during the mark-removal phase. Pan and Delphi's behavior during the mark removal, though not definitive, is consistent with the notion that they used the mirror to obtain information regarding a marked area of their body.

Discussion

Patterns of viewing time for dolphins were very different from those found with chimpanzees and orangutans. Typically, chimpanzees spend an increasing amount of time at the mirror in the first two or three 8-hour days after initial exposure. By the 3rd or 4th day there is a sustained decrease in viewing time, followed by a renewed interest in viewing themselves after seeing their marked facial areas in the mirror. In chimpanzees, the drop in viewing time around the 3rd or 4th day coincides with the transition from treating the mirror as a social stimulus, to apparent self-recognition (see Gallup, 1970).

In our dolphins there was an almost converse pattern in viewing time in the course of days, even when taking into account the differing overall length of mirror-exposure periods between our study and the primate studies. The dolphins showed an increase in viewing behavior over the 10 days of exposure. When marked, they actually spent less time at the mirror. Furthermore, the notable viewing-time decrease associated with the emergence of self-recognition in chimpanzees was not observed in dolphins, and their viewing time patterns did not correspond to the presence or absence of any specific behaviors regarded as social/aggressive, sexual, or "suggestive."

Some behaviors that Pan and Delphi exhibited during mirror exposure might suggest that they were engaged in testing the contingencies between the mirror image and their own bodies. However, the presence of apparently social/aggressive behaviors at the same time prevents us from concluding that the dolphins were in fact exhibiting self-recognition. Furthermore, the apparent mimetic behavior between a dolphin and the reflection of the companion dolphin may not be as telling as it may seem because rhesus monkeys, who do not show self-recognition, respond similarly to the mirror image of familiar conspecifics (Gallup, Wallnau, & Suarez, 1980). Even for the behavioral sequence we observed when each marked subject swam to the mirror during the mark-removal period, alternative explanations cannot be ruled out.

Methodological and conceptual issues

It is important to consider the issue of technical and conceptual replications in the context of the present study. We attempted to replicate the primate mirror studies *technically*. We were arguably successful, with a few exceptions. First, our subjects were exposed to the mirror for only 33 hours before a mark test was conducted, as compared to the 80 hours of exposure the chimpanzees received prior to marking. However, our subjects received the minimum amount of mirror exposure found necessary for chimpanzees and orangutans to shift from social responding to self-directed behaviors. Second, we utilized sham marking instead of anesthetization; sham marking is commonly used as an alternative to more invasive procedures and does not seem to have a significant impact on test outcomes (Calhoun & Thompson, 1988; Lin, Bard, & Anderson, 1992). Moreover, because the mark we used had tactile properties, the importance of the subjects' inability to detect the mark prior to mirror exposure was essentially negated. It was expected that the dolphins would use the mirror to investigate the mark that they felt on their skin.

Gallup (*SAAH3*) states, "To became a bonafide scientific fact an observation must be replicated by a number of different observers, in a number of different settings, on a number of different subjects." This statement does not refer simply to matters such as interrater reliability or increasing the n in a study, but alludes to the convergent validation that is required in order for a scientific finding to become viable. This is true of both intra- and interspecies replications, which are valid to the extent that they represent conceptual, not simply operational or technical, replications of hypothesized suppositions.

The mark test has been replicated in chimpanzees, orangutans, and, less formally, in one gorilla (see Gallup, 1975, for review; Patterson, 1984; Suarez & Gallup, 1981). Investigations of numerous monkey species, gibbons, and elephants (*Elephas maximus*) have failed to demonstrate convincing evidence of self-recognition (for reviews see Anderson, 1984; Povinelli, 1987, 1989).

Until our study, with the exception of the elephant, attempts at assessing self-recognition within this paradigm have been limited to close interspecific comparisons that have been both technically and conceptually similar. Even the elephant shares enough morphological and perceptual capacities with the primates to make technical replications of the mirror study conceptually similar to that with primates; but as we discovered, extensions of the mirror self-recognition procedure to a species as highly dissimilar from the primates as *Tursiops* spp. stretches the useful limits of a strictly technical replication and demands deeper attention to conceptual issues.

Our study of dolphin self-recognition can be examined, consequently, on the basis of the extent to which it meets the criteria for a *conceptual* replication of the primate work. The inconclusive nature of our findings with dolphins may be due to the fact that we focused on technically replicating the primate studies and may not have adequately replicated the conceptual features.

Morphology

The dolphins' lack of forearms and hands precludes their ability to exhibit observable self-directed responses in the same manner as primates. It was thought, however, that the dolphins would demonstrate self-directed behavior by posturing in ways that would "look like" self-directed behavior. The dolphins did so, but these findings were not conclusive. Whereas we did find that, during mark removal, both subjects positioned themselves at the mirror with the mark exposed, we did not find any other instances of posturing that were as unambiguous.

Motivation

The motivation to examine marks on the body may be low in species that do not have the ability to do much about them. (Dolphins do little else to remove debris from their bodies except breach or rub against the side of a pool.) For example, we found that as one of the subjects was marked, the unmarked dolphin displayed no special attention toward the mark on its lifetime companion, even though the marked dolphin looked radically different. This suggests that visual appearance may not be as important for dolphins as for primates. Chimpanzees have been observed to examine marks on cagemates, and have a proclivity for social grooming that is lacking in the natural social ecology of the dolphin.

Sensoriperceptual

Whereas the primary sensoriperceptual modality in primates is visual, the rich use of auditory information by dolphins suggests that the inclusion of auditory information may be important for future investigations of self-awareness. Dolphins have excellent visual acuity and, especially in close spatial proximity, rely on visual identification of objects. On the other hand, the way in which dolphins have evolved to identify themselves and others may be more richly represented in the auditory realm. Recent evidence regarding the reactions of bottlenose dolphins to real-time and delayed videotapes of themselves suggests they may have a more temporally dependent dynamic form of information processing (Marten & Psarakos, *SAAH* 24). This suggests not only that the more dynamic auditory realm would be better suited to future investigations of self-awareness, but that self-identity and awareness in dolphins may be different in nature than that in primates.

The questions that arise from our discussion of the differences between primates and cetaceans are far-reaching. In our search for self-awareness in dolphins are we looking for something that we recognize as humanlike? Are we prepared to recognize "it" in such a different organism as the dolphin? Can dolphins conceive of themselves? Indeed, some of the most important questions about convergence in the evolution of intelligence will only be answered by comparisons across highly dissimilar species such as cetaceans and primates. It is clear that in order to proceed we must translate our hypotheses into a species-relevant conceptual replication for the species under study.

Acknowledgments

The authors wish to thank those at Marine World Africa USA who worked on the project with us. This research was partially funded by grants from Sigma Xi and the State University of New York Benevolent Association to the first author.

References

Anderson, J. R. (1984). Monkeys with mirrors: Some questions for primate psychology. *International Journal of Primatology*, 5, 81–98.

Bradbury, J. (1986). Social complexity and cooperative behavior in delphinids. In R. Shusterman, J. A. Thomas, & F. G. Wood (Eds.), *Dolphin cognition and behavior: A comparative approach* (pp. 358–374). Hillsdale, NJ: Erlbaum.

Calhoun, S., & Thompson, R. L. (1988). Long-term retention of self-recognition by chimpanzees. *American Journal of Primatology*, 15, 361–365.

Connor, R. C., & Norris, K. S. (1982). Are dolphins reciprocal altruists? *American Naturalist*, 119(3), 358–374.

Gallup, G. G., Jr. (1970). Chimpanzees: self-recognition. *Science*, 167, 86–87.

——— (1975). Towards an operational definition of self-awareness. In R. H. Tuttle (Ed.), *Socioecology and psychology of primates*. The Hague, Netherlands: Mouton.

——— (1982). Self-awareness and the emergence of mind in primates. *American Journal of Primatology*, 2, 237–248.

Gallup, G. G., Jr., Wallnau, L. B., & Suarez, S. D. (1980). Failure to find self-recognition in mother–infant and infant–infant rhesus monkey pairs. *Folia Primatologica*, 33, 210–219.

Glezer, I., Jacobs, M., & Morgane, P. (1988). Implications of the "initial brain" concept for brain evolution in Cetacea. *Behavioral and Brain Sciences*, 11, 75–116.

Herman, L. M. (1986). Cognition and language competencies of bottlenosed dolphins. In R. J. Shusterman, J. A. Thomas, & F. Wood (Eds.), *Dolphin cognition and behavior: A comparative approach* (pp. 221–252). Hillsdale, NJ: Erlbaum.

Humphrey, N. K. (1983). *Consciousness regained*. Oxford University Press.

Jerison, H. J. (1982). The evolution of biological intelligence. In R. J. Sternberg (Ed.), *Handbook of human intelligence* (pp. 723–791). Cambridge University Press.

Johnson, C. M., & Norris, K. S. (1986). Delphinid social organization and social behavior. In R. Shusterman, J. Thomas, & F. Wood (Eds.), *Dolphin cognition and behavior: A comparative approach* (pp. 335–346). Hillsdale, NJ: Erlbaum.

Lin, A. C., Bard, K. A., & Anderson, J. R. (1992). Development of self-recognition in chimpanzees (*Pan troglodytes*). *Journal of Comparative Psychology, 106*(2), 120–127.

Mitchell, R. W. (1992). Developing concepts in infancy: Animals, self-perception, and two theories of mirror self-recognition. *Psychological Inquiry, 3,* 127–130.

Norris, K. S., & Dohl, T. P. (1988). The structure and function of cetacean schools. In L. M. Herman (Ed.), *Cetacean behavior: Mechanisms and functions* (pp. 211–261). New York: Wiley.

Patterson, F. G. (1984). Self-recognition by *Gorilla gorilla gorilla. Gorilla, 7*(2), 2–3.

Povinelli, D. J. (1987). Monkeys, apes, mirrors, and minds: The evolution of self-awareness in primates. *Human Evolution, 2*(6), 493–509.

(1989). Failure to find self-recognition in Asian elephants (*Elephas maximus*) in contrast to their use of mirror cues to discover hidden food. *Journal of Comparative Psychology, 103*(2), 122–131.

Pryor, K. W. (1973). Behavior and learning in porpoises and whales. *Die Naturwissenschaften, 60,* 412–420.

Reiss, D. (1990). The dolphin: An alien intelligence. In B. Bova & B. Preiss (Eds.), *First contact: The search for extraterrestrial intelligence* (pp. 31–40). New York: New American Library.

Suarez, S. D. & Gallup, G. G., Jr. (1981). Self-recognition in chimpanzees and orangutans, but not gorillas. *Journal of Human Evolution, 10,* 175–188.

Tyack, P. (1986). Population biology, social behavior and communication in whales and dolphins. *Trends in Ecology and Evolution, 1*(6), 144–150.

26 Further reflections on mirror usage by pigeons: Lessons from Winnie-the-Pooh and Pinocchio too

Roger K. R. Thompson and Cynthia L. Contie

In the Disney film version of Milne's (1958) *Winnie-the-Pooh*[1] there is a scene where Tigger (a tiger) introduces himself to Winnie, or Pooh bear, as being unique. "I'm the only one!" he announces proudly. Pointing to a mirror in which Tigger's appearance is reflected, Pooh responds by asking, "Then, who's that?" Tigger's response is interesting. Instead of reasoning that the mirror image is a reflected identical representation of himself – he is, after all, unique – he concludes instead that it must be an impostor. His subsequent attempts to threaten this other animal succeed only in scaring himself under a bed. Pooh's behavior toward his own mirrored reflection also is interesting. When some of the sewing on his back comes undone – he is a stuffed teddy bear – Pooh uses the mirror to guide his efforts at repairing that area of his body that he otherwise cannot see directly. After finishing this task, Pooh turns, faces the mirror directly, and says, "Thank you." Pooh's behavior suggests that he recognizes the functional significance of the spatial invariants between objects in space, including his body, and their images reflected in the mirror. However, he nevertheless perceives the image as "other" and fails to recognize that it is simply a representation of himself. Pooh doesn't grasp the fact that he is literally talking to himself (cf. Jaynes, 1976). Pooh was, after all, "a bear of very little brain" (Milne, 1958).

Neither Tigger nor Pooh seem capable of self-recognition of the kind displayed by the Ugly Duckling (Andersen, 1984) as depicted in another Disney film.[2] Rejected by his supposed siblings, the depressed young bird catches sight of his reflection off the water's surface at the edge of a lake. He gazes at his image, which is distorted from the movements of the water, and then, turning toward the movie viewer with a look of abject despair, he points at himself as if to say, "C'est moi!" Unlike Pooh and Tigger, the Ugly Duckling spontaneously labels the reflection as self in conjunction with a self-evaluative emotional response (cf. Lewis, *SAAH2*; Lewis & Brooks-Gunn, 1979). His behavior suggests a self-concept as rich as that of any normal mature human.

392

The disparate reactions of the three animated characters to reflected self-images provide useful analogies for patterns of behavior typically displayed by nonhuman species in the presence of mirrors (see Gallup, 1975, for a review). Tigger's spontaneous reaction to his mirrored image is typical of that shown by most nonhuman primate species, other mammals, birds and fish. Initially, an animal will react to its mirrored reflections as if faced with another animal. With continued exposure to the mirror, the animal either avoids the reflected information or ignores it. Behaviors elicited by the mirror image habituate, and the animal shows no evidence that it recognizes itself in the mirror. Some monkeys, however, can be trained to use a mirror as a perceptual tool. That is, they learn to use information reflected from a mirror to guide their own actions as, for example, in reaching for otherwise unobservable objects (Anderson, 1986; Itakura, 1987a,b, 1992). Nevertheless, unlike Winnie-the-Pooh, no monkey has ever been observed to spontaneously use mirrored information to guide self-directed movements toward otherwise invisible marks on its own body (Anderson, 1983; Anderson & Roeder, 1989). The general consensus has been that, apart from humans, only chimpanzees and orangutans, and possibly gorillas, have this latter ability (Gallup, 1970; Ledbetter & Basen, 1982; Lethmate & Dücker, 1973; Patterson, 1984). In this regard then, these apes, if not monkeys, are like Winnie-the-Pooh. The received wisdom, however, suggests that the Ugly Duckling is an even more appropriate analogy for these apes.

Some investigators have argued that an ape's ability to use information from mirror images spontaneously to guide responses to its own body (i.e., self-recognition) and, for some, to label self-images, like the Ugly Duckling, is evidence of a self-concept and self-awareness that is absent in other species (e.g., see Gallup, 1970; Suarez & Gallup, 1981). Epstein, Lanza, and Skinner (1981), however, claimed that pigeons as well as chimpanzees could use a mirror spontaneously, as was evidenced by the former species's ability to locate and peck at dots on their bodies that were otherwise not visible. Epstein et al. (1981) dismissed the notion that their birds' behavior was mediated by a self-concept. Instead, they argued that the mirror's use for guiding responses directed at their bodies by both pigeons and chimpanzees could be accounted for by appealing solely to a history of reinforcement contingencies.

Epstein et al. (1981) reported that the self-directed behavior of their pigeons, which they did not explicitly reinforce, emerged spontaneously after the birds had been trained first on two different tasks. In the absence of a mirror, the pigeons were trained to peck visible blue dots placed on their bodies. They were trained also to peck those locations on the walls of the test chamber where a dot had been flashed momentarily. In this latter case, a bird was required to face a mirror in the front wall of the experimental chamber. The experimenters then briefly exposed a dot behind the bird from one of several holes located in the rear and right walls of the chamber. Hence, presumably the birds did not view the dots directly, but saw only their mirror images. Only after the flashed dot's mirror image had disappeared was the bird

Table 26.1. *Range of median dot-directed responses by pigeons in the control and experimental conditions*

		No. of responses	
Study	No. of subjects	Mirror covered	Mirror exposed
Epstein et al. (1981)	3	0	4–16
Thompson & Contie (1986)	6	0	0

rewarded for turning away from the mirror and pecking the chamber wall location where the dot had been flashed. The experimenters observed the bird through the remaining transparent chamber wall.

After a pigeon had acquired the two above tasks the investigators placed dots on it, but, importantly, a bib worn by the bird around its neck obstructed any direct view of these dots. The bib covered the dot whenever the bird leaned forward. The investigators reported that in the presence of the mirror the pigeons bobbed and pecked at the position on the bib corresponding to the hidden dot's location on the bird's breast. As shown in Table 26.1, however, these same birds never pecked toward the obscured dot when the mirror was covered. Epstein et al. (1981) interpreted these results as demonstrating that their birds used the mirror to locate the position of otherwise invisible body dots, even though they had not been explicitly trained to do so. Furthermore, as noted earlier, these investigators argued that this ability resulted solely from the pigeons' prior reinforcement histories. Epstein et al. (1981) inveighed against the necessity of appealing to a self-concept as a causal explanation for any species' ability to guide self-directed responses via mirror images.

Further reflections on mirror usage by pigeons

Even if one eschews the absence or presence of a self-concept, one cannot ignore the fact that learning to use a mirror to locate things, including on one's body, depends on discovering a mirror's perceptual properties and affordances. As described by Loveland (1986), these properties include, for example, reflection of symmetrical virtual and actual layouts of surfaces in which the axis of left and right, but not the vertical relationship, is reversed; ordinal and dynamic properties (i.e., rhythm) of events are preserved also. A mirror uniquely affords perception of otherwise invisible portions of a layout, but one cannot physically act upon (e.g., grasp) reflected objects, although one may interact socially with the reflection of another organism (Loveland, 1986). By focusing on the final performance of their birds alone, Epstein et al. (1981) failed to identify precisely which of several possible perceptual relationships between objects and their mirror reflections were functionally significant for their birds.

Failure to replicate "self-awareness" in the pigeon

The emphasis by Epstein et al. (1981) on the role of reinforcement in the acquisition of mirror-controlled behavior may have documented *how* their pigeons learned the affordances of the mirror. However, they failed to identify *which* specific perceptual contingencies between objects and their mirrored reflections were learned by their birds. Our goal was not only to replicate Epstein et al.'s earlier results, but also to explore the latter issue. Our first unsuccessful attempts (Gelhard, Wohlman, & Thompson, 1982) to replicate the results of Epstein et al. (1981) were criticized by Epstein (personal communication, 1983) on several grounds, including that our strain of birds differed from his and that our experimental chamber also differed from his in several dimensions.

Subsequently, we attempted to replicate and extend Epstein et al.'s (1981) findings by using white Carneaux birds and by constructing a new chamber that met his criteria (Thompson & Contie, 1986). The front wall of the test chamber consisted of a mirror that could be either exposed or hidden by an opaque cover. The opening to an automatic grain feeder was situated on the gray rear wall of the chamber (i.e., as viewed by the bird) as were three circular holes arranged in an inverted triangular pattern. Each hole was backed by white opaque Plexiglas upon which a blue stimulus dot was mounted. These could be presented or removed from sight at each hole by sliding the plexiglas back and forth. Stimulus dots could be presented also at two signlike rectangular extensions of the left-hand side wall that were angled such that all holes were visible in the exposed mirror. The right-hand side chamber wall was clear Plexiglas through which each bird's behavior was observed and videotaped. Like Epstein et al. (1981), we trained our six pigeons, in the absence of the mirror, to peck at visible blue stick-on paper dots placed on various parts of their bodies. With the mirror exposed, we rewarded the same birds for approaching and pecking holes in walls, but only if we had first flashed a blue dot briefly at that location when the birds faced the mirror. Subsequently, we tested each bird to see if it would direct pecks toward dots that, although hidden by a bib, were reflected visibly in the exposed mirror. Each bird was so tested a minimum of three sessions.

As shown also in Table 26.1, all our birds, unlike those tested by Epstein and his colleagues, failed this transfer test. Results from additional tests further revealed that the failure of our birds to pass the transfer test was not caused by the bib inhibiting body-directed pecking. As shown in Table 26.2, regardless of whether or not they wore a bib, each bird pecked body dots that were directly visible when the mirror was covered. However, there was evidence that the mirror inhibited body-directed pecking even when body dots were directly visible. As shown also in Table 26.2, the birds were much less likely to peck visible body dots when the mirror was exposed. Two additional observations supported this evidence for discriminative stimulus control of the birds' behavior by the mirror per se. When the mirror was covered, our

Table 26.2. *Effects of bib and mirror on the number of pecks directed toward body dots that were directly visible*

| | Mirror covered | | Mirror exposed |
	Bib on	Bib off	(bib off)
Mean	13.10	15.25	4.0
SD	8.68	16.23	2.61
Median	10.5	8.5	4.0
Range	4–27	1–50	0–7

Table 26.3. *Response of six pigeons to flashed wall dots*

| | | Proportion of trials with response to dot | | Proportion of responses at correct location | |
Condition[a]	Stimulus condition	Range	Overall	Range	Overall
1	Mirror	—	1.0	.8–1.0	.90
2	Nonreflective glass	0.0–.5	.46	.4–1.0	.71

[a] On each trial (10 per subject) in Condition 1, dots were flashed at one of three rear and one side wall locations in the test chamber. In Condition 2, dots were flashed at rear and side walls located in a chamber that was a mirror replica of the test chamber.

birds often scanned their bodies for body dots, even when none were present. When the mirror was exposed the birds were more likely to turn away from it and spontaneously approach the rear and side walls, even when no wall dots were flashed.

This latter observation raised the question of whether the reflections of dots flashed on the walls in any way controlled our birds' behavior. Using a mounted bird as a model, we had originally positioned the wall holes at locations outside of a pigeon's visual field as it faced the mirror (McFadden, 1987). However, it is difficult to determine by observation exactly where a live pigeon is looking (Friedman, 1975), and therefore, perhaps our birds had detected the flashed dots directly. To test this possibility we replaced the mirror with nonreflective glass and flashed dots at the holes in the rear and side walls only when the birds faced the nonreflective glass.

The results are summarized under Condition 2 in Table 26.3. Despite the absence of reflected dot cues the birds overall detected when a dot had been flashed on about one-half of the test trials. Also, on such trials they subsequently pecked with considerable accuracy the correct wall location after the dot had disappeared. In fact, the detection of flashed dots and subsequent choice of the correct location by two birds was only marginally lower in the nonreflective glass condition than in the mirror condition. Reflection of a dot, then, was unnecessary for these two "cheater" birds. Stronger evidence for discriminative control by dot reflections, however, was obtained from the remaining four subjects. Three of these latter birds detected all reflected flashed dots in the mirror condition, and they responded to the correct hole on 9 or all 10 trials. When they faced the nonreflective glass these same birds detected only one-half or fewer of the flashed dots, but when they did so, two of the three chose the correct location to peck as accurately they did in the mirror condition. Only one of the six subjects failed to detect and respond to any of the dots flashed as it faced the nonreflective glass. The same bird, however, had responded to dots without fail, and with 80% accuracy, in the mirror condition.

Two of the six subjects had used peripheral information rather than information from reflections to detect the location of flashed wall dots. This result indicated that a glass surface alone – mirrored or not – was a sufficient discriminative cue for these birds to continuously monitor the rear and side walls for directly visible dots as best they could with their bodies oriented toward the front of the test chamber. Flashed dot reflections were effective discriminative stimuli in the case of the mirror condition for the remaining four subjects. This was particularly true for one bird. What remained unclear, however, was whether or not dot reflections alone provided sufficient information for these latter four birds to accurately determine the location of the correct wall hole. One might argue from the results that reflection of a dot anywhere in the mirror served as a discriminative cue for these birds to engage in a visual search for any directly visible dot and then, if they were quick enough, to determine the dot's location before it had disappeared from sight completely.

Sources of stimulus control on pigeon performances

We next looked at whether the performances of the four birds for whom dot reflections were discriminative stimuli depended upon their ability to visually detect a dot directly. In these tests we used an additional second chamber that was a mirror image replica of the original. The two chambers were joined together by a common clear Plexiglas front wall that replaced the mirror used in the original test chamber. When viewed from the original chamber through the Plexiglas divider, the mirror image replica chamber preserved the perceptual spatial invariances associated with the visual array of the original mirror. For example, if a bird in the original chamber were to face the new chamber and observe it through the common Plexiglas wall, then the opaque wall of both chambers would be to its left. Similarly, the Plexiglas observation wall of both chambers would be to its right. Also, for each hole in the original chamber's rear and opaque side walls there was a corresponding contralateral hole in the rear and opaque walls of the replica chamber. Hence, if a bird were to face, say, the rear wall of the second chamber, then a hole to its left corresponded with a hole to its right as it faced the rear wall of the original chamber.

We tested the four pigeons for whom dot reflections were discriminative stimuli using the mirror image replica chamber under three conditions, each of which preserved to differing degrees the information associated with a bird's reflection in the mirror condition (which was used also as a control):

1. Another live, but experimentally naive, pigeon was placed into the mirror image replica chamber during testing of the original four subjects. This manipulation preserved movement cues, if not the contingency between a subject's movements and its reflection in the mirror.
2. A mounted (i.e., "stuffed") bird was placed into the mirror image replica chamber and oriented toward the Plexiglas divider facing the original chamber. This manipulation preserved the physical features of a bird, but eliminated both contingency and movement cues associated with a subject's reflection.
3. No bird, live or mounted, was placed in the mirror image replica chamber. Hence no cues associated with self-reflection in the mirror were present in the visual array facing a subject bird.

In all three test conditions a dot was flashed at one of the wall hole locations in the mirror image replica chamber, but only when the subject in the original test chamber was facing the Plexiglas partition between the two chambers. This eliminated, therefore, the possibility of a bird seeing a dot directly at a location in the original chamber that corresponded to the location of a dot flashed in the mirror image replica chamber. Dots were never flashed in the original chamber except in the mirror condition. In each of the three test conditions described above we recorded whether the subject birds pecked a wall hole, correct or not, in the original chamber on trials whenever a dot was flashed from one of the wall hole locations in the mirror image replica chamber. In test sessions a block of 10 test-condition trials alternated with a block of 10 trials in the mirror-control condition.

The results are summarized in Table 26.4. When exposed to the mirror condition the performance of all subjects was as impressive as previously. In each test condition, however, overall responsiveness of all four birds to dots flashed in the mirror image replica chamber was no better than when the nonreflective glass was used and dots were flashed from locations within the original test chamber. No birds responded to dots flashed in the mirror image chamber when they faced the mounted bird in Condition 2. Two birds never responded to flashed dots in the mirror image replica chamber when there was no bird opposite them in Condition 3. One of these latter birds never responded when it faced the live bird in Condition 1 and, unlike the other 3 subjects, had not responded previously in the nonreflective glass condition.

One bird then responded only to dots flashed in the original chamber, and only if they were reflected in the mirror. This pattern of results indicates that for this bird a dot reflection served as a discriminative cue to engage in a visual search for moving dots that were directly visible, and then to respond to the location where a dot had just disappeared. The low proportion of trials on which the other 3 subjects responded to dots in the three test conditions are consistent with this hypothesis. Interestingly, however, on those few trials

Table 26.4. *Response of four out of six pigeons to flashed wall dots*

Condition[a]	Partition & stimulus bird	Proportion of trials with response to dot		Proportion of responses at correct location	
		Range	Overall	Range	Overall
1	Plexiglas/live bird	0.0–.2	.125	.5–1.0	.8
	(Mirror reflection)	(.8–1.0)	(.95)	(.8–1.0)	(.92)
2	Plexiglas/mounted bird	—	0.0	—	—
	(Mirror reflection)	(—)	(1.0)	(.8–1.0)	(.92)
3	Plexiglas/no bird	0.0–.2	.125	.3–1.0	.60
	(Mirror reflection)	(.9–1.0)	(.97)	(.88–1.0)	(.93)

[a] On each trial (10 per subject) of each experimental condition a dot was flashed at one of three rear and two side wall holes located in a chamber that was a mirror replica of the test chamber. In the mirror-reflection control condition, dots were flashed at rear and side wall locations.

when they did respond to dots flashed in the mirror image replica chamber, these birds often pecked the correct location in the test chamber. This result provides some evidence that these birds in fact had learned something about the invariant spatial relationship between dot reflections and specific wall hole positions in the test chamber. That the latter positions were on the contralateral side of the former flashed dot locations likely made the task even more demanding, and any successes by the birds are all the more striking. Nevertheless, although overall accuracy in the test conditions was better than chance (which was 20%), it was less than in the control mirror conditions. For these birds, then, detecting a disappearing dot directly after it initially had been signaled by the dot's reflection likely facilitated their performances.

The results leave open the possibility that responding also may have been inhibited by perceived stimulus novelty associated with temporary changes in the visual array between conditions from that of contingent motion in the subject's reflection, to movement alone with the live bird, or no movement at all with the mounted bird. If the subject birds had noticed the changes, then it would leave open the possibility of contextual control of responding to flashed dots by the contingency between a subject's movements and its reflection. However, if the subject birds had attended to the contingent movement, it is clear that they never learned that the image in the mirror was their reflection (cf. Gallup & Suarez, 1991).

Summary and conclusions

We found no evidence that our pigeons learned self-directed behaviors in the presence of mirrors as reported for other pigeons (Epstein et al., 1981).

Contrary to what those investigators reported, simply exposing our pigeons to the contingencies of reinforcement associated with a mirror was not a sufficient condition for their learning to use a mirror for guiding self-directed behaviors, let alone the sophisticated self-recognition exhibited by that water fowl, the Ugly Duckling. Our pigeons did peck visible body dots, but the mirror had an inhibitory effect on this behavior. When the mirror was exposed, the same birds correctly chose which of several wall holes to peck when a dot was flashed seemingly out of direct sight as they faced the mirror. We attribute this latter ability primarily to different levels of discriminative stimulus control by the mirror on the performances of a given bird or birds.

Two of the six subjects were "cheaters"; information from dot reflections was never used by these birds. A glass surface alone was an effective discriminative stimulus to engage in a visual search for dots that were directly visible. A similar visual search was controlled in the other birds by dot reflections and not by the mirror per se. The use of dot reflections by these latter birds was perhaps analogous to a human using a fleeting reflection to turn, and only then locate and identify what it is behind him or her. There was some evidence that in addition, three birds had learned something about the invariant relations between specific virtual dot (i.e., reflections) locations behind the mirror plane and real wall hole positions in the space lying in front of the plane (cf., Gibson, 1966; Loveland, 1986). These birds may have discovered something about the affordances of reflecting surfaces for locating objects from a geocentric perspective; nevertheless, they failed to generalize this information to an egocentric perspective involving body dots.

The discrepancies between our results (Thompson & Contie, 1986) and those reported by Epstein et al. (1981) suggest that perhaps the ability of pigeons to discover the perceptual affordances of mirrors hinges in large part on the skill of the experimenters. Perhaps with more ingenuity, if not patience, one might eventually produce in pigeons the same performance that encourages some individuals to attribute chimpanzees and children with a self-concept. This expectation, however, is overly optimistic. No amount of training will render an animal capable of mirror-mediated object location, let alone self-recognition, if it cannot detect the requisite perceptuocognitive relationships between objects and their mirror images (cf. Platt, Thompson, & Boatwright, 1991).

Requisite perceptual relationships for self-recognition

The minimum requirement for using a mirror as a perceptual tool is the ability to form conditional associations without necessarily detecting that the events instantiating the association are identical. Tests for conditional associations typically are of the "if A, then C, and if B, then D" type in which, for example, an animal is trained to respond to a circle and not a square when these shapes are presented on a red background, but to respond to the square and not to the circle when they are presented on a green background (e.g., Thomas, 1980). Sensitivity to simple associative relationships of the above type alone constrains mirror usage by animals to a level no more sophisticated than that shown by Tigger the tiger when confronted by his own reflection.

It is difficult to see how any organism, pigeon or other, could use mirrors for self-recognition and self-evaluation without the additional ability to perceive two formal properties of equivalence relations, and without having a broadly construed concept of identity relations (Premack, 1983a,b; Sidman et al. 1982; Sidman & Tailby, 1982). A more sophisticated use of mirrors than that demonstrated by Tigger and many vertebrate species (e.g., Gallup, 1968; Thompson & Sturm, 1965) dictate that, at the very least, an organism respond to the relational properties of *symmetry*. Symmetry is demonstrated when an association between two events is equally strong in both forward and backward directions, even though the animal was trained only in the forward condition. Tests for symmetry typically involve training animals first on a conditional (i.e., symbolic) matching task where, if sample *A* precedes choices *C* and *D*, then respond *C*; but if the initial sample stimulus is *B*, then respond *D*. Symmetry is demonstrated when the animal chooses *A* and not *B* when the sample is now *C*, and *B* rather than *A* if the sample is now *D*. Recall that when Winnie-the-Pooh detected the undone sewing on his back, he used the mirror image to guide his repair efforts. If Pooh, like humans, also were to look toward his body after having first seen undone sewing via the mirror, then he would have demonstrated symmetry.

Winnie-the-Pooh demonstrated that he was sensitive to the fact that his body and his reflection shared common visual attributes, but, unlike the Ugly Duckling, he did not unequivocally demonstrate *reflexivity*, another formal property of equivalence relations. Reflexivity is demonstrated when an organism spontaneously classifies events not only as sharing attributes, but as being identical. One could have concluded that Pooh demonstrated reflexivity had he not thanked his reflection, or better still, had he pointed to Tigger's reflection and exclaimed, "it's you," rather than, "then, who's that?"

Using a mirror for self-recognition with the competence of a normal adult human also requires one to have a broadly construed concept of reflexive physical and functional identity relationships that extend both within and across sensory modalities (e.g., Bahrick & Watson, 1965; Mitchell, 1993; Spelke, 1987). Pooh's ability to match visual information from the mirror with the kinesthetic feedback from his hands while he repaired his torn sewing indicates he had that capacity. If Pooh also had assigned the same label (e.g., "reflection") to mirror-images of physically disparate objects, then he would have further demonstrated his understanding that reflections were members of a common functional class (Eikeseth & Smith, 1992; Goldiamond, 1962). At that point we should concede that Pooh truly had a mirror concept, if not a self-concept.

As noted earlier, the performances of our pigeons (Thompson & Contie, 1986) could be accounted for in terms of different levels of associative stimulus control by either glass surfaces per se or dot reflections. Nothing in the pattern of results indicated that the performances of our birds were mediated by their perceiving equivalence relations or matching stimuli on the basis of physical identity. Our results are not surprising as they are consistent with most findings from studies of equivalence relations and matching-to-sample in pigeons (cf., D'Amato, Salmon, Loukas, & Tomie, 1985; Hayes, 1989; Hogan & Zentall, 1977; Lipkens, Kop, & Matthijs, 1988; Sidman et al., 1982; Vaughan, 1988). Results from these studies provide good evidence that pigeons do not demonstrate symmetry. They are predisposed to acquire narrowly construed matching rules based primarily on absolute stimulus values (i.e., match red with red) rather than on physical (i.e., match like with like) or relational similarity (cf. Wilson, Mackintosh, & Boakes, 1985a; Zentall, Edwards, Moore, & Hogan, 1981).

In contrast to pigeons, children and chimpanzees demonstrate symmetry (Daehler, Lonardo, & Bukatko, 1979; Menzel, Savage-Rumbaugh, & Lawson, 1985; Premack, 1976; Savage-Rumbaugh, 1987) and they spontaneously perceive physical and more abstract visual identity (i.e., same–different) relations (Oden, Thompson, & Premack, 1990; Tyrrell, Stauffer & Snowman, 1991). To our knowledge, however, corvids and African Grey parrots (*Psittacus erithacus*) are the only bird taxa that have broadly construed identity concepts approximating those reported for chimpanzees and children (cf. House, Brown, & Scott, 1974; Pepperberg, 1987; Premack, 1983a; Wilson et al., 1985b). Perhaps the parrot then, if not the pigeon, may one day emulate the Ugly Duckling.

In short, pigeons lack the requisite perceptual skills necessary for self-recognition in mirrors. Given the substantial perceptual differences across taxa, we conclude that the use of mirrors by pigeons reflects only a surface similarity with the performance of chimpanzee and child. This conclusion is important because it challenges the claim by Epstein et al. (1981) that mirror self-recognition in chimpanzees (Gallup, 1970) is mediated by the same mechanism as mirror usage by pigeons and does not necessarily imply self-awareness or self-recognition.

Monkeys 'n' mirrors

Might not monkeys be better than pigeons at learning to use mirrors? As reviewed elsewhere in this volume (e.g., see Anderson, *SAAH*21; Boccia, *SAAH*23; Thompson & Boatwright-Horowitz, *SAAH*22), most studies of mirror use by monkeys report that they can be trained to use mirrors for object location and recognition, but that they typically fail the traditional mark test for self-recognition (cf., Anderson, 1983, 1984; Gallup, 1977; Gallup, Wallnau, & Suarez, 1980; Platt & Thompson, 1985). What can account for this asymmetry of performance? Some have interpreted it as evidence that monkeys lack a self-concept (e.g., cf. Gallup, 1970; Povinelli, 1987; Suarez & Gallup, 1981). However, as was the case with understanding mirror usage by pigeons, we believe a more profitable approach lies in identifying which perceptual information afforded by mirrors is perceived by monkeys and then mapping out the processes that facilitate their discovering and utilizing this information as a perceptual tool (Field & Hogg, 1992; Loveland, 1986; Platt et al., 1991; Robinson, Connell, McKenzie, & Day, 1990).

Some investigators have reported mirror-mediated object recognition by monkeys (e.g., Anderson, 1986; Itakura, 1987a,b). Performance of this mirror-related task likely hinges on the independently demonstrated ability of these animals to perceive physical, and possibly reflexive, identity relationships. For example, both Old and New World monkeys can match objects on the basis of identity (i.e., match like with like) rather than absolute cues alone (i.e., match red with red). However, demonstrations in macaque and cebus monkeys of what has been labeled a "generalized matching concept" typically depends on the use of a large number of training stimuli, often trial-unique novel objects, during acquisition of the matching-to-sample task (D'Amato, Salmon, & Colombo, 1985; Wright, Cook, & Kendrick, 1989). When a small number of training objects are used, cebus and macaque monkeys are more likely to match on the basis of absolute cues rather than on identity, as is demonstrated by their failing to match novel items in transfer tests (cf. D'Amato et al., 1985; Fujita, 1982). No such procedural constraint is evident in results obtained from infant chimpanzees that spontaneously match novel objects after training with only a single pair of items (Oden, Thompson, & Premack, 1988).

Success on mirror-guided reaching tasks in which monkeys monitor the reflection of an arm or hand in relation to the reflection of, say, a food item, implies that these animals perceive the contingency between objects and their reflections. However, we see little reason to assume that performance of this task is mediated by the perception of either symmetrical or reflexive relations between object and reflection. Indeed, it would be surprising if it were so, as attempts to demonstrate stimulus equivalence with monkeys have not been unequivocally successful (cf. D'Amato et al., 1985; Hayes, 1989; McIntire, Clearly, & Thompson, 1987; Saunders, 1989; Sidman et al., 1982).

We suspect the performance of mirror-guided reaching tasks by monkeys is mediated by the same processes underlying their acquisition of video-task procedures (Rumbaugh, Richardson, Washburn, Savage-Rumbaugh, & Hopkins, 1989). Rhesus macaques learn to move a cursor on a computer monitor screen by hand manipulating a joystick directionally. As in mirror-guided tasks, the animal continuously tracks its progress in bringing a hand or cursor into contact with a still or moving target. For example, in the so-called chase task, a monkey is rewarded for moving a cursor on a video screen so as to touch a moving video target (Rumbaugh et al., 1989). Despite the physical separation between the joystick and the monitor screen, rhesus monkeys readily learn this task; they map hand movements with movement of objects in space. There is no evidence, however, that these monkeys see the cursor as an extension of their hand, arm, or body. We see no compelling reason to assume that monkeys in mirror guidance tasks are any different in this regard. Menzel et al. (1985) reported, for example, that rhesus monkeys trained to reach for food reflected in a mirror threatened the image of their own hand as it approached the food.

Gaze aversion by monkeys has been appealed to as a source of interference for monkeys in mirror self-recognition tasks (e.g., Premack, 1983a). The recent claim by Thompson & Boatright-Horowitz (*SAAH*22) that a macaque explicitly reinforced to maintain eye contact with its mirror image passed the mark test would appear to vindicate this particular argument. As with the report by Epstein et al. (1981), this is a finding investigators should attempt to replicate. Unfortunately, however, the prognosis is not encouraging. As summarized above, monkeys may have the cognitive capacities for understanding relations between reflections and objects external to their bodies, but unlike humans and chimpanzees, they lack the perceptuocognitive skills that would permit them to readily grasp relations between reflections and marks on their bodies (e.g., Anderson, 1986; Itakura, 1987a,b; Platt et al., 1991). Furthermore, the expression by monkeys of those perceptuocognitive skills that they do possess often depends on specialized training procedures. It is not surprising, then, that instances in which monkeys are reported to display self-directed responses, let alone unequivocal self-recognition, are rare, occur only after extraordinary training, and show little or no evidence of transfer to novel situations (cf. Anderson, *SAAH*21; Boccia, *SAAH*23; Thompson & Boatright-Horowitz, *SAAH*22).

The evolutionary ontogeny of Pinocchio

As indicated above, one important source of the different facility with which monkeys and chimpanzees use information from mirrors is that the latter animals have the necessary perceptuocognitive capacities for grasping the significance of the affordances associated with reflective surfaces. Equally important is that, unlike studies with pigeons and monkeys, the experimenter's expertise is not a limiting factor in studies of self-recognition in mirrors by chimpanzees. An essential difference, then, between monkeys and chimpanzees is the spontaneous nature in the way many if not all chimpanzees (e.g., Swartz & Evans, 1991, *SAAH*11) discover the affordances of mirrors and use them instrumentally. This claim has been criticized as stating only that the reinforcement history responsible for shaping the behavior is unspecified and lies forever buried in the past (Platt et al., 1991). Our emphasis here, however, with respect to spontaneity, is on the nature and degree of external intervention required for an animal to express its perceptuocognitive sensibilities (e.g., Oden et al., 1990). The issue then is one of locus of control or the nature of causal agency in the regulation of action (see Bandura, 1989).

The children's classic story *Pinocchio* (Lorenzini, 1882; see, e.g., the Disney film[3]) is instructive for understanding this difference. Initially, Pinocchio is merely a puppet; a simple S–R creature whose every movement is caused by the manipulation of his strings by Geppetto the wood-carver. Geppetto's wish for a real boy is granted in part when the Blue Fairy – call her Psyche – gives Pinocchio the gift of life. Nevertheless, Pinocchio is still not a "real boy"; he does not yet have a conscience by which he can tell right from wrong. What he does have, of course (at least, in the film), is Jiminy Cricket as his guide. We would argue that many nonhuman species are similar to Pinocchio in their limited ability to use their own initiative for discovering salient information of varying abstraction within a complex environment, and then deciding which course of action to follow given a host of options. That is, the environment, often times an experimenter, is analogous to Jiminy Cricket in pointing out those features and relationships in the world to which the animal should attend and then act upon. Eventually, having proven himself worthy, Pinocchio becomes a real boy whose choices are now self-mediated. Consequently, Jiminy Cricket is no longer needed to guide Pinocchio's actions. Infant chimpanzees are analogous to Pinocchio at this transition stage. As they develop, the agency regulating their actions becomes increasingly internalized.

Our emphasis on the role of emergent agency, or the internal locus of control, should not be construed as implying that external environmental factors, including other social agents, do not contribute meaningfully to the discovery and expression of perceptuocognitive capacities, including self-recognition, during development. For example, the chimpanzee's propensity to spontaneously explore the functional contingencies afforded by natural objects, including reflective surfaces, likely has its origins in the species natural history of extractive foraging in ways that have been described elsewhere (Gibson, 1986; King, 1986; Parker & Gibson, 1977, 1979). Studies of self-recognition in chimpanzees also point to the importance of social experience (Gallup, McClure, Hill, & Bundy, 1971). Recent evidence from studies involving human experimenters and infant chimpanzees suggests, however, that social interactions between, say, adult and infant, function not so much to reinforce learned associations, but rather to establish a particular hedonic set that sustains problem solving in general (Oden & Thompson, 1992). Social motivation of this type is likely very important also in the wild for chimpanzees, where, like humans, there is good evidence of imitative learning from adults by infants (cf. Meltzoff, 1990), but, unlike humans, no clear evidence of pedagogic instruction of the latter by the former (cf. Boesch, 1991; Premack, 1984, 1986; Sugiyama, 1990).

This propensity in chimpanzees for internally induced regulation of action, coupled with the capacity to directly perceive conceptual equivalencies, forms a powerful cognitive base from which self-recognition can emerge spontaneously in an experimental setting. In studies involving pigeons and monkeys we would argue that the environment not only provides the motivational "push," but also that its structure provides much, if not all, of the animal's competence. Unlike that for a chimpanzee, our question of a pigeon or a monkey, to paraphrase Mace (1977), should not be, "What's inside it's head?" but rather "What's it's head inside of?"

Acknowledgments

Support for the research on mirror usage by pigeons reported here was provided by Franklin & Marshall College's Committee on Grants, and the Hackman Scholar Summer Research Program administered by the college. We thank Karyl Swartz, Gordon Gallup, and the editors for their comments on this research. We extend also our appreciation to the editors for their patient cooperation. Cindi Contie is now at the Institute for Human Resources, Rockville MD.

Notes

1. *The Many Adventures of Winnie the Pooh*, dirs. Wolfgang Reitherman & John Lounsbery (Disney, 1977; trilogy comprising *Winnie the Pooh and the Honey Tree, Winnie the Pooh and the Blustery Day*, and *Winnie the Pooh and Tigger Too*).
2. *The Ugly Duckling*, prod. Walt Disney (Disney, 1939, short).
3. *Pinocchio*, suprv. dirs. Ben Sharpsteen & Hamilton Luske (Disney, 1940, full-length).

References

Andersen, H. C. (1984). *Hans Andersen's fairy tales*. Oxford University Press.

Anderson, J. R. (1983). Responses to mirror-image stimulation, and assessment of self-recognition in mirror- and peer-reared stumptail macaques. *Quarterly Journal of Experimental Psychology, 35*, 201–222.

(1986). Mirror-mediated finding of hidden food by monkeys (*Macaca tonkeana* and *Macaca fascicularis*). *Journal of Comparative Psychology, 100*, 237–242.

Anderson, J. R., & Roeder, J. J. (1989). Responses of capuchin monkeys (*Cebus apella*) to different conditions of mirror-image stimulation. *Primates, 30*, 581–587.

Bahrick, L. E., & Watson, J. S. (1985). Detection of intermodal proprioceptive–visual contingency as a potential basis of self-perception in infancy. *Developmental Psychology, 21*, 963–973.

Bandura, A. (1989). Human agency in social cognitive theory. *American Psychologist, 44*, 1175–1184.

Boesch, C. (1991). Teaching among wild chimpanzees. *Animal Behaviour, 41*, 530–532.

Daehler, M. W., Lonardo, R., & Bukatko, D. (1979). Matching and equivalence judgments in very young children. *Child Development, 50*, 170–179.

D'Amato, M. R., Salmon, D. P., & Colombo, M. (1985). Extent and limits of the matching concept in monkeys (*Cebus apella*). *Journal of Experimental Psychology: Animal Behavior Processes, 11*, 35–51.

D'Amato, M. R., Salmon, D. P., Loukas, E., & Tomie, A. (1985). Symmetry and transitivity of conditional relations in monkeys (*Cebus apella*) and pigeons (*Columba livia*). *Journal of the Experimental Analysis of Behavior, 44*, 35–47.

Eikeseth, S., & Smith, T. (1992). The development of functional and equivalence classes in high-functioning autistic children: The role of naming. *Journal of the Experimental Analysis of Behavior, 58*, 123–133.

Epstein, R., Lanza, R. P., & Skinner, B. F. (1981). "Self-awareness" in the pigeon. *Science, 212*, 695–696.

Field, J., & Hogg, V. (1992). Young children's ability to find objects reflected in mirrors. *Australian Journal of Psychology, 44*, 9–11.

Friedman, M. B. (1975). How birds use their eyes. In P. Wright, P. G. Caryl, and D. M. Vowles (Eds.), *Neural and endocrine aspects of behaviour of birds* (pp. 183–204). Amsterdam: Elsevier.

Fujita, K. (1982). An analysis of stimulus control in two-color matching-to-sample behaviors of Japanese monkeys (*Macaca fuscata*). *Japanese Psychological Research, 24*, 124–135.

Gallup, G. G., Jr. (1968). Mirror image stimulation. *Psychological Bulletin, 70*, 782–793.

(1970). Chimpanzees: Self-recognition. *Science, 167*, 86–87.

(1975). Toward an operational definition of self-awareness. In R. H. Tuttle (Ed.), *Socioecology and psychology of primates* (pp. 309–342). The Hague, Netherlands: Mouton.

(1977). Absence of self-recognition in a monkey (*Macaca fascicularis*) following prolonged exposure to a mirror. *Developmental Psychobiology, 10*, 281–284.

Gallup, G. G., Jr., McClure, M. K., Hill, S. D., & Bundy, R. A. (1971). Capacity for self-recognition in differentially reared chimpanzees. *Psychological Record, 21*, 69–74.

Gallup, G. G., Jr., & Suarez, S. D. (1991). Social responding to mirrors in rhesus monkeys (*Macaca mulatta*): Effects of temporary mirror removal. *Journal of Comparative Psychology, 105*, 376–379.

Gallup, G. G., Jr., Wallnau, L. B., & Suarez, S. D. (1980). Failure to find self-recognition in mother–infant and infant–infant rhesus monkey pairs. *Folia Primatologica, 33*, 210–219.

Gelhard, B. S., Wohlman, S. H., & Thompson, R. K. R. (1982). *Self-awareness in the pigeons: second look.* Paper presented at the Northeast regional meeting of the Animal Behavior Society, Boston.

Gibson, J. J. (1966). *The senses considered as perceptual systems.* Boston: Houghton-Mifflin.

Gibson, K. R. (1986). Cognition, brain size and the extraction of embedded food sources. In J. G. Else & P. C. Lee (Eds.), *Primate ontogeny, cognition, and social behaviour.* Cambridge University Press.

Goldiamond, I. (1962). Perception. In A. J. Bachrach (Ed.), *Experimental foundations of clinical psychology* (pp. 280–340). New York: Basic.

Hayes, S. C. (1989). Nonhumans have not yet shown stimulus equivalence. *Journal of the Experimental Analysis of Behavior, 51*, 385–392.

Hogan, D. E., & Zentall, T. R. (1977). Backward associations in the pigeon. *American Journal of Psychology, 90*, 3–15.

House, B. J., Brown, A. L., & Scott, M. S. (1974). Children's discrimination learning based on identity or difference. In H. W. Reese (Ed.), *Advances in child development and behavior* (Vol. 9, pp. 1–45). New York: Academic Press.

Itakura, S. (1987a). Mirror guided behavior in Japanese monkeys (*Macaca fuscata fuscata*). *Primates, 28*, 149–161.

(1987b). Use of a mirror to direct responses in Japanese monkeys (*Macaca fuscata fuscata*). *Primates, 28*, 343–352.

(1992). Task-specific hand preferences of two Japanese macaques on mirror-guided reaching. *Psychological Record, 42*, 173–178.

Jaynes, J. (1976). *The origin of consciousness in the breakdown of the bicameral mind.* Boston: Houghton Mifflin.

King, B. J. (1986). Extractive foraging and the evolution of primate intelligence. *Human Evolution, 1*, 361–372.

Ledbetter, D. H., & Basen, J. A. (1982). Failure to demonstrate self-recognition in gorillas. *American Journal of Primatology, 2*, 307–310.

Lethmate, J., & Dücker, G. (1973). Untersuchungen zum Selbsterkennen im Spiegel bei Orang-Utans und einigen anderen Affenarten (Studies of self-recognition in a mirror by orangutans and some other primate species). *Zeitschrift für Tierpsychologie, 33*, 248–269.

Lewis, M., & Brooks-Gunn, J. (1979). *Social cognition and the acquisition of self.* New York: Plenum.

Lipkens, R., Kop, P. F. M., & Matthijs, W. (1988). A test of symmetry and transitivity in the conditional discrimination performances in pigeons. *Journal of the Experimental Analysis of Behavior, 49*, 395–409.

Lorenzini, Carlo [Collodi] (1882). *Le avventore di Pinocchio.*

Loveland, K. A. (1986). Discovering the affordances of a reflecting surface. *Developmental Review,* 6, 1–24.

Mace, W. M. (1977). James J. Gibson's strategy for perceiving: Ask not what's inside your head, but what your head's inside of. In R. E. Shaw & J. D. Bransford (Eds.), *Perceiving, acting and knowing* (pp. 43–65). Hillsdale, NJ.: Erlbaum.

McFadden, S. A. (1987). The binocular depth stereoacuity of the pigeon and its relation to the anatomical resolving power of the eye. *Vision Research, 27,* 1967–1980.

McIntire, K. D., Cleary, J., & Thompson, T. (1987). Conditional relations by monkeys: Reflexivity, symmetry, and transitivity. *Journal of the Experimental Analysis of Behavior, 47,* 279–285.

Meltzoff, A. N. (1990). Foundations for developing a concept of self: The role of imitation in relating self to other and the value of social mirroring, social modeling, and self-practice in infancy. In D. Cicchetti & M. Beeghly (Eds.), *The self in transition: Infancy to childhood* (pp. 139–164). Chicago: University of Chicago Press.

Menzel, E. W., Jr., Savage-Rumbaugh, E. S., & Lawson, J. (1985). Chimpanzee (*Pan troglodytes*) spatial problem solving with the use of mirrors and televised equivalents of mirrors. *Journal of Comparative Psychology, 99,* 211–217.

Milne, A. A. (1958). *The world of Christopher Robin, the complete When we were very young and Now we are six.* New York: Dutton.

Mitchell, R. W. (1993). Mental models of mirror self-recognition: Two theories. *New Ideas in Psychology, 11,* 295–325.

Oden, D. L., & Thompson, R. K. R. (1992). The role of social bonds in motivating chimpanzee cognition. In H. Davis & D. Balfour (Eds.), *The inevitable bond: Examining scientist–animal interactions* (pp. 218–231). New York: Oxford University Press.

Oden, D. L., Thompson, R. K. R., & Premack, D. (1988). Spontaneous transfer of matching by infant chimpanzees (*Pan troglodytes*). *Journal of Experimental Psychology: Animal Behavior Processes, 14,* 140–145.

Oden, D. L., Thompson, R. K. R., & Premack, D. (1990). Infant chimpanzees (*Pan troglodytes*) spontaneously perceive both concrete and abstract same / different relations. *Child Development, 61,* 621–631.

Parker, S. T., & Gibson, K. R. (1977). Object manipulation, tool use and sensorimotor intelligence as feeding adaptations in cebus monkeys and great apes. *Journal of Human Evolution, 6,* 623–641.

(1979). A developmental model for the evolution of language and intelligence in early hominids. *Behavioral and Brain Sciences, 2,* 367–408.

Patterson, F. (1984). Self-recognition by *Gorilla gorilla gorilla. Journal of the Gorilla Foundation, 7,* 2–3.

Pepperberg, I. M. (1987). Acquisition of the same/different concept by an African Grey parrot (*Psittacus erithacus*): Learning with respect to color, shape, and material. *Animal Learning & Behavior, 15,* 423–432.

Platt, M. M., & Thompson, R. L. (1985). Mirror responses in a Japanese macaque troop (Arashiyama West). *Primates, 26,* 300–314.

Platt, M. M., Thompson, R. L., & Boatright, S. L. (1991). Monkeys and mirrors: Questions of methodology. In L. M. Fedigan & P. J. Asquith (Eds.), *The monkeys of Arashiyama: Thirty-five years of research in Japan and the West* (pp. 274–290). Albany, NY: SUNY Press.

Povinelli, D. (1987). Monkeys, apes, mirrors and minds: The evolution of self-awareness in primates. *Human Evolution, 2,* 493–509.

Premack, D. (1976). *Intelligence in ape and man.* Hillsdale, NJ: Erlbaum.

(1983a). Animal cognition. In M. R. Rosenzweig & L. W. Porter (Eds.), *Annual Review of Psychology, 34,* 351–362.

(1983b). The codes of man and beast. *Behavioral and Brain Sciences, 6,* 125–167.

(1984). Pedagogy and aesthetics as sources of culture. In M. Gazzaniga (Ed.), *Handbook of cognitive neuroscience* (pp. 15–35). New York: Plenum.

(1986). *Gavagai! or the future history of the animal language controversy.* Cambridge, MA: Bradford.

Robinson, J. A., Connell, S., McKenzie, B. E., & Day, R. H. (1990). Do infants use their own images to locate objects reflected in a mirror? *Child Development, 61,* 1558–1568.

Rumbaugh, D. M., Richardson, W. K., Washburn, D. A., Savage-Rumbaugh, E. S., & Hopkins, W. D. (1989). Rhesus monkeys (*Macaca mulatta*), video tasks, and implications for stimulus–response spatial contiguity. *Journal of Comparative Psychology, 103,* 32–38.

Saunders, K. J. (1989). Naming in conditional discriminations and stimulus equivalence. *Journal of the Experimental Analysis of Behavior, 51,* 379–384.

Savage-Rumbaugh, S. (1987). *Ape language: From conditioned response to symbol.* New York: Columbia University Press.

Sidman, M., Rauzin, R., Laxar, R., Cunningham, S., Tailby, W., & Carrigan, P. (1982). A search for symmetry in the conditional discriminations of rhesus monkeys, baboons, and children. *Journal of the Experimental Analysis of Behavior, 37,* 23–44.

Sidman, M., & Tailby, W. (1982). Conditional discrimination vs. matching-to-sample: An expansion of the testing paradigm. *Journal of the Experimental Analysis of Behavior, 37,* 5–22.

Spelke, E. (1987). The development of intermodal perception. In L. B. Cohen & P. Salapatek (Eds.), *Handbook of infant perception: From perception to cognition.* New York: Academic Press.

Suarez, S. D., & Gallup, G. G. Jr. (1981). Self-recognition in chimpanzees and orangutans, but not gorillas. *Journal of Human Evolution, 10,* 175–188.

Sugiyama, Y. (1990). Untitled talk and video presentation in roundtable discussions. In T. Matsuzawa (organizer), *Future of language research in primatological perspective.* Satellite symposium, 13th Congress of the International Primatological Society, Kyoto, Japan.

Swartz, K. B., & Evans, S. (1991). Not all chimpanzees (*Pan troglodytes*) show self-recognition. *Primates, 32,* 483–496.

Thomas, R. K. (1980). Evolution of intelligence: An approach to its assessment. *Brain, Behavior, and Evolution, 17,* 454–472.

Thompson, R. K. R., & Contie, C. (1986). *Further reflections on mirror usage by pigeons.* Paper presented at the annual meeting of the Psychonomic Society, New Orleans.

Thompson, T., & Sturm, T. (1965). Classical conditioning of aggressive display in Siamese fighting fish. *Journal of the Experimental Analysis of Behavior, 8,* 397–403.

Tyrrell, D. J., Stauffer, L. B., & Snowman, L. G. (1991). Perception of abstract identity / difference relationships by infants. *Infant Behavior and Development, 14,* 125–129.

Vaughan, W., Jr. (1988). Formation of equivalence sets in pigeons. *Journal of Experimental Psychology: Animal Behavior Processes, 14,* 36–42.

Wilson, B., Mackintosh, N. J., & Boakes, R. A. (1985a). Matching and oddity learning in the pigeon: Transfer effects and the absence of relational learning. *Quarterly Journal of Experimental Psychology, 37B,* 295–311.

(1985b). Transfer of relational rules in matching and oddity learning by pigeons and corvids. *Quarterly Journal of Experimental Psychology, 37B,* 313–332.

Wright, A. A., Cook, R. G., & Kendrick, D. K. (1989). Relational and absolute stimulus learning by monkeys in a memory task. *Journal of the Experimental Analysis of Behavior, 52,* 237–248.

Zentall, T. R., Edwards, C. A., Moore, B. S., & Hogan, D. E. (1981). Identity: The basis for both matching and oddity learning in pigeons. *Journal of Experimental Psychology: Animal Behavior Processes, 7,* 70–86.

Part V

Epilogue

27   Evolving self-awareness

Sue Taylor Parker and Robert W. Mitchell

Introduction

In this volume and in related publications, various authors have presented a rich and heterogeneous array of hunches and hypotheses about the possible mechanisms of self-awareness, and their ontogeny and evolution. Some of these are full-blown theories, others are presented merely as intuitions. In this last chapter we take on the task of identifying, classifying, and comparing these diverse and often contradictory models. For clarity, we organize our review and evaluation, following Tinbergen (1963) and others (Hinde, 1982; Yoerg & Kamil, 1990), according to four approaches to the study of evolved behavioral capacities: their proximate or immediate causation (physiological, information processing, or psychological mechanisms or organizing principles), ontogeny, specific evolutionary history in a particular lineage, and adaptive functions in various species. A full evolutionary explanation should embrace all four approaches.

These four approaches focus on complexly interrelated facets of behavior: The proximate mechanisms and ontogeny of behaviors in each lineage have been favored by a specific history of selective and random forces and developmental constraints operating on a set of preexisting mechanisms and developmental patterns. Specific mechanisms and their development have been favored because they met certain selection pressures present in other members of that lineage at the time these mechanisms evolved. Similar functional systems based on different mechanisms and developmental patterns have often evolved independently in distantly related species with differing histories and distinct preexisting complexes because they met similar selection pressures. Flight in bats and birds is a classic example of convergent function: In both taxa, wing structures and muscle-movement patterns facilitate motion through the air, but the details of their mechanisms and development differ in accord with the ancestral structures that were modified by selection for more efficient locomotion. Such convergent functional systems provide clues to the adaptive significance of structures and behaviors.

Each of these four approaches entails its own methodology. Studies of immediate causation entail analysis of physiological mechanisms – including information processing mechanisms – and the physical and social environments

413

in which they operate. Physiological and cognitive psychology are in this domain. Studies of the ontogeny of behavior entail analysis of the means by which more complex systems develop from earlier, less complex systems. Embryology and developmental psychology are in this domain. Studies of the evolutionary history of behaviors rely on a base of comparative studies of immediate causation and ontogeny of behavior in a group of closely related living species, supplemented by a base of comparative studies on closely related fossil species. Taxonomy, ethology, and paleontology are in this domain. Finally, studies of adaptive function and reconstruction of past selection pressures entail comparative studies of the functions of structures and behaviors across a broad spectrum of species, both closely and distantly related, who display similar adaptations. Ethology and sociobiology are in this domain.

Although each of these approaches entails its own methodology, theories in all of these domains must postulate mechanisms that can explain how the capacity came about and/or how it functions today. Such mechanisms must in all cases specify not only the necessary but the sufficient conditions to explain the phenomenon.

Immediate causation

By far the most frequent explanations of mirror self-recognition (MSR) and other aspects of self-awareness in human children and great apes address immediate causation. These explanations fall into two categories: those that purport to explain the presence of MSR in great apes and human infants, and those that purport to explain the absence of MSR in other species or the nature of other simpler self-mechanisms.

Explanations for MSR in great apes and humans

Gallup, who was the first to study self-awareness in nonhuman primates empirically, defines self-awareness as "the ability to become aware of your own existence, and mind as the ability to monitor your own mental states," and concludes as a corollary of this definition that self-awareness and mind "are all part and parcel of the same process" (Gallup, 1983, p. 500). He distinguishes the following empirical markers of mind: anthropomorphism, intentional deception, reciprocal altruism, empathy, reconciliation, and pretending. He argues that MSR in chimpanzees is made possible by the existence of a self-concept without, however, specifying the origins of the self-concept.

Lewis (SAAH2) notes that self-awareness in humans and great apes is an expression of "the idea of me," which develops in human infants in the 2nd year of life, probably in conjunction with the maturation of the frontal lobes of the cortex. He contrasts this "me" self, which knows that it knows, with the ontogenetically and phylogenetically prior "machinery of self" which underlies self-regulation and self–other differentiation in all animals without knowing that it is doing so. He notes that the "me" self is necessary for the self-conscious and self-evaluative emotions that drive socialization.

In contrast to these broad definitional approaches, other approaches focus more specifically on explaining causes of MSR. These explanations fall into various (not necessarily mutually exclusive) categories: contingency testing, imitation, kinesthetic–visual matching (KVM), and mental representation.

Several investigators have followed the lead of Lewis and Brooks-Gunn (1979), who identified contingency testing or contingency play as one of the correlates of MSR in human infants. Since then, testing of the perfect contingency in mirrors has been observed in several species. Gergely (*SAAH*4), like Lewis and Brooks-Gunn (1979), Parker (1991) and Mitchell (1993a), notes that the detection of the perfect contingency between its own and the mirror-image's body movements is necessary for MSR, and that contingency testing is observed in dolphins (Marten & Psarakos, *SAAH*24), chimpanzees, bonobos, and gorillas (Hyatt & Hopkins, *SAAH*15; Parker, *SAAH*19), and perhaps macaques (Boccia, *SAAH*23). Interestingly, dolphins exhibited more contingency play with the small-image video of themselves than with the large-image mirror, suggesting to Marten and Psarakos that the dolphins needed more information in this condition to verify the contingency of the image. One of their dolphins also showed contingency testing when observing itself in the simultaneous video condition, but not when observing a non-simultaneous video.

Several investigators point to imitation as the mechanism underlying mirror self-recognition. Parker (1991) and Mitchell (1992, 1993a) provide different but complementary perspectives on imitation. Parker, following Lewis and Brooks-Gunn, argues that the discovery of MSR is brought about by the infant's "attempt to imitate the mirror image" (Parker, 1991, p. 439), which brings about the "discovery of the contingency of the image's actions on their actions with their head and facial muscles" (Parker, 1991, p. 442). This "contingent play with facial muscles via [Piagetian] 4th or 5th stage imitation may be sufficient to allow the infant to associate his own proprioceptive facial feedback with the contingent facial action he sees in the mirror" (Parker, 1991, p. 445). MSR "depends on (1) voluntary control of the facial muscles; (2) at least 4th and probably 5th stage understanding of causality in the facial modality; (3) at least 4th stage and probably 5th stage imitation in the facial modality" (Parker, 1991, p. 442).

According to Mitchell's (1992, 1993a) inductive model (following Guillaume [1926/1971]), MSR results from the kinesthetic–visual matching involved in bodily and facial imitation, and from an understanding of mirror correspondence, that is, the understanding that most images in the mirror (other than one's own) are contingent and accurate images of things outside the mirror. Using inductive inference, the organism recognizes itself in the mirror. The finding by Boysen, Bryan, and Shreyer (*SAAH*13) that chimpanzees can rapidly "integrate incoming sensory information" from the mirror also suggests skill at KVM.

Gopnik and Meltzoff (*SAAH*10) likewise argue that kinesthetic–visual matching present in imitation is likely the basis for MSR. They argue that this capacity begins at birth and provides the foundation for later manifestations of self-awareness. They note that at 14 months of age children can not only imitate others but also recognize when others are imitating them. The imitating adult can therefore provide a mirror for the child's own behavior. Although Gopnik and Meltzoff suggest that neonatal imitation involves cross-modal skills equivalent to those involved in the recognition that one is being imitated (which occurs at about 14 months of age), they present no evidence that neonatal infants can recognize that they are being imitated.

Hart and Fegley (*SAAH*9) also view imitation as fundamental to the mental model of self necessary for MSR in late infancy, and offer evidence that infants who engage in more social imitation generally achieve MSR at an earlier age. Because very young infants are incapable of all but the simplest representational thought, they are unable to integrate the self-relevant information derived from imitation. Close to the end of the 1st year, when representational abilities appear, infants become capable of integrating self-relevant information. This explains the increase in social imitation prior to MSR.

In contrast to imitation theorists, Gergely (*SAAH*4) argues that MSR is made possible by the same inferential capacity that underlies reconstruction of serial invisible-object displacements in Stage 6 of Piaget's object concept series. To recognize itself in a mirror, an organism must also construct a generalized visual feature representation of the nonvisible parts of its body (largely the face), and recognize the "mismatch" (and match) between this visual feature representation and its own mirror image. The organism then recognizes, through conceptual inference, that to coordinate its understanding of mirrors with the discrepant information requires modification of its self-representation to include the specific facial information it sees in the mirror. Organisms that fail to recognize their mirror images have an inability for conceptual inference, and thus cannot update their self-representation. Gergely's theory parallels one of Mitchell's models (1993a) in postulating the organism's capacity for inference.

According to Mitchell's (1992, 1993a) deductive model, MSR occurs via the sort of deductive inference Piaget (1954, p. 96) attributed to the child's understanding of object permanence. His deductive theory proposes that MSR depends on an understanding of object permanence, an understanding of mirror correspondence, and objectification of body parts in the following way:

[T]he organism knows that the mirror presents an image which is contingent with the action the organism is performing, and looks like what the organism expects its own action to look like. The organism also knows that most mirror-images are images of what they are contingent with and resemble. The organism infers that this mirror-image may be an image of what it is contingent with and similar to, which is the organism's action. Therefore, the organism infers that the mirror-image is of its body. (Mitchell, 1992, p. 129)

(See also Mitchell, 1993a; Thompson & Boatright-Horowitz, *SAAH*22.)

Custance & Bard (*SAAH*12) argue that MSR arises from a generalized sixth-stage sensorimotor capacity for mental representation rather than from imitation. They base their interpretation on their failure to elicit fifth-stage imitation in a chimpanzee that had already displayed MSR, noting that longitudinal studies are needed to resolve the question.

Thompson and Contie (*SAAH*26) argue that MSR in chimpanzees depends on a spontaneous and broadly construed matching concept ("reflexive equivalence relations"), coupled with "internally induced regulation of action." In this view, when monkeys use mirrors to find objects, they do not recognize that the mirror image and the object are identical, but merely that one is a sign for the other. Apes, on the other hand, *do* recognize that mirror image and object are identical, and this recognition is reflexive (i.e., bidirectional). As evidence, Thompson and Contie note that, in visual–visual matching tasks, chimpanzees spontaneously match by identity, whereas monkeys must be trained to do so by reinforcement (Oden, Thompson & Premack, 1990). This formulation, however, fails to account for cases in which monkeys have displayed spontaneous recognition of objects in mirror tasks (Anderson, 1986), presumably through visual–visual matching.

Explanations for absence of MSR in monkeys

Gallup (1977) has argued that monkeys fail to display MSR because they lack a self-concept and the cognitive capacities on which a self-concept is based. Several other investigators have tried to specify more exactly which cognitive capacities may be lacking in monkeys.

Thompson and Contie (*SAAH*26) suggest that the processes that mediate mirror-based location of nonself objects differ in organisms that recognize themselves and in those that do not. When monkeys use mirrors to locate objects other than their own body (Anderson, 1986), they are comparing two "objects" that are "out there"; but when they observe the monkey in the mirror (i.e., themselves), there is no object "out there" to compare to themselves. Thus, the locus of the self for monkeys offers no means of objectifying that self in their bodies, and may make mirror self-recognition impossible for monkeys.

Boccia suggests that monkeys lack MSR because they are not interested in deviations from their self-image, and as a result fail to respond to deviant mirror images. Her description fits Watson's suggestion that human infants who fail to show MSR may be uninterested by the perfect contingency afforded by the mirror. By contrast, human children prior to the end of the 2nd year develop a fascination with deviations from the norm in behaviors and appearances (Kagan, 1989), which may make them more attuned to the odd mark on their face during the mark test (Mitchell, *SAAH*6).

Several investigators have speculated that monkeys fail to develop MSR because of their tendency to avert their eyes from faces (Anderson, *SAAH*21; Boccia, *SAAH*23; Thompson & Boatright-Horowitz, *SAAH*22). In a related vein, Parker (1991) has suggested that monkeys display little of the voluntary control over their facial muscles that would allow them to discover the perfect contingency between their facial movements and those of the mirror image. Both of these notions are consistent with the tendency of monkeys to threaten their mirror image as if it were a conspecific.

A few investigators have tried to imagine the kinds of self-awareness that monkeys might display in the absence of the capacity for MSR. Mitchell (*SAAH*6) suggests that they display both perceptual and rudimentary imaginal selves. Parker and Milbrath (*SAAH*7) argue that the ability of macaques to inhibit and hide certain behaviors by turning away suggests that these monkeys have some rudimentary form of self-image. Comparative developmental research on self phenomena in early infancy, for example, Watson's work on self-detection, may help to elucidate the mechanisms underlying simpler forms of self-awareness. Such studies may in turn provide models for studying self-awareness in monkeys and other species that do not display MSR.

Ontogeny of self-awareness

As the foregoing discussion clearly illustrates, many of the proximate hypotheses concerning MSR and self-awareness are based on developmental models. As we will see, some of the functional and ultimate hypotheses discussed later in this chapter are also based on developmental models. One of the reasons developmental models are generative is that they specify the order of emergence of particular abilities: Studies on the order in which specific abilities emerge in human infants and children may suggest some possible mechanisms, or at least disallow other mechanisms. Several of the alternative explanations for MSR, for example, may rise or fall partly on the basis of the order of emergence of their causal candidate relative to the emergence of MSR, for example, object permanence, contingency testing and imitation, or theory of mind in humans and nonhumans. On the other hand, studies on the order of emergence of specific abilities in other primates are also important because species may differ in the order as well as in the timing and extent of their development.[1]

Research on human infants and children suggests the following developmental sequence of self-awareness:

1. neonatal imitation (Gopnik & Meltzoff, *SAAH*10);
2. development of self-detection at about 3 months of age followed by preference for imperfect contingency in social and object interactions from 3 months (Watson, *SAAH*8) to approximately 18 months;
3. embarrassment or coy reactions to mirror image (Amsterdam, 1972);

4. development of MSR through contingency testing at about 18 months (average), coincident with mirror-assisted, self-addressed behaviors (including mark-addressed behaviors) (Lewis & Brooks-Gunn, 1979);
5. followed by development of objective MSR through feature self-recognition in photos at about 24 months, coincident with verbal self-labeling;
6. development of self-evaluative emotions of shame and pride (see Lewis, *SAAH*2), perhaps coincident with role reversal in symbolic play at about 3 years (Lewis, *SAAH*2);
7. emergence of theory of mind about self and others at about 4 years (see Gopnik & Meltzoff, *SAAH*10; Moses & Chandler, 1992).

Although more research is needed on these and related phenomena in human children and nonhuman primates, the sequence in the former provides a tentative guideline for assessing various models of self-awareness in the latter.

Developmental sequences in human infants and children

As indicated above, Gopnik and Meltzoff trace the origins of self-awareness in human infants back to the neonatal capacity to engage in facial imitation. This early form of imitation is a precursor to more elaborated forms of imitation, which they believe provide the mechanism for both identification and differentiation between mother and infant. According to their research, mutual imitation between infant and caretakers occurs as early as 14 months. The beginnings of imitation of novel schemes and deferred imitation also occur at this time.

Watson (*SAAH*8) proposes a mechanism for self-detection in infants. He contrasts the perfect contingency of proprioceptive and visual feedback from the infant's own actions to the imperfect contingency of various forms of feedback to those actions that the infant receives from others. Watson argues that by 3 months of age, human infants have learned to assume that perfect contingency indicates self. At this time they become less aroused by and interested in perfect contingency, as they begin to be more interested in seeking the *im*perfect contingency associated with the responses of others to them. Because the mirror image, in having perfect contingency with the self, is apparently assumed by the infant to be itself, it is consequently uninteresting (although it becomes interesting later on, at about 15 months).

Hart and Fegley (*SAAH*9) describe the correlation between high frequencies of imitation of social gestures by infants and the onset of MSR. This is consistent with Lewis and Brooks-Gunn's (1979) observations on the close connection between imitation and contingency testing in the infant's responses to videotapes at the time they began to display self-recognition. Much remains to be learned, however, about the relationship between MSR and the Piagetian stages in the sensorimotor series. One unresolved dispute about developmental timing centers on whether MSR depends upon sixth-stage sensorimotor intelligence and mental representation (in whatever series) as argued by Hart and Fegley, Custance and Bard, and by Gergely, or whether it depends merely on

4th- or 5th-stage sensorimotor intelligence in the imitation and causality series as argued by Parker (1991).

Gopnik and Meltzoff (*SAAH*10) trace the origins of theory of mind (ToM), as well as MSR, back to imitation. They emphasize, however, that ToM is a more complex phenomenon that only emerges at about 4 years of age. This temporal sequence suggests that the association between MSR and ToM that Gallup has posited is not a direct one. Gallup predicts that organisms that recognize themselves in mirrors should not only have a self-concept, but should use their self-awareness to intentionally deceive, pretend, and empathize in a manner suggesting that they have a theory of mind. Although it is true that these capacities are generally observed in the same species, the developmental delay between MSR and ToM suggests that factors in addition to those necessary for MSR are involved (Gergely, *SAAH*4; Mitchell, 1993a,b).

Longitudinal developmental studies of self-awareness in the great apes (see Miles, *SAAH*16; Patterson & Cohn, *SAAH*17) will provide critical data for determining whether the sequences of development are parallel in these species and in human infants. The evidence presented in this volume generally favors the notion that self-awareness in great apes develops in the same sequence as in human infants, but at a slower rate. It also seems to display a similar instability in its early manifestations (Miles, *SAAH*16). Much remains to be learned, however, about individual differences within species and subtle differences among great ape species (Patterson & Cohn, *SAAH*17; Swartz & Evans, *SAAH*11).

Social context of development of self-awareness

Many psychological theorists have emphasized the social context of the development of self in humans. The same point has been made by Gallup and his colleagues, who have demonstrated experimentally that social interaction is necessary for the development of MSR in chimpanzees (Gallup, McClure, Hill, & Bundy, 1971). At least two major questions arise in relation to the socialization of the self: By what means does it occur, and how does the development of self connect with the development of the other.

The means by which social interaction differentiates and objectifies the self are only partially understood. Baldwin (1902) emphasized the role of social imitation; Mead (1934/1974) emphasized the contribution of role-playing games; Cooley (1902/1983, p. 184) proposed the idea of the "looking-glass self," that is, that self-objectification derives from the imagined judgments of other people, the imagined expectation of how one appears to them, and an initial "self-feeling" or internal experience. It is important to keep in mind in this context, however, that the mechanisms of socialization are themselves internal to the organism and change in conjunction with development. Indeed, according to Lewis (*SAAH*2), the emergence of self-awareness (the "idea of me") initiates the self-correcting processes that allow the child to begin to socialize itself.

As indicated above, Gopnik and Meltzoff, as well as Hart and Fegley, have traced the role of social imitation in the development of MSR and ToM. Several investigators have emphasized the importance of face-to-face social interactions and facial imitation in the development of self-awareness from the neonatal period on. Social imitation during the 1st year of life provides the scaffolding needed for the development of factors such as representation of one's own facial features, and body part objectification. Parker and Milbrath, as well as Custance and Bard, note that games that encourage children to point to particular facial features provide scaffolding for the contingency needed for self-recognition. These "Where's your nose/my nose?" games between parents and children can lead to body-part objectification (Mitchell, 1993a), and build up kinesthetic–tactile–visual representations needed for facial representations (Piaget, 1945/1962).

In a slightly different approach that also centers on the role of the face, Gómez (*SAAH*5) emphasizes the role of visual attention in the development of self-awareness. According to this view, when one organism attends to signs of another organism's visual attention, the latter organism can use this attention-directed attention as a clue to its own attention. Recognition of another's attention can, as Gómez notes, allow an organism to attend to the same thing that the other attends to, and such attention may result in objectification. A similar point is made by Merleau-Ponty (1960/1982, p. 139): "I understand all the more easily that what is in the mirror is my image for being able to represent to myself the other's viewpoint on me." This view contrasts with Gallup's (1985) view that organisms who recognize their mirror image are likely to recognize mental states in others by extrapolation from their own mental states.

On the subject of the developmental relationship between awareness of the self and of others, Gopnik and Meltzoff argue that neither precedes the other, but rather that the two forms of awareness develop in tandem through the same mechanisms. The idea of developmental parallelism as opposed to developmental derivation of the concept of other from that of self, or vice versa, is also consistent with Lewis's (*SAAH*2) discussion of the self-socializing role of self-evaluative emotions (e.g., shame and pride) that emerge coincident with the first internalization of social standards.

Evolutionary history of self-awareness in great apes and humans

Reconstruction of the evolutionary history of a capacity depends upon comparisons of closely related species to more distantly related species. The cladistic method provides a systematic methodology for such comparisons. Cladistic methodology was designed to provide a logically consistent procedure for classifying species into higher taxonomic categories (genera, families, orders, etc.) strictly on the basis of the closeness of their phylogenetic relationships (Ridley, 1984; Wiley, 1981). Once these relationships are known, as they are among the primates, they can be used as a basis for reconstructing the probable character states of the common ancestors of the various reference groups. When combined with geological or other data on temporal relations, they can also be used to construct phylogenies. (See Figure 27.1 for a phylogeny of primates.)

By mapping data on various cognitive abilities onto a primate family tree and comparing their distributions, we can see which abilities are unique to particular species or species groups. Comparisons with more distantly related outgroups provide a basis for determining which of the abilities are unique to humans and the great apes, that is, which of these abilities do not occur in the outgroup (so-called shared derived character states) and which abilities occur in the outgroup as well as the ingroup (so-called shared character states).

Choosing character states for comparing species is the most challenging aspect of cladistic methodology. Paleontologists typically select an anatomical character and identify varying character states among the in-group species (e.g., see Martin, 1985, for a discussion of the distribution of enamel states in great apes). Ideally the character is part of an anatomical complex the functions of which are well understood (Clark, 1964).

Ethologists typically choose as their character states highly stereotyped (ritualized) motor patterns involved in eating, mating, parenting, and so forth (though they do not use cladistic terms) to compare the forms and functions of behaviors among closely related species. Like cladists, they use comparative data to reconstruct the behavior of the common ancestor, and the stages of evolution of various character states.

Comparative developmental psychologists who study animal mentality typically choose cognitive abilities as character states (e.g., Parker, 1991; Parker & Gibson, 1977; Povinelli, SAAH18). Use of cognitive abilities is necessary because the complexity of primate behavior and the multiple means that monkeys and apes use to accomplish the same goal renders comparison of motor patterns alone impractical and insufficient for analyzing their behaviors.[2]

When we map the distribution of full abilities for MSR, fifth- and sixth-stage understanding of object permanence, imitation of novel schemes, and the capacity for symbol use among primates, for example, we see that all occur in the great apes and humans, and none but object permanence occur in the other primates. Thus, we can see that these abilities arose in the common ancestor of the great apes (Parker, 1991, SAAH19; Povinelli, 1987, SAAH18). Once character states are selected, the rather rigorous cladistic methods render this aspect of evolutionary modeling fairly straightforward and uncontroversial (see Figure 27.2).

Povinelli notes that MSR may be polymorphic in great apes, and believes that gorillas may have lost the ability for MSR due to changes in the onset and timing of cognitive development, such that MSR in gorillas will only follow from elaborate social scaffolding provided by humans, which "turns on" the ability for MSR. However, the evidence that many chimpanzees do not exhibit MSR (Povinelli, SAAH18; Swartz & Evans, 1991, SAAH11), that

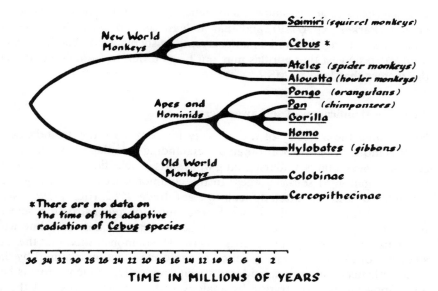

Figure 27.1. The phylogeny of anthropoid primates (from Parker & Gibson, 1977).

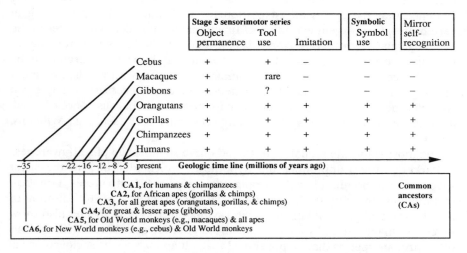

| | Stage 5 sensorimotor series | | | Symbolic | Mirror |
	Object permanence	Tool use	Imitation	Symbol use	self-recognition
Cebus	+	+	–	–	–
Macaques	+	rare	–	–	–
Gibbons	+	?	–	–	–
Orangutans	+	+	+	+	+
Gorillas	+	+	+	+	+
Chimpanzees	+	+	+	+	+
Humans	+	+	+	+	+

~35   ~22 ~16 ~12 ~8 ~5   present   Geologic time line (millions of years ago)

CA1, for humans & chimpanzees
CA2, for African apes (gorillas & chimps)
CA3, for all great apes (orangutans, gorillas, & chimps)
CA4, for great & lesser apes (gibbons)
CA5, for Old World monkeys (e.g., macaques) & all apes
CA6, for New World monkeys (e.g., cebus) & Old World monkeys

Common ancestors (CAs)

Figure 27.2. Mapping sensorimotor intelligence series, symbolic intelligence, and mirror self-recognition onto primate phylogeny to reconstruct the common ancestry of particular abilities.

chimpanzees and gorillas show many behaviors toward mirrors that monkeys do not (Parker, *SAAH*19), and that at least one gorilla without extensive human scaffolding removed a mark surreptitiously placed on her face (Parker, *SAAH*19), suggest that gorillas may not be so different from chimpanzees in the factors that lead to MSR.

Adaptive significance of self-awareness

Establishing the ancestry of the capacity for self-awareness as indicated by MSR is more straightforward than understanding the adaptive significance of self-awareness, that is, understanding what sorts of selection pressures favored the propagation of self-awareness in the common ancestor. Although functional hypotheses are often difficult to test (Hinde, 1982; Ridley, 1986; Wiley, 1981; Williams, 1966), the general procedure for reconstructing selection pressures is to look for socioecological factors that uniquely correlate with a particular structure or ability across species. The demonstration that the target structure or ability is uniquely suited to serve the function required by those socioecological factors is considered strong supportive evidence for its being an adaptation. Demonstration by analogy with optical devices that the vertebrate eye is uniquely well suited to process information available in a certain region of the electromagnetic spectrum, for example, is considered strong evidence that the eye is an adaptation for such information processing in vertebrates. Convergent evolution of functionally equivalent structures in the distantly related group of *Octopus* is considered to be independent evidence for this interpretation. This class of reasoning is known as "argument by design" (see, e.g., Dawkins, 1986; Thompson, 1981). In contrast, the argument that a given structure or ability is beneficial or even necessary in any given context is not considered to be adequate for adaptation, because once evolved, a structure or behavior may have many secondary uses (see, e.g., Williams, 1966).

The most elaborate scenario for the adaptive significance of self-awareness is Parker's (1992) apprenticeship hypothesis, which identifies in the sister species in the hominoid clade (orangutans, gorillas, chimpanzees, bonobos, and humans) the unique cluster of associated characters – MSR, imitation of novel facial and gestural behaviors, and symbol use. (These species also show object permanence and tool use, but these abilities are shared with various other primate species that do not show MSR.) It appeals to the socioecological context of this complex of abilities in chimpanzees, and notes that they are associated with apprenticeship in foraging and feeding on a variety of extractible foods, particularly apprenticeship in tool-aided extraction of these foods (Parker, in press; Parker & Gibson, 1979; Thompson & Contie, *SAAH*26). Parker notes that self-awareness and imitation are involved in the process of directed trial-and-error and insightful learning of novel sequences and orientations of behavior from a model, and in teaching by the demonstration of a technique to a naive observer (Boesch, 1991; Parker, 1992, 1993).

Comparing your own actions to those of your mentor (or conversely your pupil's actions to your own) entails awareness of yourself in relation to the other, as does adjusting your actions to the capacities of your pupil. Whereas among the hominoids only chimpanzees and humans currently are known to engage in apprenticeship in tool-aided extractive foraging, it seems likely that the common ancestor of the great ape–human clade in whom the target abilities first arose depended on this subsistence behavior.[3]

The question here is whether self-awareness is uniquely suited to learning by trial-and-error imitation of novel behavior sequences from a demonstrating model. One approach to this question is to look for parallels in distantly related species, or perhaps even in learning machines. Dolphins provide an interesting analogy to the great ape clade in that they also display the target abilities of MSR (Marten & Psarakos, *SAAH*24) and imitation of novel actions (Norris, 1974; Tayler & Saayman, 1973). Whether dolphins engage in apprenticeship in foraging and feeding through imitation and teaching, as do members of the great ape–human clade, is an interesting test case for the apprenticeship hypothesis.

Two other scenarios for the adaptive significance of self-awareness are less elaborate, but suggestive. The Machievellian intelligence hypothesis (Byrne, in press; Whiten & Byrne, 1988) suggests that, following the evolution of technological understanding, the social environment was the selection pressure that resulted in MSR, empathy, imitation, pretend play, deception, theory of mind, and teaching. This complex arose in the common ancestor of great apes and humans as an adaptation for social manipulation, such that "any increase in Machiavellian skill by one 'player in the game' will select for enhanced skill in the other, both in competitive and cooperative interaction" (Whiten & Byrne, 1988, pp. 8–9; see also Humphrey, 1976). Finally, the clambering hypothesis (Povinelli, *SAAH*18; Povinelli & Cant, 1992) proposes that the special demands of clambering locomotion of large-bodied orangutan-like ancestors in the high canopy entailed selection for the sense of self as a causal agent.

Conclusion

We have briefly outlined a variety of complementary and conflicting models for self-awareness proposed by various contributors to this volume. For clarity we have classified these models according to the four complementary kinds of approaches recognized by ethologists. It is important to emphasize that all four approaches are necessary for a comprehensive understanding of the phenomena of self-awareness in evolutionary context. It is also important to emphasize that behavioral evolution can be reconstructed only through comparative studies. The models presented in this book are a first step toward a new heuristic because they suggest new directions for future research. The fates of these and other models remain to be determined by future research and theory.

Notes

1.   Obviously, the immediate causes of behavior entail an ontogeny within which they can occur. Ontogeny itself, like proximate mechanisms, has itself been shaped by both developmental constraints and selection in a variety of contexts. Developmental constraints operate in epigenetic systems when later structures develop from earlier ones. It is impossible, for example, to develop grasping before developing the hand, impossible to develop the limb buds before the necessary three tissue layers form in the embryo. Selection on developmental sequence occurs when alternative developmental paths to the same adult outcome have different effects on survival and reproductive success. It would probably be less adaptive, for example, for a mammal to achieve sexual maturity before weaning than after weaning, because early mating might lead to incest and interfere with learning to find food and to fight, as well as competing with energy-demands of growth.
2.   The comparative use of cognitive abilities described in human infants and children by developmental psychologists as character states can be justified on the theoretical grounds that humans and great apes are sister species in a clade whose last common ancestor lived in the last 10 million years and who therefore share approximately 95% of their DNA (e.g., Weiss, 1987). The comparative use of these characters and character states can also be justified on the empirical grounds that many investigators have been able to identify these character states in both great apes and monkeys.
3.   Although a number of nonhominoid species display tool use, imitation, and even teaching, the mechanisms underlying their behaviors seem to be different from those underlying these performances in great apes (and probably dolphins). Specifically, these other mechanisms involve greater instinctual patterning and hence greater stereotypy than the more flexible and intelligent behaviors of great apes.

References

Amsterdam, B. K. (1972). Mirror self-image reactions before two years of age. *Developmental Psychobiology, 5*, 297–305.

Anderson, J. R. (1986). Mirror-mediated finding of hidden food by monkeys (*Macaca tonkeana* and *M. fascicularis*). *Journal of Comparative Psychology, 100*, 237–242.

Baldwin, J. M. (1902). *Social and ethical interpretations of social life.* New York: Macmillan.

Boesch, C. (1991). Teaching among wild chimpanzees. *Animal Behaviour, 41*, 530–532.

Byrne, R. (in press). Empathy in primate social manipulation and communication: A precursor to ethical behaviour. In G. Thines (Ed.), *Biological evolution and the emergence of ethical conduct.* Jean-Marie Delwart Foundation: Bruxelles.

Clark, W. LeGros. (1964). *The fossil evidence for human evolution.* Chicago: University of Chicago Press.

Cooley, C. H. (1902/1983). *Human nature and the social order.* New Brunswick, NJ: Transaction.

Dawkins, R. (1986). *The blind watchmaker.* New York: Norton.

Gallup, G. G., Jr. (1977). Self-recognition in primates: A comparative approach to the bidirectional properties of consciousness. *American Psychologist, 32*, 329–338.

Gallup, G. G., Jr., McClure, M. K., Hill, S. D., & Bundy, R. A. (1971). Capacity for self-recognition in differentially reared chimpanzees. *Psychological Record, 21*, 69–74.

Guillaume, P. (1926/1971). *Imitation in children.* Chicago: University of Chicago Press.

Hinde, R. A. (1982). *Ethology: Its nature and relations with other sciences.* New York: Oxford University Press.

Humphrey, N. K. (1976). The social function of intellect. In P. P. G. Bateson & R. A. Hinde (Eds.), *Growing points in ethology* (pp. 303–317). Cambridge University Press.

Kagan, J. (1989). *Unstable ideas: Temperament, cognition, and self.* Cambridge, MA: Harvard University Press.

Lewis, M., & Brooks-Gunn, J. (1979). *Social cognition and the acquisition of self.* New York: Plenum.

Lewis, M., Sullivan, M., Stanger, C., & Weiss, M. (1989). Self-development and self-conscious emotions. *Child Development, 60*, 146–156.

Martin, L. B. (1985). The significance of enamel thickness in hominoid evolution. *Nature, 314*, 260–263.

Mead, G. H. (1934/1974). *Mind, self, and society from the standpoint of a social behaviorist.* Chicago: University of Chicago Press.

Merleau-Ponty, M. (1960/1982). The child's relations with others. In J. M. Edie (Ed.), *The primacy of perception* (pp. 96–155). Illinois: Northwestern University Press.

Mitchell, R. W. (1992). Developing concepts in infancy: Animals, self-perception, and two theories of mirror self-recognition. *Psychological Inquiry, 3*, 127–130.

(1993a). Mental models of mirror self-recognition: Two theories. *New Ideas in Psychology, 11*, 295–325.

(1993b). Recognizing one's self in a mirror? A reply to Gallup and Povinelli, Byrne, Anderson, and de Lannoy. *New Ideas in Psychology, 11*, 351–377.

Moses, Louis, & Chandler, M. J. (1992). Traveler's guide to children's theories of mind. *Psychological Inquiry, 3*, 286–301.

Norris, K. S. (1974). *The porpoise watcher.* New York: Norton.

Oden, D. L., Thompson, R. K. R., & Premack, D. (1990). Infant chimpanzees spontaneously perceive both concrete and abstract same/different relations. *Child Development, 61*, 621–631.

Parker, S. T. (1991). A developmental approach to the origins of self-recognition in great apes. *Human Evolution, 6*, 435–449.

(1993). Evolved mechanisms for cognitive construction: Circular reactions and imitation as self-teaching mechanisms. *Human Development, 36*, 309–323.

(in press). *Imitation, teaching, and self-awareness as adaptations for apprenticeship in foraging and feeding.* In A. Russon, K. Bard, & S. T. Parker (Eds.), *Reaching into thought.* Cambridge University Press.

Parker, S. T., & Gibson, K. R. (1977). Object manipulation, tool use and sensorimotor intelligence as feeding adaptations in cebus monkeys and great apes. *Journal of Human Evolution, 6*, 623–41.

(1979). A developmental model for the evolution of language and intelligence in early hominids. *Behavioral and Brain Sciences, 2*, 367–408.

Piaget, J. (1945/1962). *Play, dreams, and imitation in childhood.* New York: Norton.

(1954). *The construction of reality in the child.* New York: Basic.

Povinelli, D. (1987). Monkeys, apes, mirrors and minds: The evolution of self-awareness in primates. *Human Evolution, 2*, 493–509.

Povinelli, D., & Cant, J. G. H. (1992). *Orangutan clambering and the evolutionary origins of self-conception.* Paper presented at 14th International Primatological Society meeting, Strasbourg, France.

Ridley, M. (1986). *Classification and evolution.* New York: Oxford University Press.

Swartz, K. B., & Evans, S. (1991). Not all chimpanzees (*Pan troglodytes*) show self-recognition. *Primates, 32*, 483–496.

Tayler, C. K., & Saayman, G. S. (1973). Imitative behaviour by Indian Ocean bottlenose dolphins (*Tursiops aduncus*) in captivity. *Behavior, 44*, 286–298.

Thompson, N. S. (1981). Toward a falsifiable theory of evolution. In P. P. G. Bateson & P. H. Klopfer (Eds.), *Perspectives in ethology* (Vol. 4, pp. 51–73). New York: Plenum.

Tinbergen, N. (1963). On aims and methods of ethology. *Zeitschrift für Tierpsychologie, 20*, 410–429.

Weiss, M. (1987). Nucleic acid evidence bearing on hominoid relationships. *Yearbook of the American Journal of Physical Anthropology, 30*, 41–74.

Whiten, A., & Byrne, R. W. (1988). The Machievellian intelligence hypotheses: Editorial. In R. W. Byrne & A. Whiten (Eds.), *Machievellian intelligence: Social expertise and the evolution of intellect in monkeys, apes, and humans.* Oxford: Oxford University Press (Clarendon Press).

Wiley, E. O. (1981). *Phylogenetics: The theory and practice of phylogenetic systematics*. New York: Wiley.

Williams, G. C. (1966). *Adaptation and natural selection*. Princeton, NJ: Princeton University Press.

Yoerg, S. I., & Kamil, A. C. (1990). Integrating cognitive ethology with cognitive psychology. In C. A. Ristau (Ed.), *Cognitive ethology: The minds of other animals* (pp. 273–289). Hillsdale, NJ: Erlbaum.

Author index

429

Subject index